ARCHAEOACOUSTICS III

The Archaeology of Sound

Publication of Proceedings from the 2017 Conference in Portugal

Edited by

Linda C. Eneix and Michael Ragussa

OTSF

ARCHAEOACOUSTICS III
PROCEEDINGS FROM THE 2017 CONFERENCE IN PORTUGAL
Edited by Linda C. Eneix and Michael Ragussa

Published by OTSF in 2018

Published by

OTSF

The OTS Foundation
P.O. Box 323
Myakka City, Florida 34251
www.OTSF.org

ISBN 978-0965625258

www.archaeoacoustics.org

ARCHAEOACOUSTICS III

TOMAR AND MAÇÃO, PORTUGAL

OTSF

ipt Instituto Politécnico de Tomar

INSTITUTO TERRA E MEMÓRIA

centro de geo ciências

REPÚBLICA PORTUGUESA
CULTURA

PATRIMONIO CULTURAL
Direção-Geral do Património Cultural

CONVENTO DE CRISTO

WWW.ARCHAEOACOUSTICS.ORG

Introduction

Linda C. Eneix
President, The OTS Foundation

Malta, then Turkey, and then Portugal: we have now bracketed the Mediterranean with this third international multi-disciplinary conference on Archaeoacoustics.

Building on foundations cited elsewhere in this book, a risky idea took hold a few years ago. *What if we reversed the traditional conventions of specialization and brought together people from a range of professional viewpoints to look at the subject of archaeoacoustics on an international scale?* Obviously, it was effective. It has been said that these three conferences have facilitated a fresh blossoming of interest. More people than ever before are pursuing the study of the human experience of sound in antiquity. The fresh perspective and fervent voices that we have witnessed hint at an increasing respect for the fact that the ancient world was not silent. Hearty congratulations to all who have been on the journey with us.

Personally, I can say that the meetings have been a revelation. It is no easy thing to make sure everyone has time to speak and all the equipment works as promised. More importantly, it is no easy thing to bring presenters from so many disciplines into the same room to share knowledge and enthusiasm about something intangible although intrinsic. Nevertheless, these non-profit conferences have allowed participants to share common interests, where before many were the only ones in their immediate fields with this sometimes-odd interest in the archaeoacoustic aspects of their work. They have also broadened the perspective of everyone who paid attention.

We at The OTS Foundation are extremely proud to have been an instrument for advancing study in this sensory chapter of the story of human development. With the formation of an independent International Society, the stage has now been set for the future. How this organization moves forward will depend on constituents who care about it and what they decide to do.

Moving forward will mean facing quite a few challenges. Even within this book, we see that different participants have different ideas of what Archaeoacoustics is or should be. A multi-disciplinary approach is imperative but each field has its own methodology and focus. There is an intimidating range that the subject area must encompass. We are, after all, considering a whole dimension of experience that has customarily been considered irretrievable.

Herein, you will find new research, updates & expansions on earlier presented work, methodology, interpretation, opinion, instruction and just plain food for thought. We have opted to sort them in alphabetical order by the last name of the lead author.

Sincere thanks to the authors and researchers, whose work carries us forward.

ARCHAEOACOUSTICS

www.Archaeoacoustics.org

Paul Devereux recording the sound of a fallen lithophone on the Carn Menyn ridge, Preseli, Wales, during a night session, when extraneous, unwanted, sound is at a minimum allowing for clearer recording of the stone. (Photo: S. Devereux)

When the Ancient World Got a Soundtrack

Paul Devereux
devereux.uk@btinternet.com

Summary of Audio-Visual presentation given at the Archaeoacoustics III Conference, Tomar, Portugal, on 5th October 2017.

In opening the presentation [and the conference], I gave a brief note about the symposium called by Professor Chris Scarre and Graeme Lawson and held at the McDonald Institute of Archaeological Research, Cambridge, in 2003, where the formal acceptance and naming of archaeoacoustics as an area of archaeological study took place. The book of the Proceedings was published in 2006, and was the first to use the title 'Archaeoacoustics'. I then briefly outlined three main methods of studying sound at archaeological sites:

(i) modern musical or vocal performance at sites to test their acoustic properties;

(ii) the use of electronic monitoring equipment to explore and measure the acoustics of monumental sites;

(iii) listening to the natural sounds occurring at a site, whether issuing from the structure itself, or from environmental factors such as echoes, the local sound of rushing water, wind, rain, etc.

(In actuality, some of these methods can be intermixed and expanded upon, but these are the basics.)

I gave an example of each of the first two methods, including the 1995 project that explored the fundamental acoustic resonance frequency of selected megalithic chambered monuments in southern England and in Ireland (Devereux and Jahn 1996; Jahn et al. 1996). This was the work of the ICRL (Institute of Consciousness Research Laboratories) then based at Princeton University, of which I was a part. It was as a result of this effort that a recurring narrow fundamental frequency range centring on 110 Hz in all the variously-sized chambers was noted, and this eventually led to neuroscientific research which showed this audio frequency to unexpectedly have regional brain effects (Cook et al. 2008).

But I mainly focused on method (iii), in particular explaining the cross-cultural perceived supernatural significance given to ringing rocks and lithophones in the ancient world, especially in China, India and Africa. I gave a few specific examples of such features, including ones in the British Isles and pre-Columbian ones in the USA. Finally, I described some of my own fieldwork with ringing rocks and lithophones, and referenced an archaeoacoustic study of Carn Menyn, the Welsh origin area of the Stonehenge bluestones, which I and Jon Wozencroft initiated with the Royal College of Art (www.landscape-perception.com) (Devereux and Wozencroft 2014).

We confirmed that it was a significant soundscape, and thus gave an extra possibility for explaining why Stone Age people made the huge effort to transport bluestones the approximately 200 km from South Wales to Stonehenge. Was it because sound gave these rocks a perceived mana, or spiritual power?

The presentation concluded with a very brief look at echoes related to Neolithic sites, other ways rocks can make sounds, and ways we can 'see sounds' – including kunst tube work showing how particulate matter in the air (steam, incense, etc.) illuminated by a strong light beam can be organised into visible soundwaves by acoustic influence (speculation this could have happened when the laser-like midwinter sunbeam enters the chamber at Newgrange), and my own work identifying the ceiling painting in the so-called 'oracle room' in the Hal-Saflieni hypogeum, Malta, as acoustic notation (Devereux 2009).

The ancient lithophone on the Scottish Hebridean island of Tiree. The rock is not geologically native to the island and is presumed to have been brought there by the ice sheet during the Ice Age. Whatever, the prehistoric inhabitants of the island clearly discovered the rock's ringing properties. (Photo: Author)

REFERENCES

Devereux, P., and Jahn, R.G., 1996. "Preliminary investigations and cognitive considerations of the acoustical resonances of selected archaeological sites", Antiquity, vol. 70, no. 26.

Jahn, R.G., Devereux, P., and Ibison, M, 1996. "Acoustical resonances of assorted ancient structures", Journal of the Acoustical Society of America, vol. 99, no. 2.

Cook, I., Pajot, S.K., and Leuchter, A.F., 2008. "Ancient Architectural Acoustic Resonance Patterns and Regional Brain Activity", Time & Mind, vol. 1, No.1., March.

Devereux, P., & Wozencroft, J., 2014. Stone Age Eyes and Ears: A Visual and Acoustic Pilot Study of Carn Menyn and Environs, Preseli, Wales, Time & Mind, 7(1), March. 47-70.

Devereux, P. 2009. "A Ceiling Painting in the Hal-Saflieni Hypogeum as Acoustically Related Imagery: A Preliminary Note", Time & Mind, vol.2, no.2., July.

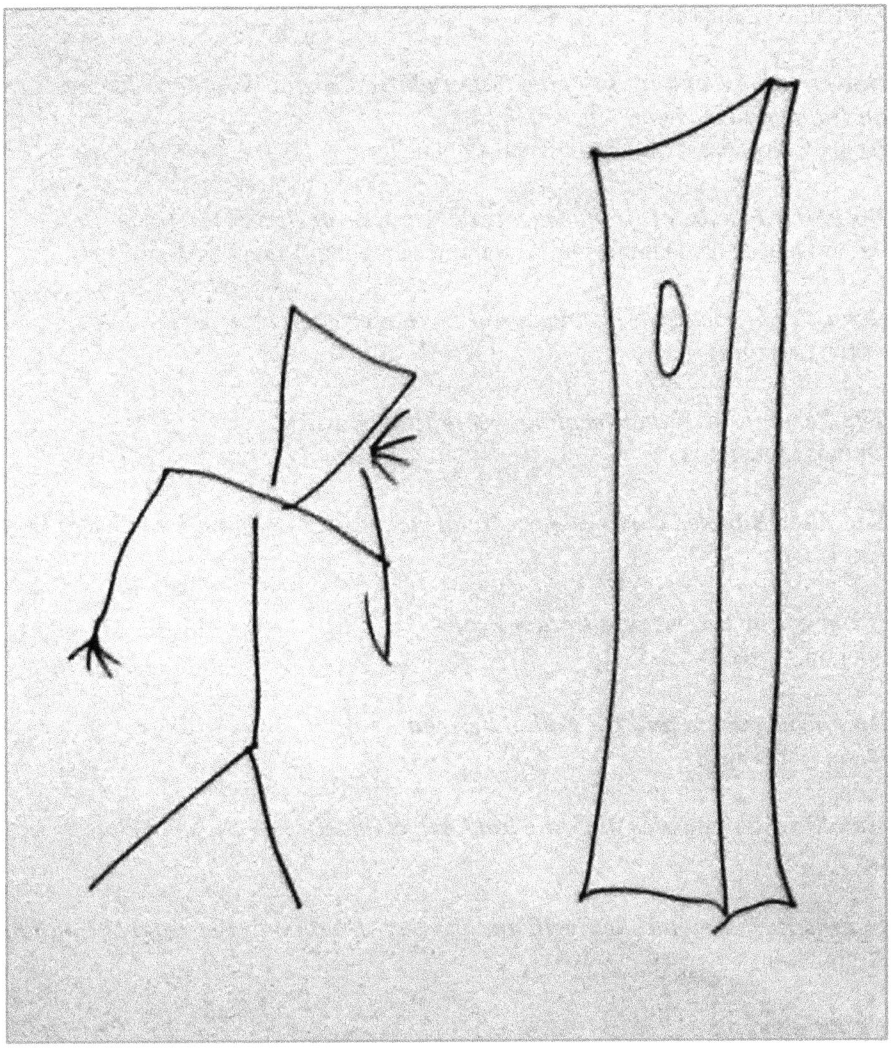

CONTENTS

CONTENTS

Dancing and Music in Rock Art

Paintings by Rosário Sousa, with Comments by Fernando Coimbra

Five paintings were displayed during the International Conference *Archaeoacoustics III,* as a kind of small exhibition, due to the connections of the represented themes with Archaeoacoustics.

Rosario Sousa is a contemporary artist that gets inspiration from rock art examples, exploring the directness that these images bring from the past. She reproduces the exact engravings, adds texture and colour to her works, transforming rock art examples into contemporary art.

rosariosousa67@yahoo.com

Neolithic paintings from Wed Mertoutek (Algeria). Acrylic, wood dust and sand on canvas. This scene is probably the result of prehistoric frenetic dancing to the sound of drums and percussion. The rhythm and the dynamic present on this image give the idea of modern dancing, where the girl in the right seems to be dancing "the twist".	Paintings from Tadrart Acacus (Libya), Late Neolithic. Acrylic, wood dust and sand on canvas. The character in the centre seems to be clapping hands, while the other two are holding tree branches, performing the sound background of a possible prehistoric ritual. The heads leaning back are probably the result of a certain kind of trance.	Petroglyphs from Saimaly-Tash, Kazakhstan. II millennium BC. Acrylic on canvas. Several characters seem to be dancing (probably with an acoustic background) and maybe clapping hands, worshipping a sun image. It's another example that it's difficult to conceive without any kind of sound or even music.	Paintings from Huashan, China. End of I millennium BC. Acrylic, wood dust and sand on canvas. This painting is a detail of one of the largest rock art panels in the world, with figures painted in red in a width of about 170 metres and 40 metres high. Some of the human figures seem to be dancing and others clapping hands, such as two characters in the first row.	Native American petroglyph, 12th century (?) AD. Acrylic, sand and plastic buttons on canvas. This image depicts the legendary character Kokopelli, present in the Hopi Indians myths, appearing here playing his flute. The contemporary artist added buttons to this scene in order to give the idea of music coming from the musical instrument.

Archaeology, Archaeoacoustics and Early Musical Behaviour

Fernando Coimbra

FERNANDO A. COIMBRA, PhD in Prehistory and Archaeology (with Extraordinary Prize from the University of Salamanca), is an archaeologist and rock art researcher. He's a Visiting Professor at the Polytechnic Institute of Tomar, Portugal and Internal Researcher of the Geosciences Centre from the University of Coimbra, Portugal. coimbra.rockart@yahoo.com

ABSTRACT: Archaeological evidence has been reporting several "sound-producing devices" such as flutes, lithophones, rasps and other since the Upper Palaeolithic, where an early musical behaviour must have occurred. In Late Prehistory, rock art displays more examples of that musical evidence, representing people playing instruments such as lyre, drums, bronze lure and horns, among other cases. Research in Archaeoacoustics allows a better understanding of the contexts where the mentioned "sound-producing devices" were used, since the reverberant environment of some caves and megalithic monuments can have developed emotions and feelings in the minds of prehistoric populations.

KEYWORDS: Musical behavior, sound, Archaeoacoustics

Introductory Note

This article analyses the contribution of Archaeology and Archaeoacoustics for understanding the early musical behaviour of prehistoric populations. However, it's not our intention to determine precisely when humans started to produce music, what can be considered as a very difficult and an almost impossible task. Furthermore, as I. Morley (2014: 147) mentioned "it is evident that no single field of investigation can address the wide range of issues relevant to answering the question of music's origins", being therefore a research with a multidisciplinary character, involving archaeology, ethnomusicology, anthropology and archaeoacoustics, among other disciplines.

It's important not to forget that since the origins of Mankind there are sounds in Nature: thunder, wind, rain and the sound of the waves in the sea, bird singing and

sounds produced by other animals were obviously familiar to the first human groups. These men and women probably wanted to mark their presence in the world by also making some kind of sound, beyond their own voices.

During the Palaeolithic, the sounds of the human voice produced inside a cave were certainly object of reverberation, what can have caused deep impression on the minds of the listeners. Together with this, the sounds produced by handclapping, foot stamping or other forms of percussion such as hitting stalactites or stalagmites with a wooden stick could have also been a reason for surprise and even joy. That's probably how an early musical behaviour started.

Archaeology and Early Musical Behaviour

As a starting point it's important to see what kind of information we can get from archaeological record, in a general way, regarding an early musical behaviour.

First of all it is known that, since the Palaeolithic there are several "sound-producing devices" (Lund, 2012: 186), such as bone and ivory flutes, lithophones, rasps (or idiophones), whistles and bullroarers. For example, bone and ivory flutes were discovered in Germany at Geißenklösterle, Hohle Fels and Vogelherd (Conard et al., 2009). Other cases were found in the French Pyrenees at the cave of Isturriz (Buisson, 1990).

Lithophones are constituted generally by stalactites or stalagmites, with marks of having been deliberately struck (Morley, 2006)[1], existing some examples at the caves of Pech Merle, Rocadour, Cougnac, Les Fieux[2], Escoural (Évora, Portugal) and Cueva de Nerja (Malaga, Spain) (Dams, 1985; Morley, 2006). More recently, lithophones were also discovered in caves from the north of Spain (Till, 2014).

A rasp "can be a piece of wood, bone or stone with grooves cut into it perpendicular to its length, which are then rubbed with another object to create a staccato vibration" (Morley, 2006: 37). There are examples from Pekarna, Moravia (Czech Republic), Abri Lafaye Bruniquel and Mas d'Azil, Ariège, both in France (Morley, 2006).

Phalangeal whistles were discovered at Isturitz and Maisières, in France, among other places. However, some examples are controversial, because they may result from carnivore activity (Morley, 2006).

A bullroarer "consists of a flat perforated piece of wood or bone on the end of a cord, which creates a whirring sound when spun in a circular motion, and is, or has been, used in a great diversity of global cultures, in both spiritual and functional contexts" (Morley, 2006: 33). Some examples were found in Aurignacian (38 000 - 28 000 years) through to Gravettian (29 000 -22 000 years) technological complexes, in caves such as Laugerie-Basse, Lalinde, Badegoule (Dordogne) and Lespugue (Haute Garonne), all in France (Morley, 2006).

Among the five types of the mentioned sound-producing artefacts, whistles and bullroarers are more suitable for communication at a certain distance than to produce any kind of early music. Whistles can be very helpful among hunter-gatherer societies to warn about an imminent danger, for example the presence of predators such as lions, bears or other dangerous animals such as rhinoceros, aurochs, boars or other, whose existence is well documented in cave art (Coimbra, 2016a). Bullroarers can have been used to establish contact between different groups of hunter-gatherers or may have been used in ritual. For example, according to Morley (2006: 33), "amongst some Australian Aborigines it is used to imitate the spirits occurring in the natural sounds of nature".

After or almost simultaneously with vocalizing, drumming must have been one of the first musical activities, since it doesn't require exactly an instrument, being possible to perform with parts of the body (handclapping, foot stamping, other).

Lithophones can be associated with a percussion activity, not constituting an intentionally built musical artefact. Therefore drumming using the body or natural lithophones may have created the origins of rhythm. The development of musical behaviour led certainly to the search of other

[1] Some of these stalactites have also paintings of red and black lines and dots. Lya Dams (1985) believes that this abstract art constitutes a type of mnemonic aid indicating striking-spots on the rock.

[2] All in the region of Lot, France.

forms of making rhythm, such as, for example the use of bone rasps[3].

Regarding what seems to be intentionally built musical artefacts, the first example identified so far is a flute made on swan radius (Fig.1) discovered at Geißenklösterle (Southwestern Germany) dated between 37 000 BP to 33 500 BP[4] (Münzel et al. 2002). In this region there are seven more flutes, being four of them made in mammoth ivory and three in bird bone (Conard et al. 2009).

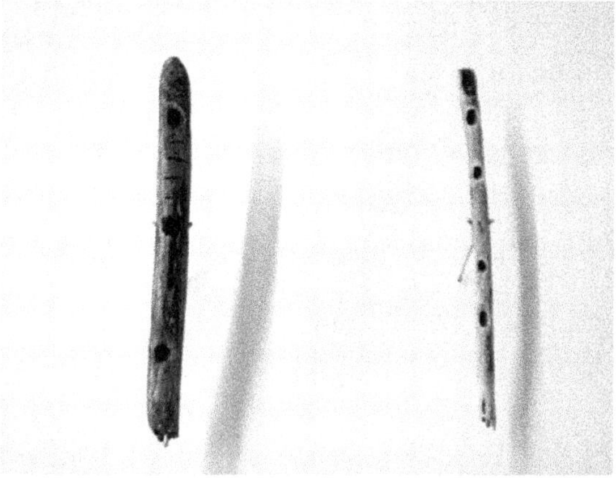

Fig.1 – Flute from Geißenklösterle (left) and flute from Hohle Fels (right), both from Southwestern Germany. Replicas of the Museum of Human Evolution, Burgos. (Photo: F. Coimbra)

However, according to E. Safa (2015) these flutes bring more questions than answers. Indeed, the presence of this kind of artefacts doesn't mean that they would have been created initially with intention of producing some kind of music. According to C. Lund (2012: 62), in the 19th century, "the shepherd in Sweden (...) blew sharp, high-pitched signals on the same type of bone flute in order to scare away his worst enemy – the wolf". For Palaeolithic peoples, this kind of sound tool may "have been considered to be a weapon against real or supernatural enemies and by no means a tool for artistic activity". (Lund, 2012: 62). Therefore, what can be at first sight interpreted as a musical instrument can also have been something else. According to C. Homo-Lechner, who analyzed musical instruments regarding their functions, they become also sound artefacts, "pouvant ainsi potentiellement revêtir une fonction musicale, signalétique, cynégétique ou rituelle, voire plusieurs d'entre elles tout à la fois'' (Safa, 2015 : 33).

After the Palaeolithic, other types of musical instruments appear, such as the Neolithic clay drums covered with animal skin, from several parts of Europe (Aiano, 2006) as for example the cases from Sjane (Sweden), Knabstrup and Garup (Denmark), Mecklenburg (Germany) and Mrowino (Poland), among other examples. From Late Neolithic there are the interesting chalk drums from Folkton (UK) displayed in the British Museum.

Still in the context of archaeological record, rock art is a privileged field for the research about cognitive expressions of prehistoric peoples, such as early acoustic abilities, constituting important case studies to understand early musical behaviours. For example, Neolithic paintings from Wed Mertoutek (Algeria) show two dancing girls, in a very dynamic scene, while another case from Tadrart Acacus (Libya) represents a dancing or a worship scene (Anati, 1994). Engravings from Saimaly-Tash (Kazakhstan), dating from the II millennium BC, show a ritual dancing in front of a man holding a sun (Coimbra, 2014: Fig.1).

All these examples seem to include an acoustic background, since it's difficult to conceive them independently from chanting or music, because, in a general way, nobody dances without any kind of sound.

[3] Rasps, as well as bullroarers, can have been made of wood, but obviously this raw material didn't "survive" in the archaeological record.

[4] We don't consider here the highly controversial "flute" from Divje Babe, in Slovenia, attributed to Neanderthals.

Fig.2 – Prehistoric pendant with dancers (original and replica). (Photo: F. Coimbra)

Fig.3 – Musical and dancing scene. (After Anati, 1994)

There are other examples of dancing in prehistoric art, being probably one of the earliest depicted on a Palaeolithic (?) pendant, kept in the archaeological museum of Argenton-sur-Creuse, in France (Fig.2)

Besides representing dancing scenes, rock art also depicts musical instruments. For example, in Bhimbetka (India) there are images of men playing drums, dating from the Bronze Age (Meshkeris, 1999).

At Wadi Harash (Negev, Israel), there's an interesting scene of two men playing an asymmetric lyre, four people dancing and a person playing a hand drum, dating from the II millennium BC (Fig.3).

At Tanum (Sweden) there are some cases of men playing a lure (Fig.4), represented in engravings dating from about 1800 to 1700 BC[5]. In Valcamonica, in the Italian Alps, there are several examples of horns depicted in carvings from Late Iron Age.

Besides rock art there are other examples of Prehistoric and Protohistoric iconography that contribute with important knowledge about the existence of musical instruments[6]. One of the earliest is a Neolithic wall painting from Çatal Hüyük (Turkey), where, in a dancing scene (?), a man is represented holding a hand drum (Mellart, 1967: Fig. 61-63).

In the III millennium BC there are the several cases of lyre that appear in Cycladic art, dating from 2500 to 2200 BC.[7]

Other types of later and smaller lyre can be seen on funerary stele from the Iberian Peninsula, dating from Late Bronze Age, on stele dating from the Iron Age Daunian Culture, in the South of Italy and on Attic pottery from Greece (Jimenez Pasalodos & Scardina, 2015).

Among complex societies from Iron Age, some "instruments" may have been associated with war of hunting, instead of producing music, such as the ceramic horns from Numancia and the carnyx, depicted for example in the Gundestrup Cauldron. This last artefact was a kind of bronze trumpet with an animal head, used by several cultures in

[5] Gerhard Milstreu, personal Information.

[6] We obviously don't aim to present an iconographic inventory about prehistoric musical instruments, which would be a task for an international team.

[7] In Egyptian art there are also many depictions of lyre, with a similar chronology with Cycladic Art.

Iron Age, probably to incite warriors to battle and intimidate the enemies.

Artefacts such as ceramic rattles that appear sometimes in European Iron Age burials may be related not to a musical behaviour but with funerary ritual, having therefore an apotropaic character. As Rustoiu & Berecki (2015: 259) stated "the sound produced by these instruments was meant to drive out the evil spirits, either in magical dances or in shamanic practices, serving also as an active protection against all sorts of perils mainly for the most vulnerable members of the communities, the women and the children".

We can summarize that archaeological evidence makes clear that in Prehistory and Protohistory there were at least three categories of what can be called early musical instruments: wind instruments (flute, lure, horn and carnyx); percussion instruments (lithophones, rasps and drums); string instruments (several types of lyre).

Fig.4 – Men playing a bronze lure, Tanum, Sweden and a replica of the instrument (Photo: Gerhard Milstreu)

Archaeoacoustics, Mind/Bodily Effects and Early Musical Behaviour

"Without music, it could be that we would never have become human"
(Ian Cross)

Archaeoacoustics, so far, has been concerning more with sound in the past than with early music. However, "an archaeological site itself may act like a musical instrument, generating sounds as a result of its own acoustical properties" (Till, 2014: 294). As a matter of fact, in the end of the 19th century, the writer Thomas Hardy reports that the wind at Stonehenge "playing upon the edifice, produced a booming tune, like the note of some gigantic one-stringed harp" (Hardy, quoted by Till, 2010: 5).

Also some Palaeolithic caves, with the sound of falling water drops, besides their reverberation, have their own soundscape (Till, 2014).

In a well known pioneering work, I. Reznikoff and M. Dauvois (1988) noticed that the location of Palaeolithic art in several caves from France was determined by the places where the sound resonance was louder[1]. This discovery seems to have been confirmed by similar research carried out more than two decades later in five caves from the north of Spain[2], where there was reported the existence of "statistically significant evidence that there are relationships between acoustical context and visual imagery in such painted caves" (Till, 2014: 298).

But, before any kind of interpretation in Archaeoacoustics it's necessary to keep in mind that "in all attempts to investigate the early past there is the risk that we first conceptualize, setting up a whole series of categories of our own construction" (Renfrew, 1994: 47). This line of thought is completed by Cross and Watson (2006: 115) arguing that "it is critical that, alongside the application of rigorous methods, acoustical investigations acknowledge the social contexts within which sound may have been experienced, and remain aware that it is easy to impose modern cultural understandings and experiences onto past societies".

We can add, as we wrote before, that "Archaeoacoustics without Archaeology is just noise" (Coimbra, 2016c: 11). In fact, in the research of Archaeoacoustics the observer is in the future regarding the object of his studies and, therefore, some subjective interpretations may happen and some mistakes may occur, with the danger of dropping to pseudoscience, if archaeological contexts are ignored.

Nevertheless, this discipline makes one be aware that most certainly early humans noticed and explored several kinds of sounds, before having a musical behaviour, which obviously did not emerge suddenly. For example, in Stonehenge the interior surface of some monoliths was intentionally polished into a concave shape in order to reflect the sound in a better way (Till, 2010). In the Hypogeum of Ħal Saflieni (Malta) the sounds produced inside the "Oracle Room" come out at great speed as it was felt by observers in the experiments carried out in 2014 (Coimbra, 2014; 2016b). The amazing reverberation produced inside this monument was impossible to be ignored by the participants in rituals carried out in its interior.

---

[1] These authors also consider caves as remarkable amplifiers of sounds, due to their inner resonance, not considered here as an echo but as the amplification of sound in intensity or in duration.

[2] The caves of Tito Bustillo (Asturias), La Garma, Las Chimeneas, La Pasiega and El Castillo, all in Cantabria.

It's not difficult to imagine that an early musical behaviour started probably by paying attention to certain kinds of sounds and then trying to repeat them or adding other sonorities. According to I. Cross (2001: 7) "a capacity for musicality (most likely, vocally expressed) must predate the construction of a musical artefact, most likely by a considerable period". However, regarding Prehistory it's necessary to distinguish "sound" from "music", which generally requires elements such as rhythm, melody and harmony that hardly could be present in an early musical behaviour, despite musical ability "can vary from context to context, within the same context, and from individual to individual" (Cross, 2001: 5)[3].

Being more related with sound than music, there are several rocks, stone blocks and slabs, from all over the world, that produce a metallic bell-like sound when struck with a pebble, being some of them "situated close to rock paintings and rock engravings" (Lund, 2009: 179). They "occur in a variety of geological contexts and are referred to as 'rock gongs', 'sounding stones', 'ringing rocks' or 'lithophones" (Kleinitz, 2004: 14), not constituting obviously a man made musical instrument, despite the possibility of having been used for the production of early music. In fact, C. Kleinitz, which studied several ringing rocks from Ishashi, in the Fourth Nile Cataract region, mentions that these sounds, in Late Prehistory, "may have been accompanied also by other forms of performance, such as singing, clapping and/or dancing, as is illustrated by spontaneous demonstrations of these percussion instruments by the contemporary inhabitants" (Kleinitz, 2004: 15). Indeed, ethnographic information from the African continent reports the use of this kind of rocks "in rites of passage, fertility or rainmaking rituals, as signaling devices or for entertainment (Idem, ibidem: 14).
Regarding the North of Europe, C. Lund (2009) listed thirty nine ringing stones in Sweden and three in Norway, having most of them cup-marks dating from Bronze Age, being therefore, in these examples, this rock art motif the result of sound production in the same rocks[4].

Recently, research in Archaeoacoustics has been paying attention to mind/bodily experiences caused by sound[5]. We must not forget that music is constituted by sounds and that sound is vibration. Interestingly, back in 1967, the anthropologist R. Needham (1967: 610) already mentioned that "there is no doubt that sound-waves have neural and organic effects on human beings, irrespective of the cultural formation of the latter. The reverberations produced by musical instruments thus have not only aesthetic but also bodily effects." For example, B. Watson (2009) argued that frenetic dancing (to the sound of instruments), which was probably used in several prehistoric rituals (Fig.5), may result in a certain trance, or maybe in altered states of consciousness, according to ethnographic parallels.

Interestingly, during excavations in Ħal Saflieni, some clay figurines were found, showing a relaxing attitude, laying on a couch (Coimbra, 2016b: Fig.2-3), as the so-called Sleeping Lady. Since this monument is essentially an underground cemetery, it's strange to imagine that people could go there just to have a sleep. Therefore, the Sleeping Lady and other people must have used the Maltese Hypogeum for other purposes. One possibility is that the Hypogeum was a kind of "a dream incubation temple" (Devereux, 2009: 226), where people could

[3] Even today, what can be considered as music by a person may be interpreted as noise by another.
[4] This fact makes crucial to revisit other cases of cup-marked stones, in order to check out if they produce a metallic sound when struck, as in the above mentioned examples.
[5] See detailed examples in Coimbra, 2016c.

go there to comfortably hearing reverberating sounds (music or chanting) and getting any kind of mind/body experience through them.

Fig. 5 - Neolithic paintings from Tadrart Acacus (Libya), where the character in the center seems to be hand-clapping (contemporary painting inspired on the original).

Already in the II millennium BC, sound vibration was known to be a liberator by the Vedanta tradition (Coimbra, 2016b). Sound has obviously a large influence on people, since prehistory, being "an integral part of human experience" (McBride, 2014: 1) and, as G. E. Williams (2012: 36) very well stated, "recent advances in neuroscience have helped expand our understanding of the cognitive aspects of many of the variables which affect the human experience; one such variable is sound". Therefore, sound turned into early music "deals with fundamental human emotional concerns" (Morley, 2014: 148).

Final Statements

The presence of "making sound artefacts" in Prehistory doesn't mean that they would have been created initially with intention of musical behaviour. Indeed, "while in Western societies music is commonly associated with leisure or culture, and considered something to be enjoyed, in many cultures music is an integral part of daily life, used to keep and transmit knowledge, to summon protection, to remember ancestors or to regulate social and economic activities" (Torres, n/dated).

Therefore in Prehistory many so called "musical instruments" must be considered to have several functions and not just one: Ritual, communication, music, hunting, other...

Archaeological evidence regarding an early musical behaviour of Mankind during Prehistory and Protohistory contributes for a better understanding of the discipline of Archaeoacoustics.

Simultaneously, Archaeoacoustics allows a better knowledge about the early musical behaviour of our ancestors and the origins of music.

Regarding the effects of some sounds and music in the human brain and sequent bodily effects, the future of the research in Archaeoacoustics has obviously to do with neuroscience. Both disciplines will much benefit from each other. Indeed, some statements of a recent Master thesis presented by a medical doctor are rather interesting:

"Music and its origins share the space with mankind, playing important individual functions. It brings pleasure, influences emotions controls anxiety, stimulates the nervous system, transcends, carries away to another place (...) Music goes beyond the individual, embracing the collective, contributing to the sharing of experiences, to the building of a social harmony" (Gomes, 2015).

Without the experience of sound and early music, human evolution probably would have been much slower.

BIBLIOGRAPHY

AIANO, L. (2006) – Pots and drums: an acoustic study of Neolithic pottery drums. *EuroRea, 3.* EXARC, Leiden: 31-42.

ANATI, E. (1994) – World Rock Art. The primordial language. *Studi Camuni, XII.* Centro Camuno di Studi Preistorici, Capo di Ponte: 160p.

BUISSON, D. (1990) – Les flûtes paléolithiques d'Isturitz (Pyrénées-Atlantiques). *Bulletin de la Société Préhistorique Française, 87, n°10-12.* Société Préhistorique Française, Paris: 420-433.

Coimbra F. A. (2014) – An Interdisciplinary Approach: the Contribution of Rock Art for Archaeoacoustic Studies. In ENEIX, L. (ed.). *Archaeoacoustics: The Archaeology of Sound.* OTS Foundation, Myakka City: 51-58.

COIMBRA, F.A. (2016a) – Archaeology, Rock Art, Archaeoacoustics and Neuroscience: What kind of relation? In ANATI, E. (ed.), *The intellectual and spiritual expressions of non-literate peoples.* Archaeopress, Oxford: 121-131.

COIMBRA, F.A. (2016b) – Neolithic art, Archaeoacoustics and Neuroscience. *in* ENEIX, L. (ed.). *Archaeoacoustics II, Proceedings of the 2015 Conference in Istanbul.* OTS Foundation, Myakka City: 15-24.

COIMBRA, F.A. (2016c) – Prologue. In ENEIX, L. (ed.). *Archaeoacoustics II, Proceedings of the 2015 Conference in Istanbul.* OTS Foundation, Myakka City: 9-11.

CONARD, N. J.; MALINA, M.; MÜNZEL, S. C. (2009) – New Flutes Document the Earliest Musical Tradition in South-Western Germany. *Nature, 460*: 737-740.
Doi:10.1038/nature08169

CROSS, I. (2001) – Music, mind and evolution. *Psychology of Music, 29 (1).* SAGE, Publishing, London: 95-102.

CROSS, I.; WATSON, A. (2006) – Acoustics and the Human Experience of Socially-organized Sound. In, *Archaeoacoustics.* SCARRE, C; LAWSON, G. (eds.), McDonald Institute for Archaeological Research, Cambridge: 107-116.

DAMS, L. (1984) – Preliminary findings at the 'organ' sanctuary in the cave of Nerja, Málaga, Spain. *Oxford Journal Archaeology, 3.* John Wiley & Sons Ltd, Chichester: 1-14.

DAMS, L. (1985) – Palaeolithic lithophones: descriptions and comparisons. *Oxford Journal Archaeology, 4.* John Wiley & Sons Ltd, Chichester: 31-46.

Devereux, P. (2009) – A Ceiling Painting in the Ħal -Saflieni Hypogeum as Acoustically-Related Imagery: A Preliminary Note. *Time and Mind, Volume 2, Issue 2.* Berg Publishers, Oxford: 225-231.

GOMES, A. F. A. (2015) – Música: correlatos bio-psico-sociais. Unpublished Master Thesis. University of Lisbon.

JIMÉNEZ PASALODOS, R.; SCARDINA, P. (2015) – The lyres on the Daunian Stelae. Towards a better understanding of chordophones in the Mediterranean iron Age". In MILITELLO, P.; ONIZ, H. (eds.), *Proceedings of the 15th Symposium on Mediterranean Archaeology.* Archaeopress, Oxford: 161-173.

KLEINITZ, C. (2004) – Rock art and 'rock gongs' in the Fourth Nile Cataract region: the Ishashi island rock art survey. *Sudan & Nubia, Bulletin No. 8.* The Sudan Archaeological Research Society, London: 12-18.

LUND, C. S. (2009) – Early Ringing Stones in Scandinavia – Finds and Traditions, Questions and Problems. In, JÄHNICHEN, G. (ed.). *Studia instrumentorum musicae popularis I.* Wissenschaft, Münster: 173-194.

LUND, C. S. (2012) – Sound Tools, Symbols or Something Quite Different? On Possible Percussion Instruments from Bronze-Age Sweden – Including Methodological Aspects of Music-Archaeological Research. *Studien zur Musikarchäologie VIII. Orient-Archäologie, 27.* Deutsches Archäologisches Institut, Rahden: 61-73.

McBRIDE, A. (2014) – The acoustics of archaeological architecture in the Near Eastern Neolithic. *World Archaeology, XXXX.* Routledge, London: 1-13.
http://dx.doi.org/10.1080/00438243.2014.909150
Accessed on 22-October-2017

MELLAART, J. (1967) – Çatal Hüyük. A Neolithic Town in Anatolia. McGraw-Hill Book Company, New York: 131-135.

Meshkeris, V. A. (1999) – Musical phenomena of convergence in Eurasian Rock Art. NEWS-95 International Rock Art Congress Proceedings. CeSMAP, Pinerolo.

MORLEY, I. (2006) – The Evolutionary Origins and Archaeology of Music. PhD Dissertation. Cambridge University.

MORLEY, I. (2014) – A multi-disciplinary approach to the origins of music: perspectives from anthropology, archaeology, cognition and behaviour. *Journal of Anthropological Sciences, 92.*
doi 10.4436/JASS.92008
Accessed on 15-November-2017

MÜNZEL, S.; SEEBERGER, F.; HEIN, W. (2002) – The Geissenklösterle flute: discovery, experiments, reconstruction. In, HICKMANN, H.; KILMER, A. D.; EICHMANN, R. (eds.), Studien zur Musikarchäologie. Leidorf: 107-118.

NEEDHAM, R. (1967) – Percussion and transition. *Man, New Series, Vol. 2, No. 4.* Royal Anthropological Institute of Great Britain and Ireland, London: 606-614.

RENFREW, C. (1994) – Towards a cognitive archaeology. In RENFREW, C.; ZUBROW, E. (eds). The ancient mind. Elements of cognitive archaeology. Cambridge University Press, Cambridge: 3-12.

Reznikoff, I.; Dauvois, M. (1988) – La dimension sonore des grottes ornées. Bulletin de la Société Préhistorique Française, tome 85, N. 8. Société Préhistorique Française, Paris: 238-246.

RUSTOIU, A.; BERECKI, S. (2015) – The Magic of Sounds. A ceramic rattle from the La Tene grave No. 1 at Fantanele – *Dâmbu Popii* and its functional and symbolic significance. In, *Representations, Signs and Symbols,* RIŞCUŢA, N. C.; FERENCZ, I. V.; BĂRBAT, O.T. (eds.). Editura Mega, Cluj: 259-274.

SAFA, E. (2015) – L'Archéologie du son: transmettre les connaissances et les savoir-faire de la France paléolithique à la France médiévale. *Polymatheia, 1.* Éditions Tourmaline, Vanves: 23-46.

TILL, R. (2010) – Songs of the stones: an investigation into the Acoustic Culture of Stonehenge. Journal of the International Association for the Study of Popular Music.
DOI: 10.5429/2079-3871(2010)v1i2.10en
Accessed on 21-November-2017.

TILL, R. (2014) – Sound archaeology: terminology, Palaeolithic cave art and the soundscape. World Archaeology, 46:3, 292-304, DOI: 10.1080/00438243.2014.909106

TORRES, J. de (n/dated) – Instruments of community: lyres, harps and society in ancient north-east Africa.
https://blog.britishmuseum.org/category/collection/african-rock-art/
Accessed on 23-October-2017.

Watson, B. (2009) – Universal visions: Neuroscience and recurrent characteristics of world palaeoart. PhD Dissertation. University of Melbourne: 47-142.

WILLIAMS, G. E. (2012) – Rock art and pre-Historic ritual behaviour: a landscape and acoustic approach. *Rock Art Research,* 29. Australian Rock Art Research Association, Melbourne: 35-46.

ARCHAEOACOUSTICS III participants visit the Caves of Lapas, Portugal
photo: OTSF

Prayer and Resonance in Paleolithic Painted Caves of Southern France: an Indigenous Science Approach

Apela Colorado, Ryan Hurd, Matthew Tucker
Worldwide Indigenous Science Network

ABSTRACT: In 2016, the Worldwide Indigenous Science Network (WISN) brought an inter-disciplinary team of scholars and indigenous practitioners to several Paleolithic painted caves in the south of France. We engaged in an evolving and collaborative methodology that incorporates archaeoacoustics, indigenous practices and narrative research to "hear" and "see" and "feel" what these Paleolithic spaces have to say to modern people. Specifically, impulse response (IR) technology was used to locate strong resonant frequencies, overtones and harmonics for each cave, creating 3D "maps" of the acoustic landscape. We engaged the caves with prayer, song, and classic Paleolithic instrumentation—conch, drum and bullroarer. Finally, we used narrative methods to document experiences as well as identify qualitative themes of researcher participation.

KEYWORDS: Indigenous Science, impulse response, Paleolithic caves, perceptual anomalies

Introduction

"Art is the way of talking to the deepest cave of the soul."
Credo Mutwa, Zulu High Sanusi, 2006

In June and July of 2016, a multidisciplinary group of researchers, scientists, artists, educators, indigenous cultural practitioners and healers from Worldwide Indigenous Science Network (WISN) embarked on a trip to the Paleolithic painted caves of southern France. The mission for the trip is best encapsulated by the aims of *indigenous science*: the place based, holistic and spiritual knowledge system and wisdom traditions of indigenous peoples. By perpetuating earth-based knowledge systems and linking it with western science to foster bio-culture resilience, the localized and isolated visions of an ecologically sustainable planet becomes a networked global reality (Colorado, 2014). Pascal Raux, Paleolithic expert and author (2004), served as guide and sacred site guardian, allowing us unparalleled access to the caves after hours. In Raux's words, "We go in the Dordogne caves to see the cave petroglyphs, rock paintings and engravings because we have lost the principal reason of life. We go into the caves to retrieve, to restore this thing" (2016).

The team included several Indigenous Cultural Practitioners (ICPs), who can be thought of as "master scientists" in native cosmologies (Han Chinese, Apache, Oneida, Maori, Australian Aboriginal, Occitan). Other participants have expertise in big cat preservation, acoustics, dream and consciousness research, phenomenology, and sacred geometry.

This presentation focuses on three levels of inquiry: indigenous practices, archeaoacoustic research, and first person narrative research. After brief site description for orientation, each methodology will be described in turn.

Paleolithic Caves of Southern France

Although we visited a half-dozen prehistoric cave sites, we are reporting on three caves in the area: Cougnac, Font-de-Gaume, and Pech Merle. This is not a formal archaeological site description, but rather an orientation to the most salient aspects of the caves' features for this presentation.

The Grottes de Cougnac are located in Payrignac in the valley of the Dordogne. The caverns of Cougnac have many fine concretions known as soda straws that also function as lithophones. Paleolithic art from Upper Paleolithic cultures (Gravettian [33-24000 B.P.], Solutrean [22-17000 B.P.] and Magdalenian [17-12000 B.P.]) in this location includes line drawings of many now extinct animals (such as the ibex and megalosaurus) as well as human figures and geometric images that are quite similar to those found in Pech Merle (Lawson 2012, 326).

Font-de-Gaume is another cavern in the Dordogne, located on the outskirts of Les Eyzies. Based on stylistic attributes, the polychromatic figures in this sinuous limestone cave span the Solutrean and Magdalenian cultures (Rosengren 2014, 66). Mammoth, bison, woolly rhinoceros, cave lions and horses are depicted as well as geometric designs.

Pech Merle is an extensive cavern system in the Occitania lot of France, located near the medieval village of Cabrerets. Paleolithic art in Pech Merle includes both the Gravettian and Magdalenian cultures (Lawson, 374). Previous collaborative research conducted with indigenous trackers from Namibia analyzed human footprints found in Pech Merle, determining the age, sex, gait and even mood of the individuals who walked across the mud 17,000 years ago (Pastoors et al 2016).

Spirited Embodiment

We entered the caves not with a hypothesis to test, or necessarily a single intention, but rather through an indigenous opening ceremony where we asked for permission to enter, and a sense of reverence and appreciation. In this way, our approach is first and foremost healing work that necessarily gives voice to the land and caves as having "spirit." In the West, land and earth itself are not special or "spirited," allowing for commodification and commercialization, and the problems of climate change, despeciation, and loss of complexity in living systems. Also, the destruction of native cultures and voices has gone hand in hand with industrialization and globalization (Mander and Tauli-Corpuz, 2006). These realities informed the prime motivations for entering the painted caves: not just to extract information about the past for its own sake, but to reconnect modern people to ancient lifeways for the sake of awakening and revitalizing sustainable ways of living. In this way, the approach of indigenous science is congruent with transformational western disciplines such as ecopsychology (Fisher 2013), applied anthropology (Kedish and Van 2005), action research (Bradbury 2015), and activist archaeology (Atalay et. al 2014).

Necessarily, the role of presence and subjective experience is highlighted in this discussion. Science as a practice often falls short of its principles when it comes to questions that demand embodied participation to collect information. The "taboo of subjectivity" has reached far into the human sciences and humanities as well, especially concerning questions about the nature of consciousness and perception (Wallace 2000). Ontologically, subjectivity is impossible to truly root out when it comes to topics that demand participation like archaeoacoustics: to understand how humans used, altered and were altered by the sonic landscapes they inhabited in the past (Till 2014). Full embodiment in the caves includes our corporal bodies, our voices, and our soulful presence, as well as interior and intersubjective elements of consciousness. This approach resonates with ecofeminism (Roszak 1995) as well as the anthropology of the extraordinary, in which researchers include their own personal experiences as relevant sources of data—including anomalous perceptions, altered states of consciousness, and dreams (Young and Goulet 1994).

The relevancy of the extraordinary in archaeoacoustics is that many acoustic landscapes were crafted by people who live or lived in cultures that value information from trance, dreams and other ecstatic states. In this context, recent experiential work that integrates the subjective experiences of archaeoacoustical researchers, including non-normative states of consciousness, into valid research protocols has begun to chip away at the invisible barrier between objects and subjects (Devereux 2001, Hurd 2011, Lindstrom and Zubrow 2014).

We will now present the methods used and some of the data collected in the painted caves, including first person experiences narratives, and discuss emergent themes.

Archaeoacoustic Research in Cougnac, Pech Merle and Font-de-Gaume

During site visits to the painted caves of Cougnac, Pech Merle and Font-de-Gaume, audio engineer Matt Tucker provided audio recording and generated impulse responses using easily transportable equipment that essentially makes for a mobile sound lab. The equipment used consisted of:

- 2 small diaphragm, matched pair, condenser microphones, set to an ORTF recording pattern
- 1 large diaphragm condenser microphone, set to an omnidirectional polar pattern
- A full range bluetooth wireless portable speaker
- An audio file of an impulse response, taking 15 seconds to go from 20Hz to 36,000Hz, a 15ms white noise and pink noise burst
- A Zoom f8 field recorder set to 192kHz and 24 bit wave files
- A pair of Sennheiser HD380 isolating headphones, so that any sound coming into the headphones would not "bleed" into the microphones at close proximity

The data recorded from the caves was analyzed with a spectrograph and a 3D sonograph to fully illustrate the harmonic content of the recordings. Note that the following acoustic results are not a formal, or comprehensive acoustic profile of all the caves' interior spaces, but a pilot sampling with attention to anomalous and curious acoustic features that warrant more research.

<u>Pech Merle</u>

In general, some interesting sonic manipulation can be detected near the paintings known as the "Panel of the Spotted Horses," from the Gravettian culture [carbon dated to 24,640 +/- 290 B.P. (Lawson, 379)]. Recording took place near a monolith known as "black-face" (Raux, personal communication). The sonograph shows an area between 40Hz and 600Hz with intense harmonic amplification. Not only do the standing waves produce strong resonances, but the harmonies compounding the complexity of the frequencies are particularly reactive. This creates a massively large, booming sound from a small sound source, a similar effect as noted by Rezinkoff as the "Bison effect" (2013, 35). If someone sings or chants, one only need to do so in a soft voice, as the cave enhances the song tremendously. Should the singer increase the performance, the sound vibrates through the listener, and surrounds them, engulfing their experience with sound.

Figure 1: Impulse response recording of Pech Merle, showing harmonic resonance from 40Hz-600Hz. Image by Matthew Tucker © 2017

Cougnac

One noticeable aspect of the Cave of Cougnac was the display of high frequencies within the back chamber, adjacent to the Principal Frieze [Gravettian, dated to between 25,000 and 19,000 radiocarbon years ago (Lawson, 330)]. From 1.3kHz to 10kHz, the first order and second order frequencies traded off. As the sound emanated from the speaker, the odd shape of the cave allowed certain frequencies to bounce off and phase out other frequencies. Like waves crashing against a shore, the harmonics rung out and dissipated in unique patterns, as the frequencies bounced back from the cave walls and collided with the other frequencies. When a listener is present in this location, and a high-pitched voice or instrument is played, such as a bone flute or whistle for a long enough duration, the effect is like a whirring or whooshing pressure on the eardrum. This affect, if experienced long enough, could most likely induce a trancelike state (Winkelman 2010, 133).

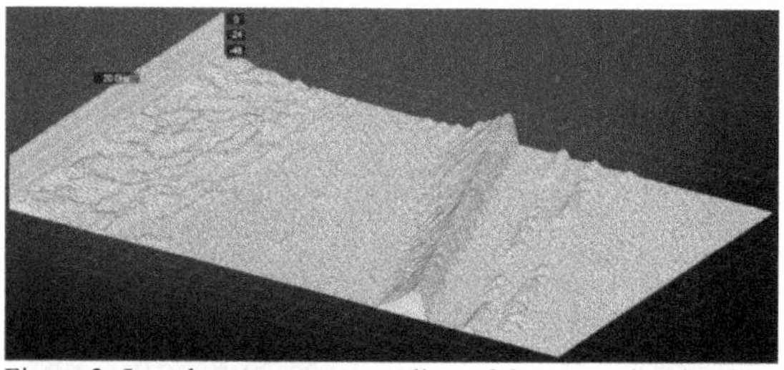

Figure 2: Impulse response recording of Cougnac showing harmonic resonances "trading off" from 1.3kHz-10kHz. Image by Matthew Tucker © 2017

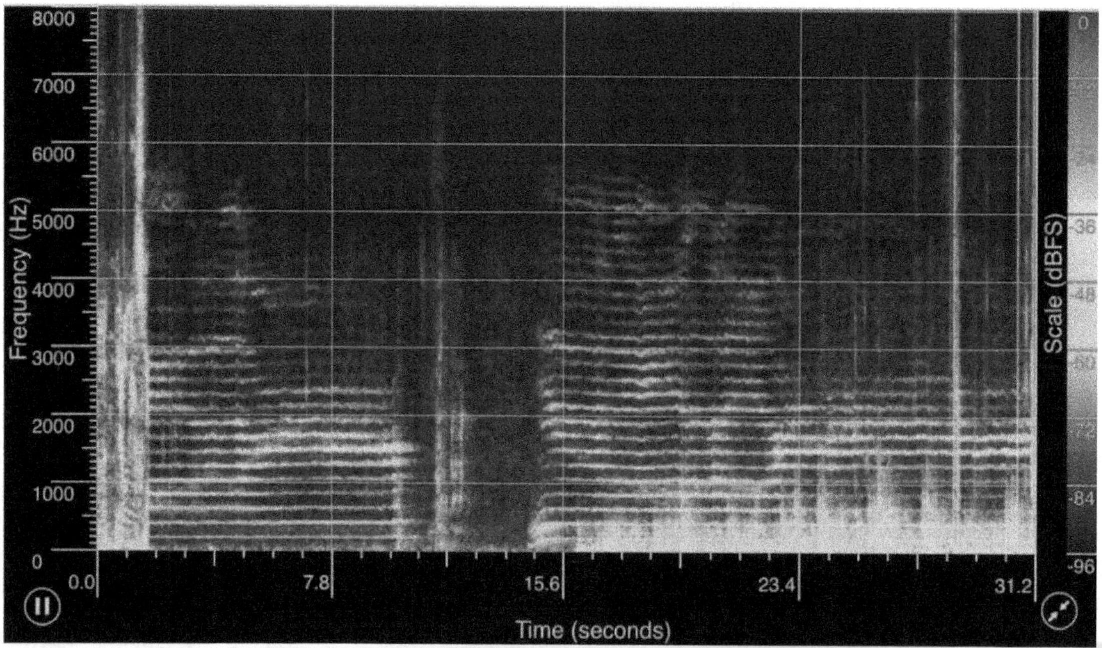

Figure 3: Spectrograph recording at Font de Gaume. While the root note does not change in pitch, the harmonies significantly waver in pitch. Image by Matthew Tucker © 2017

Font de Gaume

Font-de-Gaume had even more mysterious acoustic properties. We recorded near the "Grande Galerie des Fresques," the main panel of Paleolithic art that includes images of bison, mammoth and abstract forms, painted mostly in red and black. With its narrow walls and high ceiling, the sound ricochets off the walls nearest to the listener, travels up to the chambers top and comes crashing down to be met with new frequencies. This disorienting phasing of the sound is similar to Cougnac's back chamber. However, unlike Cougnac, more phasing of frequencies across the spectrum was evident, imparting harmonic shifts in pitch. The spectrograph illustrates that, while the root note does not change in pitch, the harmonies significantly waver in pitch.

Personal Experience Pech Merle:
Matt Tucker

The first principle of indigenous science for people re-entering holism is to put ourselves in the process. As such, as part of my work and research with WISN, I must participate in personal experiences.

What happened to me in these caves is unlike anywhere else in the world I have been. Particularly in Pech Merle, I experienced first-hand a situation that is still confounding me. In the process of recording, there are certain steps an audio engineer takes to maintain consistency and accuracy of the recorded material. The first thing I need to do is make sure everything is plugged in properly, before I start to receive signal from the microphones. Once everything is plugged in, I need to "arm" the tracks being recorded. This means that I need to engage, or turn on the channel to receive signal from the microphones. When I armed the tracks to my three microphones, I immediately heard the sound of a low frequency flutter, similar to the proximity effect of blowing into a microphone. Before I began, due to its small size, there wasn't much room for the sound to dissipate. The whole team left the cave. Any subtle noise, (shoes shifting, jackets rubbing, people moving, etc.) would ruin the recording. Also, the generators that powered the lights needed to be turned off, as they produced a significant low frequency hum. So, in that moment, with everyone gone and the lights shut off, again, I heard a sound directly similar to the sound of someone blowing into the microphones.

Alarmed, I disarmed the tracks, and went through my signal flow troubleshooting methods to make sure everything is connected properly, and that I am not causing damage to my microphones or mixer. After assessing that everything was indeed functioning and connected properly, I proceeded to rearm the tracks. Once more, I heard the fluttering, blowing sound. With a growing sense of unease, I disarmed the tracks again. Shaking with nervousness, I rechecked all my connections, made sure there wasn't any moisture on the microphones, and that no damage was occurring. I gave myself a few moments to calm down, rearmed the tracks for a third time, and waited a few seconds. I didn't hear the fluttering, so I hit record, captured the acoustics, and ran out.

As I stood outside the cave trying to make sense of what had happened, a wave of panic assailed me. What if my equipment was failing? We had more filming and recording to do; it was imperative I find the source of the sound. I couldn't work if my equipment was failing, I had to isolate the issue. Again, I went through the signal flow, trying to find the problem. After 10 or 15 minutes, I had a spark of realization: I had heard the fluttering in stereo. Each ear heard a different intensity and pattern in the headphones. As I powered up my field mixer, I saw that, in fact, my mixer channels were set to mono. I had heard a stereo signal through the microphones on a mono system—a technological impossibility.

Honoring Subjective Experience in Natural Places and Spaces

Once we decide that the subjective and the anomalous is worth studying, we can begin to track it and compare notes more formally. In this regard, archaeoacoustic researchers are ahead of the curve compared to their archaeological colleagues in other sub-disciplines and are often informed by intuition and soma during research. We all owe a debt to Paul Devereux for his call for "deep listening" at sacred sites, a method he calls *being and seeing* (1992). More formally, Ros Bandt has highlighted the importance of embodied presence in her definition of *sonic archaeology* as "methods for re-hearing the past founded on the interdisciplinary intuitive response through experiencing sound itself – through the direct connection of sound made in its fragmentary past – in situ" (2014, 87-88). In the spirit of continuing to develop more precise language and methodologies for tracking personal experiences in natural spaces, we would like to present a simple environmental observation method. This simple meditation supports the recognition of our own "ecological self," which we can loosely define as the identity we each have in relationship with the natural world. (Roszak, 1995). Cobbled together from ecopsychology, somatic psychology and indigenous practices, this field technique allows for clearer sensing, especially when sitting in locales known to affect shifts in consciousness (Hurd 2008).

Preparation

When entering a sacred site or landscape, or just prior, take a moment for yourself and check in with your body. Ask, "What is going on with me right now?" This practice, known as *focusing*, helps with minimizing inappropriate mental projections while clearing the path for more authentic projections that co-arise in communication with others with human and the more-than-human realm (Gendlin 1978; Fisher 2002). Sometimes, by inviting these feelings forth and naming them, a *felt shift* can sometimes occur that can be likened to a discharge of energy. Other times, a few moments of this somatic attention can produce a clearer frame of mind that is more "here and now."

Deep Listening and Participation

Next, shift your attention to include to world around you. This practice of nature awareness was popularized by naturalist Jon Young (1996) although its roots are in indigenous North America. In brief, this is a quiet sitting posture with alert eyes but soft gaze, also highlighting hearing, smell, and bodily reactions. Thoughts are noted but attention remains on the here and now. The focus on the body is crucial as a step away from "in your head" analyses that undercut direct, lived experience. The stage is set for the cave or resonate structure, as well as its acoustics and art, to be observed in its "natural habitat," which includes humans participating in the environment, and *belonging* there.

This practice is also phenomenological, drawn from continental philosophy and its roots in Western insight traditions. In the moment, the research attitude is to practice the *èpoché*: a pure moment of perception in which sense impressions self-arise and are clearly reflected in the mind (Depraz, Valera and Vermersch 2000). The practice is an invitation. Our own intentionality reaches forth and meets the world halfway. In phenomenology, this is the bridge to all perceptions. Similar to Eastern meditations, when you have biases or abstractions that arise and take you away from the moment and its perceptions, you acknowledge the bias and return to open reflection. No type of experience is privileged over another. This is the practice of intuition, a key concept in transcendental phenomenology which can be defined as being open to all that presents itself (Moustakas 1994, 32).

In the field, this work allows for subjective and intersubjective themes to arise amongst the researcher, the group, and the perceived non-human environment. For our purposes today, this practice can be seen to allow for uncanny and anomalous perceptions to arise and not be swatted away automatically by the filter of the rational/conceptual mind.

<u>Analysis of Personal Journals as Participant Data</u>

Phenomenology offers not only a meditative practice but a simple method for analyzing personal journals as well. So take good notes, as soon as possible, after visits to sacred sites while you are still steeped in the energy of the place. Your journal notes comprise a living document written while you were still in relationship with the place. In these moments, you are contextualized, quite literally a different person.

Afterwards, the journal entries can be typed and then explored like any narrative in ethnography. In this project, one of us (Ryan Hurd) chose to use a loosely modified phenomenological method. Like traditional textual description, phrases were reviewed impartially (horizontalization) and then condensed into themes.

As these were personal experiences, this practice is best considered a heuristic-phenomenological method, or auto-phenomenology (Moustakas 1990). For this reason, the bracketing of expectation is not intended to reject interior flashes of truth, but rather to consider them as valid insights into the data. Biases, in this sense, are important because they provide entry to the phenomena itself. The validity springs not from external facts but from internal sense of *authenticity* – and the ability to communicate that. Unlike formal phenomenology, this method cannot discover invariants of perception – or collective experiences. Rather, everything must be considered a variant (personal), until integrated with other similar data sets. In other words: we need more archaeological phenomenologists.

Personal Themes from Being-in-the-Caves:
Ryan Hurd

One of the major themes revealed from my personal notes is how entering the caves, and especially, entering them with an attitude or sacred respect in community, had a profound effect on my perceptions.

I noted the following perceptual anomalies in my visits in the caves:

Time dilation. When guessing how long we had been in the cave—often I was off by 30 minutes or more. I ordinarily take pride in guessing the passage of time correctly usually within a few minutes.

Strong emotional responses. Not only awe and profound feelings of "communitas" Communitas or 'communitas' but also at times bitter loneliness, anger, jealousy, and despair. These occurred in social settings in the caves as well as while in relationship with the cave art and the cave in general.

Sustained sensed presence. At one point, I felt I was standing next to someone while looking at an image on the wall—the feeling lasted until I looked over to say something to my companion, no one was there.

Auditory/Kinesthetic rushing. During singing/prayer, a specific note would create an auditory feedback that was a whoosh sound in my ear as if pressure in the room has suddenly dropped. It was similar to the hypnagogic gating on the edge of sleep paralysis

(Mavromatis 1987). Other times, chills from nape of neck down the shoulders and down the spine occurred in rapid floods. As I am susceptible to Autonomous Sensory Meridian Reaction (ASMR), I can state unequivocally that these effects were strongly heightened during acoustic events in the caves.

Enhanced perception. Upon leaving the cave, including sight and hearing. Everything felt "fresh" and "clean," as if the world was flooded with stimuli, most likely due to the sensory deprivation in the caves.

Discussion

A meta-theme that emerges from both Matt Tucker's and Ryan Hurd's personal experiences while in the painted caves is the *heightened presence of anomalous perceptions*. Given Tucker's experience of clearly hearing stereo sound through a mono-sound set up as well as Hurd's odd perceptions of sonic rushing that often precedes hypnagogic trance, we could more precisely describe the meta-theme of *aural multidimensionality that lends to the perception of uncanny presence*.

In general, these experiential clusters point towards a relaxation of self/other. Also known as "magical thinking," we suggest that consciousness in the caves allows for heighted access to anomalous experience including: *thoughts* that appear to cause action at a distance, *perceptions* that ordinary objects that are infused with power, and uncanny somatic *feelings* of being in the presence of something/someone larger than life.

Of course, many writers have noted that caves, with their sensory deprivation and unfamiliar settings, have a propensity to alter consciousness using classic techniques such as drumming or chanting (Winkelman 2000). Most specialists today support the idea that at least some of the work reflects shamanistic worldview and practices (see Clottes and Lewis-Williams 1998; Whitley 2009). The ancient images of animals, geometric designs, and human/animal hybrids reveal a worldview tied deeply to the animal world but also a bizarre cosmos that is more familiar to dreamers than scientists. Sometimes an image emerges from a crack in the stone, as if the cave is not just a symbol for this other world, but the actual portal.

Regardless, we feel that the caves are not just inert physical places that are valuable only because of they happened to archive the mysterious scrawlings of Paleolithic peoples, but a living container that still invites multidimensionality and more-than-human presence.

Re-Enchantment at Pech Merle

More important than these theoretical questions—which will never fully be resolved—an unspoken question revealed itself during our visits to the painted caves: are the caves still relevant to modern humanity? And if so, what can we learn? We approached the caves with a reverence for their antiquity, exquisite art, and the history of our ancestors held within, but, to our amazement, we discovered that the caves are still alive.

This theme became clear during the visit to Pech Merle, where our team was greeted by an enthusiastic young staff of tour guides. The co-mingling of our research group and the staff created an unanticipated occasion of performance, such as when we broke with norms and had a respectful silence and time for reflection before entering the cave. Deep in Pech Merle, Maori healer Timoti Bramley began a long series of singing chants and prayers to the caves and ancestors in front of the frieze "Panel of the Spotted Horses." Our large group fell into a deep and seemingly-shared reverent state that hushed all the previous giggles and bickering that accompanied the group in the beginning of the trip. The members of the staff had never spent as much time in the cave as this particular visit and our practices had left an impression. One young staff member said she felt she had been "initiated." Others excitedly said they planned to incorporate aspects of the WISN approach to entering the cave respectfully. They said they now felt they were more than workers, but were "guardians" to the sacred site. They had been re-enchanted with the world.

Given that hundreds of tourists stream through Pech Merle every week, this small shift in consciousness may affect change in the way that archaeology tourism is perceived, at least at this one site in the south of France. The shift of consciousness occurred in this group by inviting others to change their status from passive observers of "art" to embodied participants in respectful relationship with the living nature. Deep participation is necessary for this shift to occur.

Conclusion

From an Indigenous Science perspective, sharing methods of deep participation in order to re-enchant the world is of upmost importance. At a time of great global uncertainty, the Paleolithic painted caves can be seen as sacred sites that invoke multidimensionality to those who have gone to sleep due to ecological denial or despair. It's our hope that we can continue to work toward identifying, preserving—as well as participating with—sacred sites in order to inspire and re-enchant one another.

To listen to recordings we made in the Paleolithic caves, visit this webpage on the World Wide Web: https://wisn.org/cavesounds2016

REFERENCES

Atalay, Sonya, Clauss, Lee, McGuire, Randall, and Welch, John, eds. *Transforming Archaeology: Activist Practices and Prospects*.

Walnut Creek: Taylor and Francis, 2014. Accessed January 4, 2018. ProQuest Ebook Central.

Bandt, Ros. Sonic archaeologies: Towards a methodology for "rehearing" the past. In Linda Eneix (ed.) *Archaeoacoustics: The archaeology of sound. Publication of the 2014 conference in Malta*, 87-97. Myakka City: OTS Foundation.

Bradbury, Hilary. *The SAGE handbook of action research, 3rd edition.* Los Angeles: SAGE Reference, 2015.

Clottes, Jean and David Lewis Williams. *The shamans of prehistory: Trance and magic in the painted caves.* NY: Harry N. Abrams. 1998.

Colorado, Apela. Scientific Pluralism. *The Pari Dialogues: Essays in Indigenous Knowledge and Western Science*, Volume 2, edited by F. David Peat. Pari, Italy: Pari Publishing, 2014.

Colorado, Apela, and Nargiza Ryskulova. *Shamanism in Central Asian Snow Leopard Cultures.* Snow Leopard Trust, forthcoming.

Depraz, Natalie, Valera, Frances and P. Vermersch. The gesture of awareness: An account of its structural dynamics. In M. Velmans (Ed.) *Investigating phenomenal consciousness*, 121-136. Amsterdam: John Benjamin, 2000.

Devereux, Paul. *Earth memory: Sacred sites, doorways into earth's mysteries.* St. Paul: Llewellyn Publications, 1992.

Devereux, Paul. *Stone Age Soundtracks: The Acoustic Archaeology of Ancient Sites.* London: Vega, 2001.

Gendlin, Eugene. *Focusing.* New York: Everest House, 1978

Hurd, Ryan. Nature observation as a field technique: The relevancy of the ecological self. Presented at *the Annual Proceedings of the Society for the Anthropology of Consciousness* in New Haven, CT, March 21, 2008.

Hurd, Ryan. Integral archaeology: Process methodologies for exploring prehistoric rock art on Ometepe Island, Nicaragua. *Anthropology of Consciousness,* 22, no.1 (2011): 72-94.

Fisher, Andy. *Radical ecopsychology: Psychology in the service of life, 2nd edition.* Albany: SUNY Press, 2013.

Kedish, Satish and Van, Willigen. *Applied anthropology: Domains of application.* Westport: Praeger, 2005.

Lawson, Andrew. *Painted caves: Paleolithic rock art in Western Europe.* Oxford: Oxford University Press, 2012.

Lindstrom, Torill and Zubrow, Ezra. Fear and amazement. In Linda Eneix (ed.) *Archaeoacoustics: The archaeology of sound. Publication of the 2014 conference in Malta*, pp. 255-264. Myakka City: OTS Foundation.

Mander, Jerry and Tauli-Corpuz, Victora, Eds. *Paradigm wars: Indigenous peoples' resistance to globalization.* San Francisco. Sierra Club Books, 2006.

Mavromatis, Andreas. *Hypnagogia: The unique state of consciousness between wakefulness and sleep, 2nd edition.* London: Thyros Press, 2010.

Moustakas, Clark. *Heuristic research: Design, methodology and applications.* Thousand Oaks: Sage Publications, 1990.

Moustakas, Clark. *Phenomenological research methods.* Thousand Oaks: Sage Publications, 1994.

Pastoors, A., Tilman Lenssen-Erz, Bernd Breuckmann, Tsamkxao Ciqae, Ui Kxunta, Dirk Riepe-Zapp, Thui Thao (2017) Experience based reading of Pleistocene human footprints in Pech-Merle. *Quaternary International*, Volume 430, Part A, 12 February 2017, Pages 155-162.

Raux, Pascal, 2004. *Animisme et Art Premier, Nouvelle Lecture de l'Art Préhistorique.* Grenoble: ThoTs Editions.

Raux, Pascal. Interview by Apela Colorado. Dordogne, France, June 2016.

Reznikoff, Iegor. The acoustic dimension of Paleolithic painted caves. *Pleistocene Art of the World: Proceedings of the International Federation of Rock Art Congress*, 45-56. Tarascon, France: IFRAO, September 2010.

Rosengren, M. *Cave art, perception and knowledge.* New York: Palgrave Macmillan, 2014.

Roszak, Betty. The spirit of the goddess. In Theodore Roszak, Mary Gomes and Allen Kanner (Eds.) *Ecopsychology: Restoring the earth, healing the mind.* New York: Sierra Club Books, 1995.

Till, Rupert. Sound archaeology: An interdisciplinary perspective. In Linda Eneix (ed). *Archaeoacoustics: The archaeology of sound. Publication of the 2014 conference in Malta*, pp. 23-32. Myakka City: OTS Foundation.

Whitley, David. *Cave paintings and the human spirit: The origins of creativity and belief.* Amherst: Prometheus Books, 2009.

Winkelman, Michael. *Shamanism: A biopsychosocial paradigm of consciousness and healing, 2nd edition.* Westport: Praeger.

Young, David, and Jean-Guy Goulet. (Eds.) *Being changed by cross-cultural encounters: The anthropology of extraordinary experience*. Petersborough, Ontario: Broadview Press, 1994.

Young, Jon. *Seeing Through Native Eyes: Understanding the Language of Nature*. Shelton, WA: Owl Media, 1996.

Definitive Results of Archaeoacoustic Analysis at Alatri Acropolis, Italy

Paolo Debertolis, Daniele Gullà, Natalia Tarabella, Lorenzo Marcuccetti

PAOLO DEBERTOLIS, Department of Medical Sciences, University of Trieste, Italy, President of Super Brain Research Group institution [1]
DANIELE GULLA', forensic researcher, Vice-President Super Brain Research Group institution, Bologna, Italy(*)
NATALIA TARABELLA, architect, bursar of Super Brain Research Group institution, Florence, Italy (*)
LORENZO MARCUCCETTI, historian, member of Super Brain Research Group institution, Florence, Italy (*)

ABSTRACT: Our research group has used archaeoacoustic methodology over the last seven years. Archaeoacoustics has enabled us to explain some of the enigmas of ancient archaeological sites that were not previously possible to explain through other methods. Our hypothesis suggests the exposure to certain non-audible vibrations could have a significant effect on the psyche of those who came for prayer or rituals, facilitating access into a mystical state. Archaeoacoustic methodology was utilised to study Alatri acropolis in Italy. The cathedral of Alatri is located at the highest point in the town of Alatri which sits on top of a Cyclopean temple. We sought to understand why this temple was built in this location. Using a number of protocols we discovered very strong and significant low vibrations (seismic waves) continuously emitted originating from below the ground. Even though ancient people did not possess the same equipment we have today, they would have been aware of the conditions required to achieve such a mystical state, perhaps by simply sensing they were closer to God in a given location. The seismic waves appear to arise from a geological fault located on the side of the hill where the town has stood since ancient times. The presence of such seismic frequencies increases the effect of any rituals by enhancing the psyche of the participants due to their influence on human brain waves. This suggests the builders of this temple had some knowledge of their effect and offers a possible explanation as to why the temple was built on this particular hill and not on any of the surrounding hills.

KEYWORDS: archaeoacoustics, Alatri, polygonal walls, low frequency sound, infrasound

Introduction

Natural sound phenomena were utilised by some civilizations to enhance their rites and ceremonies, indeed some ancient structures were modelled in a certain way to enable the vibrations produced there to directly influence the mind towards a particular state of consciousness. [6,7,8,10,11,12,13,14,15,16,17,18,19,20,21,22,23,24,25,26,27,34].

In earlier research, SBRG demonstrated a relationship between mechanical or natural

[1] Note. Super Brain Research Group (SBRG) is an international and interdisciplinary team of researchers, researching the archaeoacoustic properties of ancient sites and temples throughout Europe and Asia (www.sbresearchgoup.eu).

vibrations originating from resonance phenomenon at some temples and brain activity. Natural low vibrations with an absence of high pressure can have a positive influence on human health and some people can perceive very low-frequency sounds as a sensation rather than a sound. Infrasound may also cause feelings of awe or fear in humans and given it is not consciously perceived, it may give the appearance that strange or supernatural events are taking place [33]. It is therefore possible to hypothesize that wherever there is a concentration of natural low vibrations, ancient populations considered these sites to be "sacred" [7]. Through archaeoacoustic analysis, it is possible to demonstrate there was some knowledge of acoustic phenomena in the past, which could for example have been used to enhance ancient rituals [6,7,8,10,11,12,13,14,15,16,17,18,19,20,21,22,23,24,25,26,27,34].

The historical town of Alatri was analysed from this point of view and with preliminary results published in 2015[15].

In this article the definitive results of this four year research project are presented along with three different research methodologies.

The Ancient Town of Alatri, Italy

Alatri is a small town located in the Frosinone province district, Lazio (Fig. 1). The city of Alatri was built around a small hill, surrounded by megalithic walls and whose remains are mostly visible today. The acropolis is at the heart of the historic centre, on the top of the hill (Fig. 2). The internal perimeter of the acropolis is defined by huge limestone walls forming a polygonal shape which measures 500 meters in length and a maximum of 15.4 meters in height. In addition, the external walls draw a perimeter of 2km made of different stone blocks layers that can be up to 3 meters in height [32].

What is striking is about the huge polygonal "*Cyclopean walls*" is the way the megalithic blocks fit together without the use of mortar, it is impossible to insert a sheet of paper between the joints. The exact date the acropolis was constructed is unknown: some believe the walls were built by the Romans or Latins, others that its origin dates to a pre-roman period, there is however no consensus [29].

Fig. 1 – The location of Alatri, Italy.

Fig. 2 – Alatri town as viewed from a neighbouring hill, the cathedral is clearly visible.

Don Giuseppe Capone was the first to explore the hypothesis that Alatri and its acropolis were built following geometrical and astronomical lines. In the 1980s, the local monk who served the church-seminary of Alatri for many years, studied further the archaeo-astronomical origins of the town. His observations were confirmed by Antony F. Aveni, professor of Astronomy and Anthropology at Colgate University (USA) in an article in collaboration with Don Capone in 1985 [1].

The fundamental concept inspiring the monk had its roots in the religious culture of the ancient Indo-Europeans. They used the sun as a point of reference by which they fixed certain points on the horizon. This principle appears to have been followed by the builders of the Cyclopean walls. Capone identified a "privileged point" which is located behind the northern wall, at the center of the acropolis. Using this point, he was able to establish that the North-East corner of the acropolis outlines a direction which indicates the rising sun at the summer solstice and that the eastern and western sides are parallel, oriented North-South. In addition, the monk noticed that several doors surrounding the city were located in strategic points and in mutual relation to one another as well as with the map of the city [29].

The Cyclopean walls extend along a North-South axis centered on a solitary rock outcrop, a sacred space reserved for religious purpose located a little higher than the acropolis, a veritable "altar of sacrifice". This rock outcrop (the so-called "*navel*" of the town) appears to be in the center of the ancient city, and is now located outside of the cathedral which was built over the remains of the ancient Temple of the Sun [1,2,3,4]. Perhaps this outcrop was used as a reference point during the construction of the city? This appears to be a plausible as several astronomical observations support this: in the early morning during the summer solstice, it is possible to see the sun rise from the rocky outcrop facing north; the corner of the Cyclopean walls casts a shadow that points directly to the outcrop. Additionally, the east side of the wall turns out to be a fundamental unit of linear measure used in the construction of the acropolis; all the gates and archways in the external wall (with one exception) lay equidistant from the outcrop at three times the length of the eastern inner wall (Fig. 3).

Fig. 3 – The map of Alatri with its astronomical and geometrical indications (Capone, 1982).

Furthermore, the builders of the acropolis divided it into quadrants centered on the outcrop. At the summer solstice sunset, the sun illuminates the city gate of the superior northwest quadrant and its shadow heads directly toward this outcrop. Another gate, "*Porta Minore*" located in the right inferior quadrant attracts further attention: it is a trilithon doorway also called the "*Gate of three phallus*" for a symbol carved over it. On the morning of the equinox in March and December the sun lights the steps that led to the door drawing a perfect rectangular shape outside the gate [1,2,3,4] (Fig. 4).

Fig. 4 – The Minor Gate at the equinox, with the sun lighting up the stairs, drawing a perfect rectangle on the stones in front of it. Over the gate are three phallus carved on the stone architrave. (Courtesy of Ornello Tofani, Italy).

Based on our archaeoacoustic experience throughout Europe [6,7,8,9,10,11,12,13,14,15,16,17,18,19,20,21,22,23,24,25,26,27,34] we decided to analyze this site starting from the rocky outcrop. Our purpose was to discover why the acropolis was built on that particular hill and not on another of the surrounding hills. Maybe this hill met certain criteria of sanctity for the architect, if so why did he consecrate the acropolis with such characteristics?

Materials and Methods

This research carried out over a four year period, used three different investigative methodologies: *full audio spectrum recording*, a *geologic device* (GeoBox SR04S3 Datasheet) to confirm the audio recording results in the infrasound range and *TRV technology* to analyse the effect of vibrations on the brain. Our research group has used this multi investigatory method previously at various archaeological sites. This experiment was carried out over four visits in different seasons between 2013 and 2016.

The audio recording was performed following the SBRG Standard for archaeoacoustics – SBSA [9]. In this case the equipment consisted of a high range dynamic recorder, extended in the ultrasound and infrasound field with a sampling frequency rate of 192 kHz (Tascam DR-680); Condenser microphones with a wide dynamic range and flat response at different frequencies (Sennheiser MKH 3020, frequency response of 10Hz to 50,000 Hz) with shielded cables

(XLR Mogami Gold Edition) and gold plated connectors.

The microphones were placed at a number of different locations around the acropolis and in the surrounding area to detect any vibrations present. Low frequencies or infrasounds (seismic waves) are non-directional and because they are not easily absorbed by the soil, they travel long distances.

Fig. 5 – The recording operation in the Acropolis's "navel"

A second technology in the form a digital sensor GeoBox SR04S3 Datasheet (Fig. 6) from Italian firm SARA, used accelerometers to acquire audible frequencies specifically in the seismic range. This devise is usually used for seismological and geophysical surveys such as the Horizontal/Vertical Spectral Ratio - HVSR. The SR04 GeoBox is designed especially for recording ambient seismic noise, but it can also record earthquakes and artificial vibrations.

TRV technology (Variable Resonance Imaging Camera), is a methodology we have used for five years. The following is a simple explanation as to how it's applied in our research and how it might be applied within the wider archaeoacoustic field. First off there is a direct correlation between emotional and functional states of the human body with precise parameters of controlled motion reflection. Until fairly recently, quantitative parameters and efficient information of the movement of the human body were not established.

Bernstein and Mira Lopez (psychodiagnostic miokinetics) [30] studied the micro-mobility of the human body and found that it represented a sophisticated mathematical problem. For example, it has been shown that the vertical balance of the human head is controlled by the vestibular system, described as a reflex function, but the balance of the head is also considered an extension of locomotor activity (micro-mobility of the head) controlled by this system. The analysis of this and other types of mobility reflexes provide a lot of

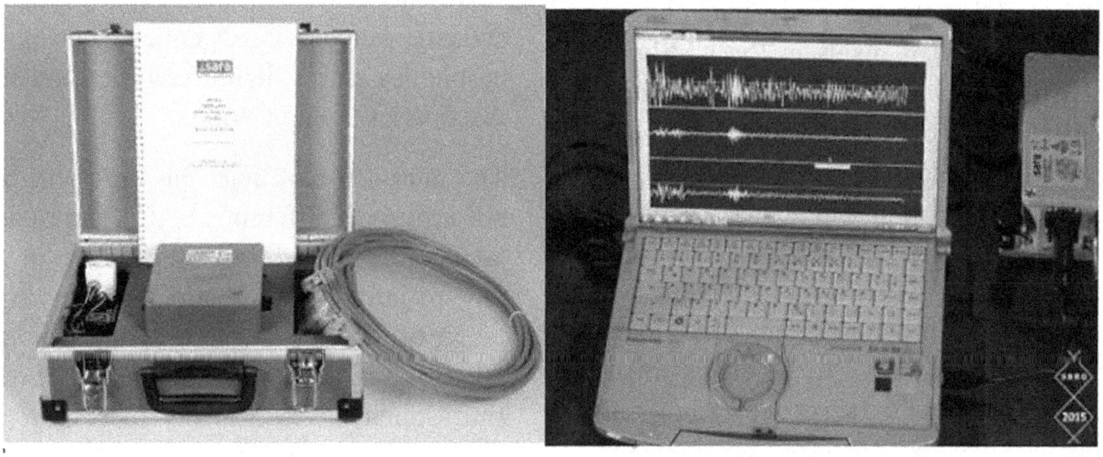

Fig. 6 – Left: GeoBox SR04S3. Right: the GeoBox connected to the computer to investigate vibrations in the acropolis.

information on the state of consciousness of the subject. From a physical point of view, the mechanical oscillations of the head, is a vibrational process, the parameters of which provide a quantitative correlation between energy and mobility of the object. Such information can be obtained using video analysis Variable Resonance Imaging Camera (TRV) technology, which provides quantitative information of the periodic movements of any part of the imaged object.

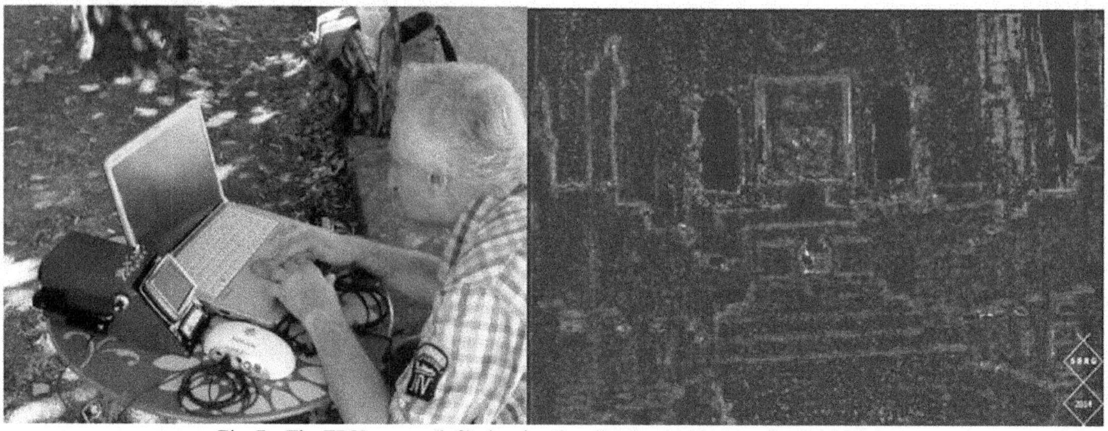

Fig. 7 – The TRV camera (left) showing the vibration inside the cathedral (right).

In order to make visible the dispersion effects of air vibrations, a TRV camera (Variable Resonance Imaging Camera, known in Italy as Merlin camera or Defend X system in Japan) was used, along with Vibraimage Pro 8.3 software. TRV video analysis technology, provides quantitative information of the periodic movements of any part of the imaged object. The software can process small vibratory changes in air movement between different video frames, by highlighting the movement and change of chromaticity of pixels in the UV band (Fig. 7). To achieve this, low resolution video is used (640x480) to prevent overloading the computers processing capabilities. Frames are collected and reassembled (standard deviation or STD) to generate a composite video. This technique published in the scientific literature is capable of identifying low vibrations in the environment and has been used to detect deep vibrations caused by the movement of underground water that affects overlying areas [13, 15, 19, 20, 24, 26].

The TRV system was used to visually confirm the subsonic vibrations detected on two previous occasions, as well as to evaluate the emotional state of the volunteers on top of Alatri.

Results

In June 2015 our final archaeoacoustical survey at Alatri took place, this confirmed acoustic data previously collected using microphones and a digital recorder and new discoveries by the Geobox.

The ultra-sensitive microphones were connected to two different TEAC Tascam digital recorders. The results showed a loud volume of infrasound vibrations present in the range of 8-9Hz that affected the whole of Alatri hill in the form of non-audible sound (infrasound). Additionally there is a frequency of around 32Hz in the audible field band, recorded during each of our four visits. The volume was between -38db and -42db (Fig. 9) and most likely represents a harmonic of the main vibration. Before recording, a spectrum analyzer (Spectran NF-3010 from the German factory Aaronia AG) was used to search for any electromagnetic phenomena which could have influenced the results (Fig. 8).

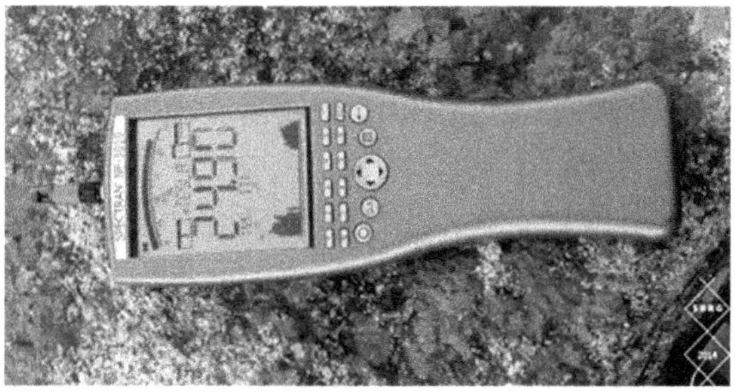

Fig. 8 – Spectran NF-3010 from the German factory Aaronia AG.

It is important to discuss these measurements. A distinction between using decibels to measure the sound pressure level (SPL) and using them to the measure signal level needs to be made. SPL is a measurement of the air pressure caused by sound, which results in physical force against the eardrum (or the diaphragm of a microphone). In the acoustic environment this translates to sound volume. Measurements of this type are usually expressed as dB SPL (decibels of Sound Pressure Level). A rock concert typically has a 100db of sound or an average conversation 60-70db of SPL. Low volumes in this field include whispered speech at 20-30 db or a residential area ambient noise at 40-50db. The threshold of human hearing is 0db. When dealing with signal level as opposed to SPL, decibels are used differently; in this case 0 dB is the highest signal level achievable without distortion. All signal levels below distortion are then represented as negative numbers. A volume fader may be labelled with a “0”, or a “U” (for unity), part way up to mark the point at which that fader is neither boosting nor attenuating the signal. So a level of -38-42db as found in Alatri acropolis must be considered very high, like a strong ambient noise and can be heard very easily, especially at night in the absence of general environmental noise.

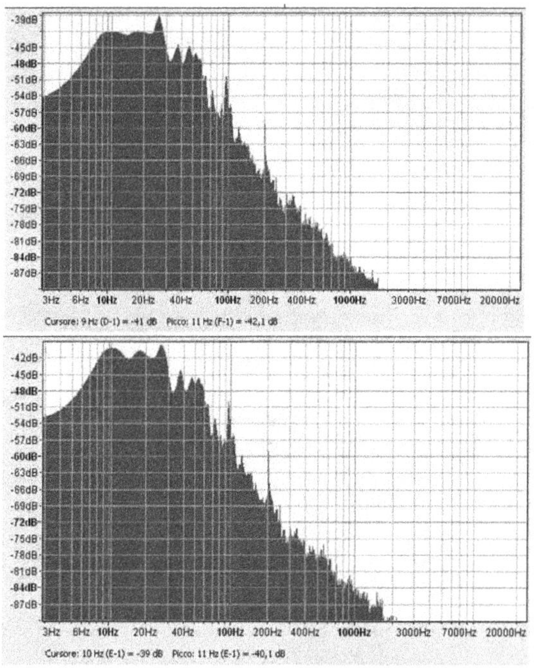

Fig. 9 – Plots of the underground vibrations taken at different times inside Alatri Cathedral. The foundations of this church are dedicated to St. Paul and made from the walls of the ancient pagan temple. In these graphs the frequency peaks at around 8-11Hz and 30-32Hz.

Because of its sensitivity in geologic explorations (using accelerometers instead of microphones), the Geobox confirmed all the frequency peaks found using audio recordings (8Hz and 32Hz), due to the sensitivity of the accelerometers a frequency of 4Hz (non-audible sound) was also discovered. The Geobox made another important discovery, a big cavity below the Acropolis. Such a cavity would function as a giant resonance box amplifying any underground vibrations (Fig. 10).

With the TRV Camera, the infrasound peaks of 4Hz and 8Hz as initially measured by microphones and GeoBox were confirmed. These vibrations appear to affect the entire acropolis and the cathedral in particular. It also appears that there are simultaneous peak frequencies below 4Hz capable of generating vibratory fields in the air (Fig. 11 and 13).

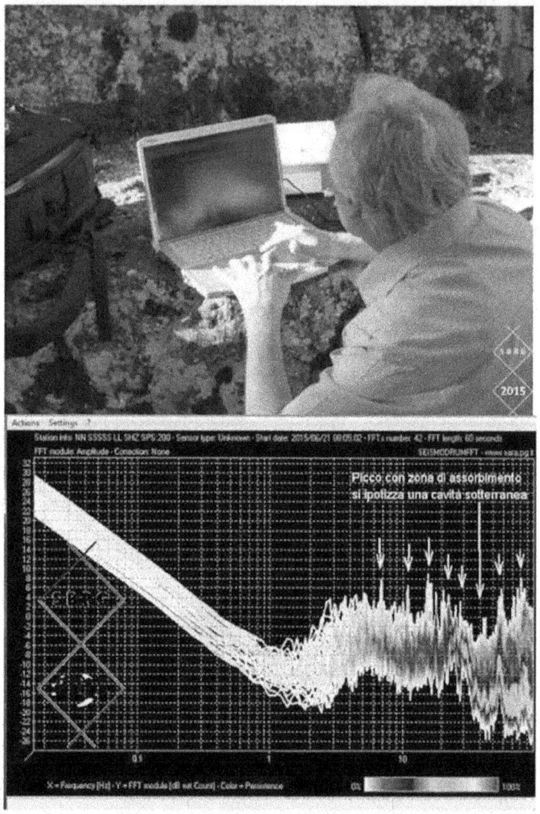

Fig. 10 – The Geobox measurement taken at the "acropolis's navel" (above) and the peaks of absorbing vibration made by the cavity below the hill (below).

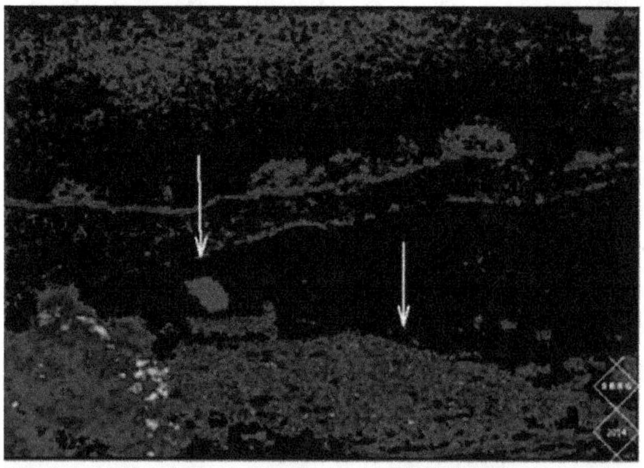

Fig. 11 – View of Alatri as taken by the TRV Camera. The arrows indicate the resonance field of 2 Hz above the acropolis and in particular over the Cathedral.

The vibrations were made visible in the UV spectrum using the TRV Camera, which was located at Acropolis's "navel" area, which is deeply immersed in the outcrop rock hill on which the pagan temple was built. This rock works as a transducer for the underground vibrations transmitting them perfectly inside the Cathedral. It is interesting to note these pulsing vibrations are not transmitted to the blocks inside the original pagan temple, which now forms the church basement. The image below shows the blocks of the ancient temple appearing black as taken by the TRV Camera. This indicates a lack of vibration, which is contrary to the underlying rock and due to the fact that the blocks are fitted together, without cement. This actually dampens the underground vibrations and confirms the seismic character of this structure that is still standing after thousands of years and many earthquakes (Fig. 12).

Fig. 12 – Image of the original pagan temple blocks (without cement) positioned to the north of the Acropolis (left). The same image captured by TRV Camera. These blocks appear black in colour because they do not transmit the vibrations coming from below the ground (right).

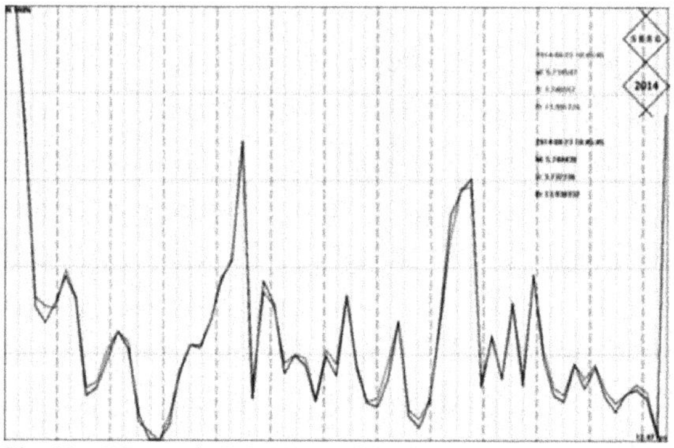

Fig. 13 – The graph (0.1-12Hz) of the low frequency (subsonic) peaks detectable in the navel area of the Acropolis by TRV camera.

Using the TRV camera on the volunteers, these subsonic vibrations do not appear to create any problem for them. Rather, as with other sacred places, it can be assumed their existence is precisely the reason why the Acropolis and the temple were built in that location as opposed to the neighbouring hills. Anyone who undertakes prayer or meditation inside the church has the potential to feel the effect of these subsonic vibrations, which could ultimately lead to altered states of consciousness, or mystical experiences (usually only experienced after many years of training as with Buddhist monks).

To test this claim, we proceeded to test the depth of meditation that can be reached in a short timeframe with a small number of volunteers (six people), seated on the Acropolis's "navel" (inside the church), and the outcrop rock at the center of the acropolis. Part of this rock protrudes from the walls of the church basement, and part is located deep within the hillside, as such it superbly transmits the subsonic underground vibrations.

The depth of relaxation reached was examined using the TRV camera, whereby the subtle body vibrations were measured, specifically the vestibular organ (inner ear), which regulates balance and spatial awareness. If the subject in question is stressed some imperceptible body vibrations increase and can be immediately detected by the TRV camcorder. However, if the subject is relaxed its vibrations diminish to become imperceptible even to the equipment. This last state is reached only in a state of deep meditation or in the state of vigil prayer.

After a few minutes of meditation, the subject begins to vibrate at a slower frequency (less than one Hz) indicating a deep relaxed state (Fig.4). At this point, it becomes difficult for the TRV camera to distinguish them from the rock. This is known as *entrainment*, a phenomenon in which two or more independent rhythmic processes synchronize with each other [17].

Fig. 14 – After a few minutes, the volunteer's vibrational frequency is so low, it becomes indistinguishable from the rock when viewed through the TRV Camera.

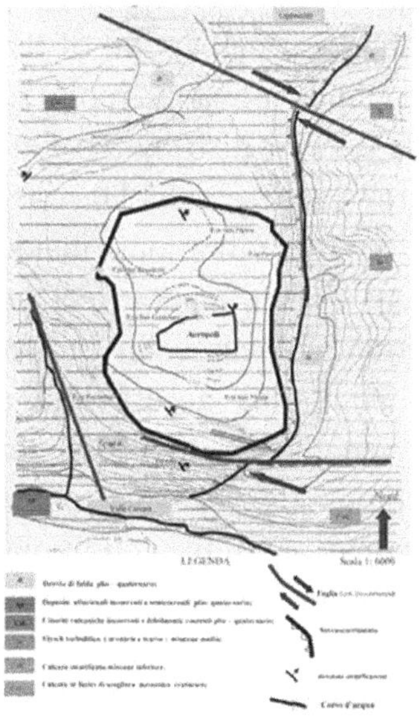

Fig. 15 – The geological map of the area around Alatri shows significant movement of the geological faults (in red) which may explain the vibrations measured at the Acropolis (in brown). The arrows placed close to the faults show the tectonic movement direction likely producing the vibrations recorded at the top of the acropolis (map made by geologist Dr. Rocco Torre).

The sound seems to be concentrated solely in the navel of the acropolis and fades into nothing when moving away from it. It is likely the vibrations are coming from the geological faults (Fig. 15) that are very close to Alatri, with their vibrations channelled via some unknown mechanism to the top of the hill.

Conclusion

When looked at alongside research on the effect of acoustics on the human body, archaeoacoustics can be viewed as a method of analysing ancient sites from another point of view. Indeed, its study presents a chance to recover "ancient knowledge" that affects the emotional sphere of human consciousness, as well as to broaden our understanding of the ancient world.

Our methodology using three different approaches has confirmed the preliminary results which were published in 2015 [15]. This focussed on the findings of the TRV camera, which showed some frequencies present at the acropolis capable of entraining the volunteer subjects into an altered state of mind in this case a positive sense of relaxation. The objective findings observed by our devices represent something already detached from subjective perception of the people considered and variously reported with subjective sensations also by the protagonists of our previous research. The continued exposure to the vibrations inside Alatri acropolis has a significant effect on the psyche of those who came for prayer and meditation, facilitating access into a mystical state. Even though they did not have the same equipment we have today, ancient people were aware of the conditions required to achieve such a state, perhaps by simply sensing that in that place they were closer to God. We have also to consider the important discovery of a cavity below the top of the hill working as a musical box for amplifying the natural vibration coming from below.

Archaeaoacoustics is an interesting method of analysing ancient sites to re-discover a forgotten technique that affects the emotional sphere of human consciousness. Ultimately the devices used confirm that a "*mystical*" state can be reached after a few minutes by those who are subjected to the vibration phenomenon inside the acropolis.

ACKNOWLEDGEMENTS

SBRG are grateful to Department of Medical Sciences at the University of Trieste (Italy) for supporting this research and in particular to the Director, Professor Roberto Di Lenarda. We would like to thank Don Antonio Castagnacci for his availability to grant us the opportunity to make recordings, including inside the Basilica - Cathedral of St. Paul (Cathedral of Alatri) and for his help, also to his collaborator Mr. Sisto Macciocca. We also thank in particular the independent researcher Ornello "Paolo" Tofani for the documentation and the extraordinary support provided for our research for more than four year. Without him, none of this would have been possible. A sincere thank you to our scientific assistant, Nina Earl, for her support in the drawing up of this paper.

REFERENCES

[1] Aveni A, Capone G : "Possible Astronomical Reference in the Urbanistic Design of Ancient Alatri, Lazio, Italy", Archeoastronomy, Vol VIII, n 1-4, 1985: pp. 12-15.

[2] Boezi G.: "In Caput Angeli" (1st edition), Associazione Culturale "Le Mura", 2013, Alatri.

[3] Capone G.: L'orientazione solstiziale dell'antica città di Alatri. Rivista di Archeologia. Suppl. n°9 (Archeologia e Astronomia), 1989: pp. 60-65.

[4] Capone G.: "La Progenie Hetea, annotazioni mitico-storiche su Alatri antica", Tofani Litografo in Alatri, 1982, Alatri.

[5] M. Clayton, R. Sager, U. Will: "In time with the music: the concept of entrainment and its significance for ethnomusicology", European Meetings in Ethnomusicology, 11, 2005: pp. 3–142.

[6] P. Debertolis, H.A. Savolainen: "The phenomenon of resonance in the Labyrinth of Ravne (Bosnia-Herzegovina). Resultsof testing" Proceedings of ARSA Conference (Advanced Research in Scientific Areas), Bratislava (Slovakia), December, 3 – 7, 2012: pp. 1133-36.

[7] P. Debertolis, N. Bisconti: "Archaeoacoustics in ancient sites" Proceedings of the "1st International Virtual Conference on Advanced Scientific Results" (SCIECONF 2013), Zilina (Slovakia) June, 10 - 14, 2013: pp. 306-310.

[8] P. Debertolis, N. Bisconti: "Archaeoacoustics analysis and ceremonial customs in an ancient hypogeum", Sociology Study, Vol.3 no.10, October 2013: pp. 803-814.

[9] P. Debertolis, S. Mizdrak, H. Savolainen: "The Research for an Archaeoacoustics Standard", Proceedings of 2nd ARSA Conference (Advanced Research in Scientific Areas), Bratislava (Slovakia), December, 3 – 7, 2013: pp. 305-310.

[10] P. Debertolis, N. Bisconti: "Archaeoacoustics analysis of an ancient hypogeum in Italy", Proceedings of Conference "Archaeacoustics: The Archaeology of Sound", Malta, February 19 - 22, 2014: pp. 131-139.

[11] P. Debertolis, G. Tirelli, F. Monti: "Systems of acoustic resonance in ancient sites and related brain activity". Proceedings of Conference "Archaeoacoustics: The Archaeology of Sound", Malta, February 19 – 22, 2014: pp. 59-65.

[12] P. Debertolis, A. Tentov, D. Nicolić, G. Marianović, H. Savolainen, N. Earl: "Archaeoacoustic analysis of the ancient site of Kanda (Macedonia)". Proceedings of 3rd ARSA Conference (Advanced Research in Scientific Areas), Zilina (Slovakia), December, 1 – 5, 2014: pp. 237-251.

[13] P. Debertolis, D. Gullà, Richeldi F.: "Archaeoacoustic analysis of an ancient hypogeum using new TRV camera (Variable Resonance Camera) technology", Proceedings of the "2nd International Virtual Conference on Advanced Scientific Results" (SCIECONF 2014), Žilina (Slovakia) June, 9 - 13, 2014: pp. 323-329.

[14] P. Debertolis, F. Coimbra, L. Eneix: "Archaeoacoustic Analysis of the Ħal Saflieni Hypogeum in Malta", Journal of Anthropology and Archaeology, Vol. 3 (1), 2015: pp. 59-79.

[15] P. Debertolis, D. Gullà: "Archaeoacoustic analysis of the ancient town of Alatri in Italy", British Journal of Interdisciplinary Science, September, Vol. 2, (3), 2015: pp. 1-29.

[16] P. Debertolis, M. Zivić: "Archaeoacoustic analysis of Cybele's temple, Imperial Roman Palace of Felix Romuliana, Serbia", Journal of Anthropology and Archaeology, Vol. 3 (2), 2015: pp. 1-19.

[17] P. Debertolis, D. Nicolić, G. Marianović, H. Savolainen, N. Earl, N. Ristevski: "Archaeoacoustic analysis of Kanda Hill in Macedonia. Study of the peculiar EM phenomena and audio frequency vibrations", Proceedings of 4th ARSA Conference (Advanced Research in Scientific Areas), Zilina (Slovakia), November 9 – 13, 2015: pp. 169-177.

[18] P. Debertolis, D. Gullà, "Anthropological analysis of human body emissions using new photographic technologies", Proceedings in Scientific Conference "The 3rd International Virtual Conference on Advanced Scientific Results (SCIECONF-2015)", Slovakia, Žilina, May 25-29, 2015; Volume 3, Issue 1: pp. 162-168.

[19] P. Debertolis, L. Eneix, D. Gullà: " Preliminary Archaeoacoustic Analysis of a Temple in the Ancient Site of Sogmatar in South-East Turkey", Proceedings of Conference "Archaeacoustics II: Second International Multi-Disciplinary Conference and workshop on the Archaeology of Sound", Istanbul Technical University, Taşkışla Building, Istanbul, Turkey, 30, 31 October and 1 November, 2015: pp. 137-148.

[20] P. Debertolis, D. Gullà: "New Technologies of Analysis in Archaeoacoustics", Proceedings of Conference 'Archaeoacoustics II: The Archaeology of Sound', Istanbul (Turkey), Oct 30-31 Nov 1, 2016, pp. 33-50.

[21] P. Debertolis, D. Gullà: "Preliminary Archaeoacoustic Analysis of a Temple in the Ancient Site of Sogmatar in South-East Turkey. Proceedings of Conference 'Archaeoacoustics II: The Archaeology of Sound', Istanbul (Turkey), Oct 30-31 Nov 1, 2016, pp. 137-148.

[22] P. Debertolis, N. Earl, M. Zivic: "Archaeoacoustic Analysis of Tarxien Temples in Malta", Journal of Anthropology and Archaeology, Vol. 4 (1), June 2016, pp. 7-27.

[23] P. Debertolis, D. Gullà: "Healing aspects identified by archaeoacoustic techniques in Slovenia", Proceedings of the '3rd International Virtual Conference on Advanced Scientific Results' (SCIECONF 2016), Žilina (Slovakia), June 6-10, 2016, pp. 147-155.

[24] P. Debertolis, D. Gullà, F. Piovesana: "Archaeo-acoustic research in the ancient castle of Gropparello in Italy", Proceedings in the Congress "The 5th Virtual International Conference on Advanced Research in Scientific Areas" (ARSA-2016) Slovakia, November 9 - 11, 2016: pp. 98-104.

[25] P. Debertolis, N. Earl, N. Tarabella: "Archaeoacoustic analysis of Xaghra Hypogeum, Gozo, Malta", Journal of Anthropology and Archaeology, vol.1 no. 5, June 30, 2017: pp. 1-15.

[26] P. Debertolis, D. Gullà: "Archaeoacoustic Exploration of Montebello Castle (Rimini, Italy)", Art Human Open Acc J 1(1): 00003, DOI: 10.15406/ahoaj.2017.01.00003.

[27] P. Debertolis, D. Gullà, H. Savolainen: "Archaeo-acoustic Analysis in Enclosure D at Göbekli Tepe in South Anatolia, Turkey", Proceedings in Scientific Conference "5th HASSACC 2017 - Human And Social Sciences at the Common Conference", Slovakia, Žilina, September 25-29, 2017: pp. 107-114.

[28] C. De Cara: "Gli Hethei-Pelasgi", Vol. 3, Chap. XVII, Tipografia Academia dei Lincei, Roma, 1902: 289-302.

[29] G. Magli: "Polygonal walls and the astronomical alignments of the Acropolis of Alatri, Italy; a preliminary investigation", Nexus Netw J, 2006: p. 1.

[30] E. Mira y Lopez: "Psicodiagnostico Miokinetico (P.M.K)", Editorial Paidos, 1979, Buenos Aires.

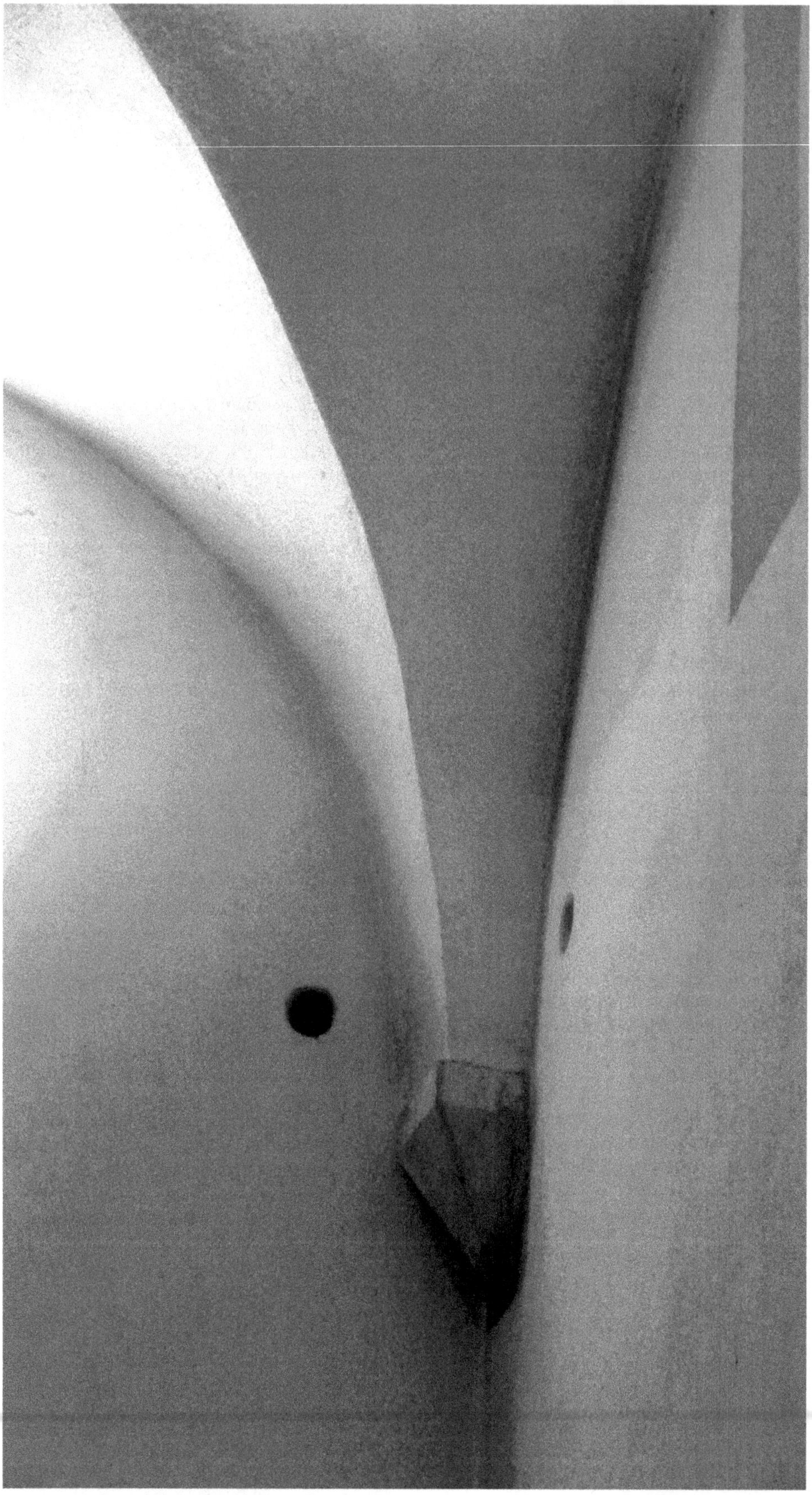

Ceramic jars placed in the walls of the ancient synagogue of Tomar to boost acoustics

Acoustics of Medieval Architectural Heritage – Research Methodology

Zorana Đorđević

ZORANA ĐORĐEVIĆ, Ph.D. Institute for Multidisciplinary Research, University of Belgrade, Republic of Serbia, zoranadjordjevic.arch@gmail.com

Introduction

The question of methodology in archaeoacoustic research was one of the main issues discussed on the conference Archaeoacoustics III in Tomar. Due to the numerous approaches of the published archaeoacoustic studies, there is a need to systemize them and acknowledge their advantages and disadvantages. This paper questions what kind of sound and in what way is related to sacral architecture. In other words, it considers the possible methodological approaches, suggesting their systematisation based on the type of sound they employed. As an example that illustrates those various approaches, this paper presents the case of architectural heritage of medieval Serbia[1].

The Orthodox Christianity was very closely related to the Serbian medieval Monarchy (XII – XV century). During the rule of Nemanjić dynasty (from XII to end of XIV century) and their successors (until the fall under Ottoman rule in 1459), monumental sacral architecture had the most significant place. That close relation of church and state is reflected in the fact that each monarch founded at least one monastery. Also, monarchs were often proclaimed a saint after death. The monastery complexes were located strategically, representing both the state power and spiritual commitment of the monarchs and the people. Consequently, they were the most sophisticated expression of building culture that incorporated the cosmological symbolism. Therefore, as a fine connection between material and spiritual worlds, the sound of these spaces is very important for the religious practice that has lasted in an intact form from medieval times until today.

Research Methods

When we research acoustics of various historical places, we use different types of sound – sounds of nature, human made sound, and sound signals. Dividing the

[1] More elaborative research results on acoustics of sacral architecture in medieval Serbia were presented on the exhibition *Archaeoacoustics: Sacral Architecture of Medieval Serbia*, authored by Zorana Đorđević (history of architecture), Dragan Novković (acoustics) and Marija Dragišić (ethnology – anthropology). The exhibition took place in the Museum of Science and Technology in Belgrade (November 8th – December 10th, 2017).

research methods upon the type of sound used in the archaeoacoustic research of architectural heritage, there are:

(1) Live sound - based on the still living, more or less intact soundscapes and religious practice;

(2) Harmonic sound – based on the inaudible relation of acoustic principles and the geometry and proportions of architecture;

(3) Impulse response – based on the in situ measurements or virtual model examination in specialized software.

Although archaeoacoustic research is multidisciplinary, each of these approaches has a primer scientific discipline it originates from: musicology for live sound, history of architectural proportions for harmonic sound and acoustics for impulse response. Nevertheless, none of these approaches excludes the others. In order to show they are complementary and point out their good and bad sides, these three approaches are further considered on the example of Orthodox monasteries and monastic churches of medieval Serbia.

Live Sound

Live sound can be considered as a soundscape of the monastery and as a divine service in the church. Both of them are preserved more or less intact from medieval times.

Soundscapes of Medieval Serbian Monasteries

Soundscape consists of a specific acoustic environment created by a combination of sounds of nature and human activities. In the case of medieval Serbian monasteries, the soundscape was very important as a recognizable overture to a sacred place. It provides a strong impression when entering a monastery, thus preparing a believer for the prayer.

Written medieval documents testify that the soundscape was considered as an important issue during the selection of the site for the future monastery. Hagiographers noted that the site had to have specific acoustic properties that induce the sense of mystery, holiness and strong connection with nature in a believer. The monks ought "to be taught by the rustling of trees and chirping of birds" (Теодосије 2009, 15). Hence, the monastery's surroundings had to highlight the depth of relationship between the man and the nature, excite the believers and summon the mightiness of God and orthodoxy, "so that everyone who sees it [the church] may think it's heaven on earth" (Теодосије 2009, 82). Those places were described as "desolate hunting grounds for beasts" (Мирковић 1939), which glorify the peace and God-given beauties of nature.

Although the medieval soundscape could not survive intact through centuries, some recognizable parts remain. These are human made sounds produced by various percussion instruments and used for calling into the temple for prayer. Klepalo (wooden board), klepalce (small klepalo) and bilo (metal plates) were incited with a hammer, and bells with a pendulum. Klepalo is exclusively used on certain occasions, for example, as a sign of mourning in the period from Good Friday to Easter, and after that on Easter bells would be heard, which "ring in joy" because of the Christ's resurrection (Павићевић-Поповић 2007, 155). The first bells were brought to Slavic countries in the 12th century and became inseparable from prayers in church. It is assumed that bell originated from hollowed out bilo which sounded better that way, and that the newly created bells were first knock on with hammer, and then later they got pendulums attached on the inside. Bells were manufactured at the Adriatic coast (early 14th century), and since the 15th century on the territory of Serbia as well (Станојевић 1933, 86–90). After Serbia fell under Ottoman rule, the bells went silent for the next five centuries when instead of bells people used klepalo and bilo. Thus, these instruments became recognizable characteristics of monastic soundscapes. Over time the bells also became a symbol of freedom, and after centuries of silence they rang for the first time when the First Serbian Uprising broke in 1804 (Аксић 2014, 154–56).

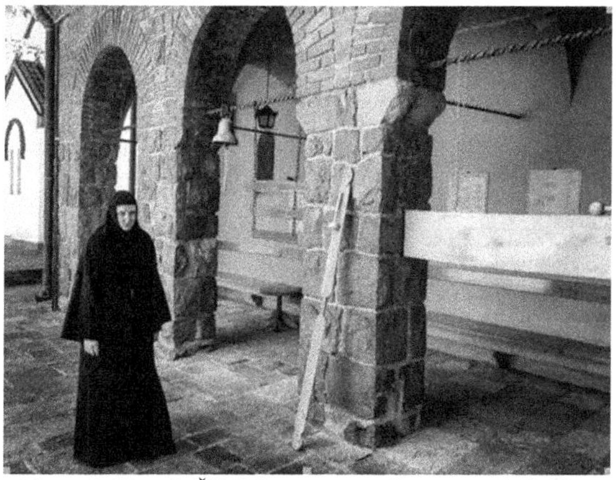

Figure 1: Sister in Žiča Monastery standing next to bell, bilo, klepalce (leaning on the column) and klepalo

Divine Service and Byzantine Chanting

The other aspect of researching live sound in architectural heritage considers the performance in the form it had during the examined historical period, in order to find out what characteristics of acoustic field is required to support the performance. Therefore, we need to understand the performance from the point of performers and from the point of listeners (spectators, believers) as well.

In the case of orthodox sacral heritage of medieval Serbia, the most important act is divine service, as an expression of the inner religion, gratitude and wish for unification with God. The main acts in the divine service are prayer - a pious conversation with God - and chant - a prayer expressed by song, bringing the heart closer to God. Medieval hagiographers highlighted the importance of sound and chanting that opens the link with heavens, also describing the mysterious acoustic events that announced the sainthoods (see Свети Сава 2009).

Orthodox church music is vocal and monophone. In Byzantine chant, used in the temples of medieval Serbia, melody is not separated from the lyrics and each spoken word is given quality of a song. Theoretic basis of this music relies on Pythagorean teaching on music intervals and antique division of scales. The principle of chanting in eight modes was present in Byzantium since the earliest times. It was adopted and developed as the system of "octoechos" in which each of the voices "rules" the divine service for a week and expresses particular feelings, in accordance with liturgical cycle, by chanting (Попмихајлов 2004). Based on recognizable musical formulas, melodic motion is mostly gradual, while the rhythm flows and depends on the lyrics. The melody has "glorious simplicity, naturalness, importance and moderation appropriate for the divine service" (Мирковић 1965, 279). There is also ison - a drone voice when one or several chanters accompany the melodic chant by singing the same note, and than shortly change it to another note following the main melody. In one of the Monastery Chilandar manuscripts, ison was described as "comprehensive sign, beginning, middle and end of all the signs in musical art… Without it, the voices cannot be performed. It is also called the voiceless since it has no voice and therefore, as a voice it does not count (Перковић-Радак 2007b, 306)."

The continuity of Byzantine chant is considered to be uninterrupted in Serbian tradition (Перковић-Радак 2007a, 34). It has been transmitted for centuries from one generation to the next and every community of chanters had a certain level of musical individuality (Пено 2014, 141–45). Simplified melodies and songs were first transferred orally, and then via manuscripts written in Byzantine neume notation. The first Serbian composers - Nikola Srbin, Stefan Srbin and Isaija Srbin - appeared during the period of Moravska Serbia (15th century) and completely adopted the Byzantine chant mastery.

After the fall of the entire Balkan Peninsula under Ottoman rule in the middle of the 15th century, construction of churches, as well as all the aspects of church art, including chanting, slowly began to die out. This is how the otherwise very complex notation became simplified (Попмихајлов 2004) and did not follow the development of analytical notation that took place in 19th century. Therefore, today is important to acknowledge the authentic Byzantine chanting when we research acoustics of medieval Serbian churches. Although it is primarily in the field of ethnomusicology, it affects our understanding of acoustic requirements of the time.

According to Byzantine tradition, chanting was performed from the choir, so two choirs were places in the apses of the Serbian medieval monastic churches. In order to achieve dynamics and variability of divine service, chanting of psalms is combined with Bible reading and common prayers – ectenes (Мирковић 1965, 212). Psalms, and especially Holy Scripture, are read at the divine service in order to convey the teaching; hence, it is important to read them clearly, intelligibly and piously. After that comes the sermon, as an extension of the Christ's teaching (Мирковић 1965, 294). From the ambo – the elevated stand opposite of the Royal Doors in the naos (or today, occasionally from the southern choir) – gospels are read and ectenes are spoken. During the divine service the believers in the Orthodox church stand facing the altar, occasionally bowing, kneeling and lifting their arms and eyes to the sky (Мирковић 1965, 304). It is of the outmost importance for them to hear well and understand the chant and the reading.

Figure 2: Northern chanter apse in the church of Ravanica Monastery

Harmonic Sound

The research of harmonic sound is closely related to the ancient understanding of the world as a whole (essentially expressed through analogies), deeply inbuilt in the foundation of European culture. It relates the musical intervals (and their ratios of great numbers) to the architectural proportions or certain building elements.

The Orthodox Church adopted the idea that beauty is immanent to the world; therefore, the temple building, as a testament of heavens, reflects the image and the beauty of the Universe with both its outer and inner appearance. This also refers to sound. In that way, the temple influences the soul of the believer so that it can participate in the church prayer more fully, easily receive the evangelist teaching and implement it in life (Патријарх Павле 1995, 17–19). Here we will consider this relation of sound and proportional principles used in church building in medieval Serbia, as well as the inbuilt acoustic vessels that follow mathematical and musical laws.

Architectural Proportions

Medieval temples express the cosmic order, so their design has always been the most sublime task. Ever since the period of the catacombs, sacral space has had three main parts: altar, naos and narthex. The Orthodox temple is entered from the West, through narthex, dedicated to catechumens and penitents, leading to naos – the inner chamber intended for the believers who participate in the divine service. On the northern and the southern wall of the naos there are places for chanters. Above the naos there is one or five domes, representing the heavens, Universe, perfect sphere. On the east side is the most sacred part of the temple - the altar, spiritual sun. It is separated from the rest of the naos by altar barrier, which had no icons until 14th century, but rather a horizontal beam on piers.

These main architectural elements were composed using geometric proportion, as a main mathematical tool for designing. Proportioning scheme was "usually obtained using a calliper, length measure as unit, while in space it was measured using geometric progression" (Vasiljević 1956). However, today we do not have reliable knowledge on the proportional diagrams, building mastery and acoustic intentions, because throughout the Middle Ages those were strictly guarded secrets of the guild. Therefore, it is important to reconstruct proportional diagrams, based on the known proportioning principles and knowing that medieval builders were searching for balanced shapes and harmony of parts with the whole.

Quadrature and triangulation were used as geometrical patterns for proportioning of plans and cross-sections. Quadrature represent the function of collapsed diagonals, producing $\sqrt{2}$, $\sqrt{3}$ and $\sqrt{5}$ lengths. The principle of triangulation is more complex, providing dynamic symmetry. It starts from equilateral triangle and a rectangle drawn around it Θ. This rectangle is obtained when rectangle ABCD, whose sides are in ratio $\sqrt{3}$:1, is divided by reciprocal diagonal BE (perpendicular to the diagonal AC) so the rectangle AFED represents rectangle Θ in which the equilateral triangle AFG can be drawn. Perpendicular diagonals AE and EP express reciprocity, i.e. they demonstrate that between two similar figures there is a gradational relationship, which gives the inner dynamics. The ratio between the sides of two concatenated rectangles Θ (AF:FP=DE:EO) is 4:3 (Vasiljević 1956). This ratio 4:3 also corresponds to the interval of the perfect fourth in Pythagorean theory of music. Marcus Vitruvius Polio (1st century BC), mentions the connection between geometry and music as "squares and triangles of the fourth and the fifth" (Vitruvije 2000, 17). Analogy of the proportion of harmonic ratios with musical theory of intervals is one of the first systems in the theory of proportions (Petrović 1972,

21). Its application in the Serbian medieval architecture points to "the remains of the basis of traditional knowledge... which passed on "from father to son" as if in a guild..." (Milosavljević 2013, 44). In certain churches, it was discovered that the heights of proportional images of squares and triangles, which can be detected during reconstruction of proportioning of Serbian medieval temples, have the same ratio as the great musical intervals. Although inaudible, in this way a direct connection was established between the sound and the space, built in the proportions of the temple (Ђорђевић 2016).

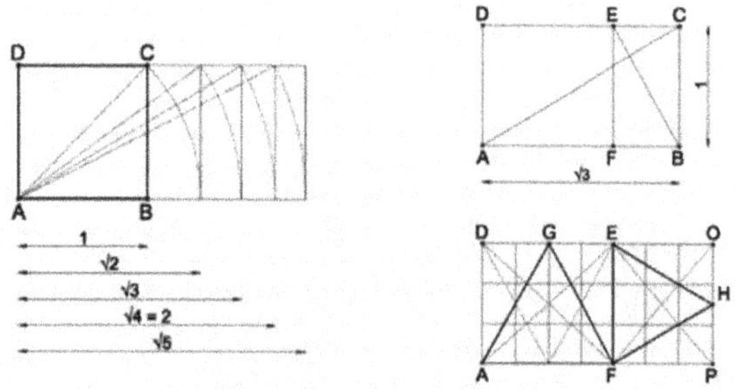

Figure 3: Quadrature - principle of proportioning based on collapsing diagonals (left) and Triangulation – principle of proportioning based on equilateral triangle and rectangle √3:1 (right)

Acoustic Vessels

The oldest surviving description of acoustic vessels - installed in accordance with the musical intervals and mathematical principles - gave Vitruvius in Ten Books on Architecture. Without elaborative description on how they work, he stated that these vessels increased the clarity of sound. Ceramic acoustic vessels were installed in sacral buildings throughout the Middle Ages. However, their real contribution to the acoustics of interior spaces was questioned even then. The archaeoacoustic research showed a relation between the type and the number of discovered vessels in a certain space with the volume, height of the church space, construction style, position and manner of placing the vessels in the walls, etc. This implies that medieval builders had empirical knowledge and they understood that the effect of acoustic vessels is cumulative (Valière, Palazzo-Bertholon, and Polack 2013, 75–79).

In other words, the number of vessels increased with the volume of church, remaining in the frequency range in which the reverberation time is generally the longest. Resonant frequencies are in the range between 100 and 500 Hz, thus they contribute to the shortening of reverberation time at those frequencies that are most often excited by human voice. A tendency to adapt the effect of the vessels by adjusting their dimensions to the absorption of low frequencies, for which the absorption of church furniture, such as iconostasis, elements made of wood and textile, is less efficient, was also noted. Two types of vessels were often used in churches whose resonant frequencies have the ratio of the fourth and the fifth musical intervals, which indicates the connection with the musical theory and tuning of vessels mentioned by Vitruvius (Valière and Palazzo-Bertholon 2014).

Other than the efforts to determine the effect of the installed vessels on correction of acoustic properties in the interior of the churches, there are also some theories on their symbolic application. This standpoint is based on the antique philosophy and belief that musical tones elate human soul (Arns and Crawford 1995). In that sense, acoustic function of the vessels can be considered to be the builders' secondary intention, while the primary idea was to express the comprehensive representation of the celestial sphere, where each echea represents a single planet and manifests its characteristic noise. Seen in this way, the system of acoustic vessels represents a part of a more general theory of spheres, which developed in the Antiquity and achieved special importance during the Middle Ages (Poulle 2000).

Acoustic vessels inbuilt in the sacral architecture were found in fifteen Serbian medieval churches. They belong to the pot, pitcher or jug type of ceramic vessels. In most cases, this was the secondary use of these vessels for acoustic purposes. The vessels built in each of the churches are similar in shape, and only appear in one or two sizes. Their height is between 20 and 50 cm (Đorđević, Penezić, and Dimitrijević 2017), which matches the findings from Western Europe (Valière and Palazzo-Bertholon 2014). The vessels were installed in line or a triangle, in pendentives, tholobate, in the zone of blind arches, apse, etc. They were installed in horizontal position in two ways: with the mouth opening facing the inside of the church and the bottom facing the wall, and with the bottom with a hole facing the inside of the church and the mouth facing the wall (Đorđević, Penezić, and Dimitrijević 2017).

Impulse Response

Objective analysis in acoustics is based on the measuring of impulse response of the space. By placing a microphone at a previously defined point in the space, it is possible to record the impulse response of the room to a purposefully generated excitation on all frequencies of interest - the hearing range of human ear. Using a mathematical tool (convolution), an information is obtained from the recorded material that completely describes the acoustic system of the analysed space in the point of measurement. Impulse response can be measured in situ or derived from the acoustical model generated in a specialized software. The analysis of impulse response provides various acoustical parameters – quantifiers that enable the comparison of different spaces' acoustics. These parameters define energy-time relationship of the acoustic energy at the point of measurement in various ways, and they are divided into several groups. A group of parameters, which are important for the analysis of church acoustics, are the parameters that define intelligibility of speech, overall experience of the sound field, relationship between frequencies, the ease of following the musical notation and all other things that can help us understand and describe the extremely complex natural phenomenon of the sound. The main spatial-acoustic parameter used in the modern acoustics is the reverberation time, which defines the time in which the acoustic energy remains in the analysed space after the sound source stops.

Figure 4: The interior view of the central dome of the church in Žiča Monastery. Under the arch on the left is a visible hole of an inbuilt acoustic vessel

The analysis of impulse responses measured in situ showed that centrally placed dome of the Orthodox Serbian churches represents acoustically coupled space with high reverberation. The central dome is a dominant form in both architectural and acoustic sense. This concave acoustic element generates an unusual acoustic environment, reflecting acoustic waves towards the imaginary focus of the dome and thus creating an interesting play of delay. Therefore, the listener is not able to define a unique source of sound, but he gets the feeling that the sound is coming from all around, which additionally increases the sense of mystery and holiness of the divine service (Ђорђевић 2016).

Auralisation

Besides the objective analysis of acoustic parameters, it is also important to include the subjective experience in the archaeoacoustic research. The distance between the temples makes the comparison of their acoustics a difficult task. However, the modern technology offers a solution in the form of auralisation of space - the reconstruction of the sound in a chosen space that can be heard using subsequent reproduction.

Firstly, the impulse response needs to be measured in the points of interest (under the dome, in naos or narthex), while the sound source is placed in the southern chanter apse. Secondly, it is important to record a relevant sound (a chant, Bible reading, etc.) in a studio, without a spatial component of the sound (echo, etc.). Afterwards, using a specialised software, this dry recording is combined with the previously obtained (from the impulse response measurement) spatial component of the sound for the chosen spaces. The result of this process is placing of a recording of speech, chanting or a song in a desired space. By exciting the space in this way, with the same acoustic content, it is possible to evaluate a subjective experience and compare the sound in different spaces.

Conclusion

Study of sound in building heritage requires a multidisciplinary approach, encouraging dialogue and combing knowledge from various scientific disciplines in an additive process. This paper suggested three approaches based on the type of sound used in archaeoacoustic research – live sound, harmonic sound and impulse response.

Live sound is considered for soundscapes and for the performances corresponding to architectural spaces. Harmonic sound is the inaudible relation of acoustic principles and the geometry and proportions of architecture. It is used for reconstruction of designing process and thus revealing the building logic when it comes to sound and beauty. Technological development provided the possibility of simple impulse response measurement, thus obtaining various acoustical parameters for objective analysis of sound field. In this way obtained spatial components of sound can also be used for auralisation and subjective comparison of sound in various spaces that are excited with the same acoustic content.

These three complementary approaches imply both objective and subjective analysis of sound field. Each in its own way helps us understand the historical dynamics between architecture and its acoustics, perceive acoustical knowledge of the builders and compare acoustic images of built heritage.

ACKNOWLEDGEMENT

The paper is the result of scientific research on the project Theory and practice of science in society: multidisciplinary, educational and intergenerational perspectives (number 179048), financed by the Ministry of Education, Science and Technological Development of Republic of Serbia.

REFERENCES

Arns, R, and B Crawford. 1995. "Resonant Cavities in the History of Architectural Acoustics." Technology and Culture 36 (1): 104–35. http://www.jstor.org/stable/3106343.

Đorđević, Zorana, Kristina Penezić, and Stefan Dimitrijević. 2017. "Acoustic Vessels as an Expression of Medieval Musical Tradition in Serbian Sacral Architecture." Musicology 22: 105-132, http://www.doiserbia.nb.rs/img/doi/1450-9814/2017/1450-98141722105D.pdf

Milosavljević, Predrag. 2013. "Zlatni presek i filozofija prirode". Doctoral dissertation, book II, University of Belgrade.

Petrović, Đorđe. 1972. Kompozicija Arhitektonskih Oblika. Beograd: Naučna knjiga.

Poulle, Bruno. 2000. "Les Vases Acoustiques Du Theatre de Mummius Achaicus." Revue Archeologique 1: 37–50.

Valière, Jean-Christophe, and Bénédicte Palazzo-Bertholon. 2014. "Le Dispositif Acoustique Du Caveau de La Cathédrale de Noyon (Oise)." In Paper Presented at the CFA 2014, Poitiers, France, April 22-25.

Valière, Jean-Christophe, Bénédicte Palazzo-Bertholon, and Jean-dominique Polack. 2013. "Acoustic Pots in Ancient and Medieval Buildings : Literary Analysis of Ancient Texts and Comparison with Recent Observations in French Churches." Acta Acustica United with Acustica 99 (November 2012): 70–81.

Vasiljević, Slobodan. 1956. "Naši Stari Graditelji I Njihova Stvaralačka Kultura." Zbornik Zaštite Spomenika Kulture VI–VII: 1–36.

Vitruvije, Marko Polio. 2000. Deset Knjiga O Arhitekturi. Beograd: Građevinska knjiga.

Аксић, Нина В. 2014. "Друге зове, себе не чује – звоно у српској народној традицији." Philologia Mediana VI (6): 151–66.

Ђорђевић, Зорана. 2016. "Принципи и историја односа архитектуре и акустике." Doctoral dissertation, University of Belgrade.

Мирковић, Лазар. 1939. Списи Светог Саве И Стефана Првовенчаног. Београд: Државна штампарија Краљевине Југославије.

———. 1965. Православна литургика или наука о богослужењу Православне источне цркве. Београд: Издање Светог архијерејског синода Српске православне цркве.

Павићевић-Поповић, Радмила. 2007. "Звоно, с освртом на жичка звона." Повеља 37 (2): 153–59.

Патријарх Павле. 1995. "О грађењу православног храма." In Традиција И Савремено Српско Црквено Градитељство, edited by Борислав Стојков and Зоран Маневић, 15–21. Београд: ИАУС.

Пено, Весна. 2014. "О истраживањима црквеног појања у средњовековној Србији." Музикологија 16: 131–54.

Перковић-Радак, Ивана. 2007a. "Стара Музика." In Историја Српске Музике - Српска Музика И Европско Музичко Наслеђе, 29–62. Београд: Завод за уџбенике.

———. 2007b. "Црквена Музика." In Историја Српске Музике - Српска Музика И Европско Музичко Наслеђе, 299–329. Београд: Завод за уџбенике.

Попмихајлов, Никола. 2004. "Црквена Музика И Богослужење." Истина 9–11. http://www.spc.rs/sr/crkvena_muzika_bogosluzhenje.

Свети Сава. 2009. Сабрана Дела. Београд: Антологија српске књижевности.

Станојевић, Станоје. 1933. "Била, клепала и звона код нас." Глас Српске Краљевске Академије CLIII: 79–90.

Теодосије. 2009. Житија. Антологија српске књижевности.

Fig. 1: The anthropomorphic vase from Parța

The Parţa Vase: Experimenting in Neolithic Acoustics

Dragoş Gheorghiu

DRAGOŞ GHEORGHIU, Professor, is an anthropologist and experimental archaeologist whose studies focus on the process of cognition and material culture of the prehistoric societies of South Eastern Europe. He is editor and co-editor of multiple books on archaic technologies and semiotics of material culture, and has sustained publication activity on prehistoric Europe. email: gheorghiu_dragos@yahoo.com

ABSTRACT: It is well known that ceramic vases, due to the resonance character of the material, were used in the past to enhance the acoustics of different buildings. But their phonic role was not limited to this. Ethnographic data and experiments show that ceramic vases were used to enhance the human voice, and to reproduce and amplify the sound produced by animals, during wolf hunting for example. The text presents an experiment conducted using a replica of a Chalcolithic vase from the Vinča tradition whose form (a human with both hands cupped around the mouth, suggesting the attitude of shouting), and resonance capacity, could have been used for animal calling, among other functions.

KEYWORDS: Vinča ceramic vase, anthropomorphic, howlers

Introduction

The study of archaeoacoustics is a recent sub-discipline in archaeology (Devereux 2001; Scarre and Lawson 2006) and relies mostly on experiments. Through experiments the acoustics of prehistoric megalithic monuments (Watson and Keating 1999), and architectural spaces (Jahn, Devereux and Ibison 1996) was tested.

Except for the experiments carried out in Palaeolithic painted caves (Reznikoff and Dauvois 1988; Scarre 1989), iconography was not utilized in the study of archaeosounds. This is the reason a prehistoric anthropomorphic vase visualizing a screaming human could be considered to represent a genuine archaeoacoustic document.

The object to be discussed belongs to the 5th millennium BC, which corresponds to the Neolithic in south-eastern Europe: an epoch when a series of phenomena of acculturation occurred between the newcomer agriculturalists and the local hunter-gatherers that led to the emergence of large cultural traditions. One of them was Vinča, which occupied the north-western part of the Balkan Peninsula up to the Danube's tributary to the east (Whittle 1996: 83 ff). Vinča could be characterized as a complex Neolithic tradition, with a mixed economy made of horticulturalists, gatherers and hunters, and a strategy of dwelling with hamlets composed of semi-subterranean houses and villages with large houses protected by wooden palisades.

The religion in the many Vinča sub-traditions, like for example, the Banat Culture (Lazarovici and Draşovean 1991), demonstrates a high level of spirituality, as shown by some buildings containing assemblages of ritual objects, identified by archaeologists as cult buildings (Lazarovici and Lazarovici 2013). At Parţa (Vinča A3-B1; Mantu 2000: 98), in south-western Romania, within the context of such a building, an anthropomorphic vase [FIG. 1] was discovered, whose shape represents a human character with the hands positioned around the open mouth, in a gesture of shouting. Its function has not been yet deciphered.

Anthropomorphism

Anthropomorphic vases are characteristic for the Balkan - Danube Chalcolithic traditions; in Vinča (Lazarovici and Lazarovici 2006: 306, fig. IIIb), or in Gumelniţa-Karanovo-Kodjadermen (Andreescu and Popa 2003: 61-63), the multi-meaning symbolism of the human body is present in the production of the ceramic containers.

In the Vinča tradition many types of vases display anatomical human features: detailed human faces, arms, feet and gender attributes (Nicolić and Vuković 2008: 58, fig. 6). This means that vases were perceived as representing entities analogous to humans. Since vases shared physical traits with humans, one can infer that part of prehistoric vases were endowed even with voice, or in other words, were in some relationship with acoustics, being utilized to distort the human voice.

Pots for Hunting

In diverse regional Vinča sub-traditions, many of the ceramic vases, of relatively small dimensions, including the anthropomorphic ones, displayed perforated protuberances or ceramic rings for being suspended with threads or transported in this way. Such a means of manipulating ceramic vases could be associated to a dynamic part of the economy, which is the activity of hunting.

Notwithstanding the emergence of the traits characterizing complex societies, the Neolithic economy was still favouring hunting instead of animal husbandry, as demonstrated by the osteological inventories in some settlements (El-Susi 1996: 150).

Examples of the use of ceramic pots for hunting are to be found in pre-modern times (Sadoveanu 1926). Today all instruments for hunting call or howlers are produced from industrial materials like metal or plastic[1]. But two to three generations before, in many parts of Europe the hunting method to call wolves using ceramic pots was current. The most recent reference about this kind of hunting was about a "shamanic" character with magic powers (*Vâlva lupilor*, *The Fairy of the Wolves*), who lived until recently in the Western Carpathian Mountains, who used to call wolves with a clay pot[2]; a fact which made me consider the Parţa vase also from a ritual and magic perspective. The special shape of this vase, compared to the rest of the Vinča portable pots, supports the idea that it could represent an object with a complex symbolic significance.

Ceramic vases fired in open fires were used as howlers because the burned clay with low sintering (i.e. temperatures under 900°C) softens higher sound frequencies and favours the medium ones, filtering the human voice and producing a final sound similar with the wolf's howl (M. Florian pers. com.).

Experiments

Based on these ethnographical premises, an analysis of a howler vase should verify its

[1] https://www.huntwolves.com/store/wolf-howlers

[2] http://cybershamans.blogspot.ro/2010/11/valva-lupilor.html

potential to produce alterations of the human voice and its portability. For this purpose, a reproduction of the vase (a globular shape with a short neck and opening under the symbolic form of a human face with the mouth open) was manufactured and burned at a temperature under 900°C, specific for Neolithic pyro-technology, to explore its acoustic potential, and the ergonomics of its portability. The two rings positioned on each side of the human figure allow the vase to be hung by the neck and easily used for shouting inside it. [FIG. 2 a, b]

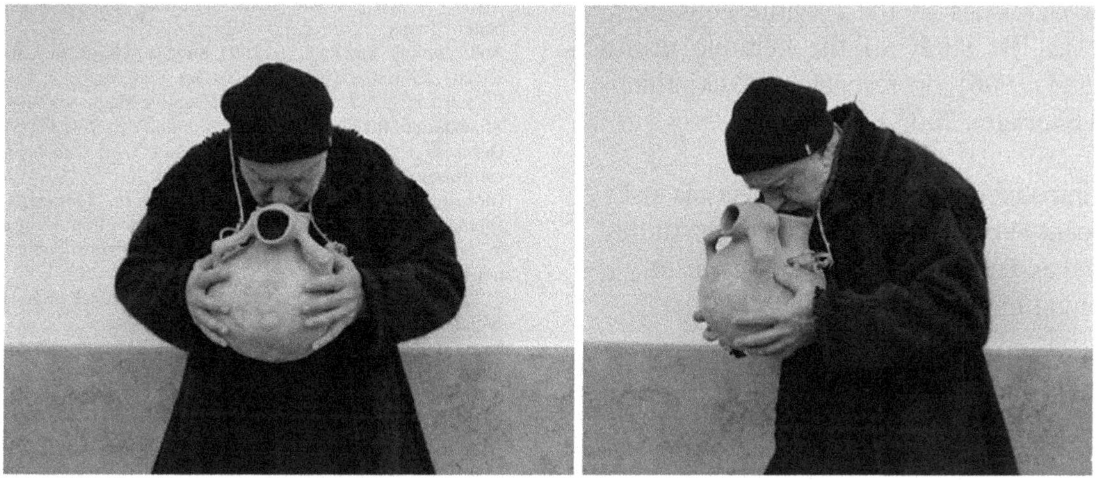

Fig. 2a, b: Experimenting the ergonomics and acoustics of the vase.

One result of the experiments was that vases burned in open firings with temperatures below 900°C, with incomplete sintering of the clay matrix, produced sounds with a lower tonality than the (more recent) vases burned in kilns at temperatures higher than 900°C (Gheorghiu 2002), that lead to a high degree of sintering.

An additional result was to notice the relationship between the shape and the acoustics of the vases: the sounds produced are also influenced by the dimension of the opening of the vase: the narrower the opening, the better the filtering of the human voice. Experiments to produce sounds with vases with narrow openings inferred that not all Vinča portable vases may have been employed as efficient callers for hunting as a second function of utilization.

A conclusion of the experiments with the Parţa vase, made of a low-burned clay that could filter the human voice, with narrow openings, with a general shape that permitted an easy manipulation, and with a symbolic image that shows an act of crying, could qualify for the function of howler.

Conclusion

One can confidently stress that this Neolithic cult instrument (see Devereux 2001: 42-43; Kováks 2014: 220-21) had a well-defined acoustic role, functional and symbolic at the same time, whose symbolic and ritual context of discovery could relate it to a ritual and magical activity of hunting (see Carneiro 1970), or to a shamanistic one. It is well-known that in the shamanic ritual the performer mimics the natural sounds, to cite only a case from Altai, where in the Sayat shamanic ritual the howling of the wolf is imitated as being the voices of helping spirits (Diószegi 1968: 203).

In both cases the role of the vase was to hide (by means of alteration) the human voice, and consequently the identity of the performer.

The Parţa vase contributes to a new direction of study in experimental and experiential archaeology: that of "animation" with sounds of the prehistoric anthropomorphic vases, and consequently to the study of archaeoacoustics of the ceramic containers, just as the study of the ceramic drums (Aiano 2006) or resonators (Zakinthinos and Skarlatos 2007.)

By introducing the ceramic objects as a significant subject of archaeoacoustic studies, a vast and novel soundscape of prehistory is opening for archaeology.

ACKNOWLEDGEMENTS

The author thanks Ms. Linda Eneix and Dr. Fernando Coimbra for the kind invitation to contribute to the Archaeoacoustics conference in Tomar, to Mr. Mircea Florian for valuable information about acoustics, and to Dr. Zoia Maxim for the image of Parţa vase. Many thanks as always to Mr. Bogdan Căpruciu and Ms. Cornelia Cătuna, for helping to improve the clarity of the text.

REFERENCES

Aiano, L., 2006, Pots and drums: an acoustic study of Neolithic pottery drums, *euroREA* 3: 31-42.
Andreescu, R., and Popa, T., 2003, Sultana Malu-Rosu. Catalog selectiv, *Cercetari Arheologice* XII, pp. 67.
Carneiro, R. L., 1970, Hunting and Hunting Magic among the Amahuaca of the Peruvian Montaña, *Ethnology*, 9(4) 331-341.
Devereux, P., 2001, *Stone Age Soundtracks: The Acoustic Archaeology of Ancient Sites*, London: Vega.
Diószegi, V., 1968, *Tracing shamans in Siberia. The story of an ethnographical research expedition.* Translated from Hungarian by Anita Rajkay Babó. Oosterhout: Anthropological Publications.
El Susi, G., 1996, *Vânători , pescari şi crescători de animale în Banatul mileniilor VI î.Ch. –I d.Ch.*, Timişoara.
Gheorghiu, D., 2002, Fire and air draught: Experimenting the Chalcolithic pyroinstruments, pp. 83-94. In D. Gheorghiu (ed.), *Fire in archaeology*, British Archaeological Reports 1089, Oxford: Archaeopress.
Jahn, R. G., Devereux, P, and Ibison, M., 1996, Acoustical resonances of assorted ancient structures, *Journal of the Acoustical Society of America* 99(2): 649-658.
Kováks, A., 2014, Ritual pots from European Neolithic and Copper Age sanctuaries, pp. 196-227. In J. Marler (ed.), *Fifty Years of Tartaria excavations*, Suceava: Lidana.
Lazarovici, Gh., and Draşovean, F. (eds.), 1991, *Cultura Vinča in România (Origine, evoluţie, legături, sinteze)*, Timişoara.
Lazarovici, Gh., and Lazarovici, C-M., 2006, A Home altar at Gura Baciului, *Analele Banatului* SN I: 103-111.
Lazarovici, Gh., and Lazarovici, C-M., 2013, "Sacred houses" and their importance for the reconstruction of architecture, inner furnishing and spiritual life, pp. 503-520. In Andres, Al., Kulcs, G., Kalla, G., Kiss, V., and Szabo, G. (eds.), *Prehistoric Studies I. Moments in Time*, Budapest: L'Hartmattan.
Mantu, C. M., 2000, Relative and absolute chronology of the Romanian Neolithic, *Analele Banatului*, VII-VIII, 1999-2000.
Nicolić, D., and Vuković, J., 2008, Vinča ritual vessels: Archaeological context and possible meaning, *Starinar* LVIII: 51-69.
Reznikoff, I, and Dauvois, M., 1988, La dimension sonore des grottes ornées, *Bulletin de la Société Préhistorique Française* 85(8), pp. 238-246.
Sadoveanu, M., 1926, *Tara de dincolo de negura*, Bucharest: editura Cartea Romaneasca.
Scarre, C., 1989, Painting by resonance, *Nature* 338: 382.
Scarre, C. and Lawson, G. (eds), 2006, *Archaeoacoustics,* Cambridge: MacDonald Institute Monographs.
Watson, A. and Keating, D. 1999. Architecture and sound: an acoustic analysis of megalithic monuments in prehistoric Britain. *Antiquity* 73, 325-36.
Whittle, A., 1996, *Europe in the Neolithic. The creation of the new worlds*, Cambridge: Cambridge University Press.
Zakinthinos, T. and Skarlatos, D., 2007, The effect of ceramic vases on the acoustics of old Greek orthodox churches. *Applied Acoustics* 68. 1307-1322. 10.1016/j.apacoust.2006.07.015.

Sounding Situated Knowledges: Archaeoacoustics as Sonic Knowledge Production

Annie Goh

ANNIE GOH: PhD candidate at Goldsmiths University of London, Department of Media and Communications funded by CHASE/AHRC and as a Stuart Hall PhD fellow, where she is also an Associate Lecturer. She holds an MA in Sound Studies, an MFA in Generative Art and a BA (Hons) German & European Studies and has previously taught at the University of Arts Berlin and Humboldt University Berlin. Email: a.goh@gold.ac.uk

ABSTRACT: Whilst archaeoacoustic's primary concerns lie within expanding archaeological knowledges, it additionally proposes intriguing epistemological challenges to Western ocularcentric conceptions and traditions of knowledge. In this paper, which draws on my current PhD research, I propose that it is pertinent to the field's advancement to critically analyse processes of knowledge production in archaeoacoustics in terms of *sonic knowledge production*. From the interdisciplinary field of sound studies, Steven Feld's notion of "acoustemology" and Julian Henriques' notion of "sonic logos" will be visited to suggest how theories of sonic knowledge from archaeoacoustics might be developed. I contend that a methodology of "sounding situated knowledges" by grounding the situatedness and embodiedness of Donna Haraway's "Situated Knowledges," will help attend to crucial epistemological, political, and ethical questions of sonic knowledge production in archaeoacoustics.

KEYWORDS: epistemology, sonic knowledge, embodiment, Steven Feld, Julian Henriques, Donna Haraway

Introduction

Archaeoacoustics studies the "role of sound in human behaviour, from earliest times up to the development of mechanical detection and recording devices in the nineteenth century" (Scarre and Lawson 2006, vii). Since the most commonly cited early fieldwork in the 1980s (Reznikoff 1987; Reznikoff and Dauvois 1988), to the first international meeting at the McDonald Institute of Archaeology in Cambridge in 2003, to the first international conference organized by the Old Temple Study Foundation (OTSF) in Malta in 2014, the field has progressed substantially; yet in 2017 it still finds itself at a formational, "pre-paradigmatic stage" where "there are no generally accepted theories...methodologies...or data yet" (Zubrow 2014, 9). As the field continues to progress, my paper, drawing on my PhD research of knowledge production in archaeoacoustics, hopes to highlight a few methodological issues central to its continuing development.

Whilst almost all work grouped under the rubric "archaeoacoustics" hitherto generally aims to further archaeological knowledges of specific sites and cultures of particular eras, my research is based upon the premise that the field additionally holds fascinating promises for related epistemological questions, namely around what I am calling "sonic knowledge." By examining the processes of knowledge production in archaeoacoustic research in terms of *sonic knowledge production*, I propose we can identify how archaeoacoustic work can not only contribute to the discipline of archaeology, but furthermore ascertain how it might provide enriching insights into broader questions around "knowledge" as such.[1]

My position as a presenter at the Archaeoacoustics III conference is somewhat unique in that I am researching the field itself, rather than seeking to contribute directly to specific or general archaeological knowledges.[2] Correspondingly, this paper sits somewhat askew from other papers in this collection, in that it is concerned more with asking questions and interrogating the questions which are being asked, than it is with answering questions. Nevertheless, this is done so in the hope of contributing productively to the field's future development.

Firstly, I will locate the emergence of archaeoacoustics as a discipline historically and in alignment with the larger interdisciplinary field of sound studies to describe the stakes of the term *sonic knowledge*. Secondly, I will describe Steven Feld's notion of "acoustemology" and Julian Henriques' concept of "sonic logos" as existing theories of sonic knowledge within sound studies to suggest how these might be useful for archaeoacoustics. Thirdly, I will introduce Donna Haraway's 1988 essay "Situated Knowledges: The Science Question in Feminism and the Privilege of Partial Perspective" to demonstrate how feminist epistemologies from science studies can be productively applied to archaeoacoustics. This intervention can be enacted on two discernible but interrelated levels: firstly, on the level of subjectivity, representation and identity, and secondly on an ethico-onto-epistemological level. Finally, I conclude by proposing *sounding situated knowledges* as a methodological tenet which vitally grounds embodiedness and situatedness within archaeoacoustics research.

Sensory Studies, Sound Studies, and Sonic Knowledge

Where is archaeoacoustics situated? Most obviously perhaps in archaeology, given its aims to examine the role of sound and listening in human behaviours in archaeological contexts. Yet the emergence of archaeoacoustics is paralleled by broader currents in cultural theory and philosophy in the late twentieth century, including the formation of the interdisciplinary field of sound studies. Archaeoacoustics is an interdisciplinary field, bridging archaeology and acoustics as well as engaging various other disciplines

1 Please note I will occasionally use "epistemology" and "epistemological" to relate to a theory of knowledge interchangeably with questions around knowledge production for stylistic reasons. However, I wish to highlight the Greek roots of the word *epistêmê* as a theoretical or scientific knowledge, developed in contradistinction to other Greek words relating to knowledge such as *technê, gnősis, phronēsis, mētis* amongst others, which various depict forms of knowledge relating to practical, spiritual, experiential, cunning or other skill-based activities. Therefore, the bias towards only one (though not necessarily unified) form of knowledge as *epistemological* amongst many is contradictory to the fundamental questioning of the limited nature in such a conception of knowledge which the term "sonic knowledge" aims to open up.

2 Currently, as part of my PhD research I have undertaken around twenty interviews with researchers (many of whom were present at the Tomar conference). However, this paper will not incorporate the interview material but instead trace the theoretical underpinnings of my PhD project, including highlighting which factors could feasibly positively contribute to the field's future development.

such as music archaeology, anthropology, musicology, ethnomusicology, music psychology, cognitive neuroscience, amongst others. Sound studies is also an interdisciplinary field, bridging diverse fields such as "acoustic ecology, sound and soundscape design, anthropology of the senses, history of everyday life, environmental history, cultural geography, urban studies, auditory culture, art studies, musicology, ethnomusicology, literary studies, and STS" (Pinch and Bijsterveld 2012, 7). So far, archaeoacoustics has not yet been substantially incorporated into the field of sound studies, aside from in the work of Barry Blesser and Linda-Ruth Salter on aural architecture where it has served as part of a useful historicisation in studies of acoustic space (Blesser and Salter 2006, 2012). Larger questions about what else archaeoacoustics can offer for sound studies are yet to be extensively explored.

In studies of Western culture and philosophy where sensory regimes are explicitly discussed, it is common to address the dominant position of vision and seeing in the modern West. A "hierarchy of the senses" is often traced back to the Ancient Greeks. Don Ihde's important phenomenology of listening begins by noting, for example, Heraclitus' claim that, "eyes are more accurate witnesses than ears" and Aristotle's that, "above all we value sight … because sight is the principal source of knowledge and reveals many differences between one object and another" (Ihde 2007, 7). Everyday language offers a plentitude of visual metaphors in pertaining to "seeing" *qua* "knowing," for example, "having insight," "becoming enlightened," "using your mind's eye," "seeing the light," and many more. Such is the "ocularcentrism" of Western culture that a crisis can be observed, at least in philosophical texts, in the growing scepticism around visuality since the late nineteenth century (Jay 1993). The relation between sensory experience and knowledge has been strongly shaped by such a hierarchisation of the senses, though the consequences of this are hugely complex and should resist any simplistic reduction, as Ihde in his work cautions against (Ihde 2007, 8-9).

Both archaeoacoustics and sound studies are positioned against the ocularcentric grain in their recent emergence in the past few decades. Archaeoacoustics has a more directly positivistic gain at stake in that it posits that the practical and theoretical neglect of considerations of sound and listening in the archaeological method, has led to "oversights" in the production of archaeological knowledge historically. By sounding and listening at archaeological sites, be this using the voice or other instruments or by doing acoustic tests and taking measurements with electronic equipment, better understandings of past human behaviours appear possible. Researchers in this budding field are well aware of its potential, even if research funding bodies and colleagues in archaeology and heritage are currently less aware of why it might be important. Sound studies, as a more diffuse collection of writings and practices, has a similarly antagonistic relationship to the dominance of visuality but with a less positivistic approach to knowledge production; its aims are looser and less clearly defined. However, I propose that bringing these two interdisciplinary fields together around the question of sonic knowledge could be hugely productive.

Historically, the emergence of both fields can be traced back to the widespread uptake of phenomenology through the work of Maurice Merleau-Ponty, with predecessors in the philosophies of Martin Heidegger and

Edmund Husserl who provided key impetuses in increasingly scholarly attention to perception, the senses, experience, and the body. Phenomenological experience, across sensory modes, foregrounded the relation between human experience and the production of knowledge in a way which opened up probing new questions. In archaeology, the turn to phenomenology was influential in the later formation of archaeoacoustics. Chris Tilley's *A Phenomenology of Landscape* placed understandings of human experiences of space and place at the basis of archaeological methodologies (Tilley 1994), which made a comparable contribution to archaeology as Ihde's aforementioned groundbreaking book *Listening and Voice: Phenomenologies of Sound* (Ihde 1976) made in the development of the field of sound studies. Today, in the early twenty-first century, such an emphasis on phenomenology and embodied experience might seem like a common-sense approach, however, it is worth remembering the deep-rooted historical ideological baggage attached to mind-body dualisms which has led to these being only fairly recent trends.

Complementarily to developments in phenomenology, anthropologists have turned to non-Western cultures to investigate less ocularcentric relational modes. Sensory anthropologists in particular have examined how cultures outside of European Enlightenment rationalist traditions use sensory perception and experience within everyday life. Some notable studies deal for example with the significances of smell and taste in the cultures of the Umeda of New Guinea, the Amazonian Desana and a variety of Melanesian peoples (Classen 1993; Classen, Howes, and Synnott 1994; Howes 1991, 2005). Classen and Howes use a schema which depicts a hierarchy of the senses in the West, in which sight is commonly denoted as the 'noble' sense, and hearing – due to the importance of speech – is also closely associated with the intellect and the ability to reason, whilst the other senses of touch, taste and smell are considered to be "lower down" in the schema (Howes and Classen 2014, 1–3). This model usefully links Western (masculinist, colonialist, imperialist, capitalist) histories with a sensory hierarchy based on Cartesian mind-body dualism, which postcolonial and decolonial thought as well as feminist theories of the body have refuted as a universal model of understanding all cultures (Said 1985; Spivak 1988; Mignolo 2000; Irigaray 1985; Grosz 1994; Butler 2011).

These ethnographic studies which attempt to understand how different sensory regimes are found in relational modes contrary to a dominant Western understanding can usefully be theorized as producing different knowledge regimes, as "sensory knowledges" or "epistemologies of perception" (Howes 2003, 54, 58). Detailed evidence of other "sensory knowledges" existing in living cultures in anthropological studies throws into relief the limited nature of typical characterisations of how knowledge can be produced. Therefore, insofar as archaeoacoustics explores sensory-knowledge regimes which deviate from the dominant ocularcentric traditions of Western post-Enlightenment culture, different ways of knowing and doing through practices of sound and listening can possibly be theorized, and in some cases, evidence of these is beginning to be collected.

The aforementioned work of anthropologies of the senses has had considerable influence on both sound studies and archaeology. In sound studies, the work is often cited and found in curricula. Archaeologists have drawn on this same work to begin working on "archaeologies of the body" (Hamilakis, Pluciennik, and Tarlow 2002). Recently, steps towards establishing multisensory archaeology as a fledgling subfield of archaeology have been taken, within which archaeoacoustics appears as a core example, especially where other sensory modalities such as smell, touch and taste leave more fragile and destructible material traces (Day 2013). However, crucially, there is yet to be an extensive and sustained engagement with what the implications are for engaging in such sensory knowledges in producing understandings which challenge dominant Western knowledge paradigms. In this sense, archaeoacoustics provides an ample productive range of case studies where such investigations might prove extremely fruitful.

However, simplistic binary models which pit conceptions of visuality against those of the auditory are a seductive trap which researchers would do well to be cautious of, particularly where these are treated as ahistorical and become idealized as universal; this risks misreading what sound and listening meant or could have meant for past cultures, as cogently described by Jonathan Sterne's critique of the "audiovisual litany" (Sterne 2003, 15). The term "sonic knowledge" is intended to indicate the potential for other modes of relation between sensory experience and knowledge production not currently recognized by Western, rationalistic, post-Enlightenment knowledge regimes. Therefore, the term urges caution in treading between overly simplified divisions between "visual" and "sonic" knowledges, as if these or the corresponding processes of knowledge production, are or can ever be clearly separated from one another, given the complex intermingling of sensorial experiences (Serres 2008) and corresponding processes of knowledge production.

Turning to archaeoacoustics to address questions of sonic knowledge, therefore, offers the potential of uncovering different ways of knowing through sound and listening which existed in other cultures at other times. Whilst researchers may more consciously be involved in advancing archaeological knowledges, they are also – often implicitly – engaged in questions of sonic knowledges. Aside from, and despite our ocularcentric methods of knowledge production in Western modern science, it may be possible to provide evidence for modes of *sonic knowledge production* which occur along non-Cartesian, non-ocularcentric lines, particularly where the definition of knowledge is brought into question itself in such an investigation.

Steven Feld's "Acoustemology" & Julian Henriques' "Sonic Logos"

As the field of archaeoacoustics progresses with its primary aims and purposes, turning to two existing notable studies of sonic knowledge may help in honing the questions which are at stake for archaeoacoustic's potential contribution to theories of sonic knowledge. Steven Feld's term "acoustemology" – a neologism of "acoustics" and "epistemology"– and Julian Henriques' 'sonic logos' both demonstrate how knowledge acquired significantly via hearing depart substantially from European, ocularcentric, rationalistic traditions of knowledge. Sonic knowledge – rather than indicating an exclusivity in how sound and listening relate to practices of knowing – merely suggests and challenges previously ocularcentric conceptions of knowledge in ways still relatively underdeveloped.

The term "acoustemology" emerged from an anthropology of sound which Feld developed in his ethnographic work over decades. His famous case study of the Kaluli people in Papua New Guinea from fieldwork undertaken in the 1970s-1990s, describes their "ear- and voice-centred sensorium" in the rainforest people which provided him with an example of a society where acute hearing, as part of a complex interplay of the senses, plays a fundamental role. Feld's careful anthropological and linguistic work describes in detail a society whose systems of knowing function upon fundamental ly different propositions to Western society's. For example, the metaphorical construction of "lift-up-over-sounding" both prescribes and describes the sonic form for Kaluli people. In the dense sound world of the rainforest, the term evokes how a sound constantly shifts between figure and ground so that its inherent fluidity is sensed auditorally, kinaesthetically and sensually. "This way of hearing and sensing the world is internalized as bodily knowledge" (Feld 1996, 100). Yet this motif of "lift-up-over-sounding" relates to diverse "things" such as "nature, music, body-painting, choreography, costume, and choreography... [it] metaphorically unites Kaluli environment, senses and arts" (ibid.: 101).

Revisiting his concept, after widespread usage within the field of sound studies in recent decades, Feld is keen to make clear that the usage of 'epistemology' is not intended to invoke the type of formal knowledge into philosophical "truth" i.e. epistemology with a capital 'E' (Feld 2015, 12), but that it "engages acoustics at the plane of the audible – *akoustos* – to inquire into sounding as simultaneously social and material, an experiential nexus of sonic sensation" (ibid.:12). As Feld comments, his daily lessons from the child and adult inhabitants of the Bosavi, showed him "routinized, emplaced hearing as an embodied mastery of locality" (ibid: 18). The knowledge Feld himself was not party to fascinated him as "bodily, powerful and gripping." Instead of the most typical paradigm of subject-object relationships, Feld draws on Bosavi acoustemology to propose that relations are "in fact, more deeply known, experienced, imagined, enacted, and embodied as subject-subject relationships" (ibid.: 19). He carefully updates its position within a discourse of relational ontology, with emphasis on actors plus relationships – actants (in Bruno Latour's language) which are "variously human, non-human, living, non-living, organic or technological" (ibid: 15). Foregrounding sound and sounding as inherently situated, the relationality of sound is readjusted as core to acoustemology, linked both retrospectively backwards to the original fieldwork in the Bosavi rainforest, and forward into contemporary debates by comparing the human/avian relations, or relations between humans to water or insects to Donna Haraway's notion of 'companion species.' As a theory of sonic knowledge, acoustemology denotes a radical re-thinking of how aural experiences relate to knowing and demands in its updated form a notion of listening with a fundamental reconsideration of relational modes.

For Henriques, the term "sonic logos" akin to sonic "ways of knowing" denotes thinking through sound based on an in-depth case study of Jamaican reggae sound system culture. It offers a dynamic model of thinking, inextricably tied to corporeal practices, where auditory propagation is a model of sociocultural as well as corporeal and material processes (Henriques 2011, xviii). A deeper understanding of the interactions between sound system crew members, selectors, MCs, audio engineers, dancers, the audio technological system of the sound system itself, demonstrates the limitations of rational and cognitive models of understanding. Drawing on Bourdieu's "logic of practice', 'practical ways of knowing' as 'sonic ways of knowing' see a merger of the knower and the known, "thinking through the vibrations of sounding helps dissolve the gulf between viewer and viewed that the visual metaphor is invariably used to describe, and indeed to justify" (ibid.: 227). According to Henriques, the 'sonic logos' articulates a fundamental relationality of sonic knowledge, diffusing the boundaries between subject and object, as well as doing and thinking.

The sonic logos, for Henriques, explores ways of knowing via auditory and embodied means to offer a deep criticism of the representational meaning and linear causality of visual ways of knowing (ibid.: 246). Whereas traditionally the "logos" represents formal knowledge or the "know-what" of the *epistême*, Henriques triangulates this type of knowledge in the sonic domain against two other ways of knowing, based on a method of thinking through sound. These two other ways of knowing, namely *techné* and *phronēsis* define respectively a "skilfulness and proficiency" and "wisdom and judgement", and both these demonstrate the lack of limiting considerations of knowledge merely to the "scientific skills" of the *epistême* (ibid.: 244-246). For the successful functioning of the system of the dancehall sound system in question, the *epistême*, i.e. scientific skills play a meagre role compared to the various skills as *techné* of the sound engineers, selectors, MCs and the *phronēsis*, for example, of the Stone Love Movement boss Winston "WeePow" Powell in managing the meta-level of the sound system's continued existence. Although the notion of "sonic logos" is clear not to claim the existence of an "alternative auditory regime" (ibid.: 248) as entirely independent of the visual world, the radical potential implied with the concept is as a method of understanding ways of knowing, and accounting for multiplicity and variation which visual accounts can only paltrily describe. This critical re-construction of the 'logos' as a 'sonic logos' lies far apart from positivistic, scientific ideas of "objective" or "pure" knowledge, and with its emphasis on practical, embodied ways of knowing, sets itself as the antithesis to traditional Western constructions of knowledge as individual, conscious, abstract and representational. Similarly, Henriques draws on Michael Polanyi's idea of "tacit knowing" as appropriate for the logic of sound practice as an alternative notion to describe "ways of knowing" which are not directly representational (Polanyi 1962; Henriques 2011, 233). These projects, Feld's and Henriques', strive towards expansion of the very concept of what constitutes "knowledge."

Feld and Henriques' work on acoustemology and sonic logos or sonic "ways of knowing" respectively, form examples from which theories of sonic knowledge can be furthered. Both case studies stem from contemporary living examples of cultures where sound and listening play rich and pivotal roles in the production of culture-specific knowledges. Different challenges stand ahead if one attempts to find comparable examples in archaeological case studies, with the large time periods which have passed imbuing the process with its speculative nature. Nevertheless, these models are extremely valuable, as in complement to the theorisation of ocularcentrism in philosophy, they provide concrete evidence of existing socialities where knowledge acquired via sound plays a large role which when applied to archaeoacoustics, open up fascinating new possibilities of sonic knowledges.

Donna Haraway's *Situated Knowledges*, Feminist Epistemologies, and Archaeoacoustics

Donna Haraway's argument for situated and embodied knowledge claims in the 1988 essay *Situated Knowledges*, provides a further crucial impetus in a proposal for theories of sonic knowledge. Presented in the context of feminist science studies, the essay usefully delineates the limitations of traditional knowledge production, particularly in comparison to the archetypal scientific laboratory as the site of such activity. The "god-trick" view from above of the effectively disembodied and purportedly disinterested observer, is vehemently rejected as an inheritance of Cartesian mind-body dualism and a correspondingly staid traditionalist conception of knowledge; *Situated Knowledges* instead foregrounds responsibility and accountability in knowledge production (Haraway 1988, 590). The scientist since the invention of rational experimental science in the seventeenth century is a self-invisible, unmarked observer who can take on the authority of establishing transcendental truths, "this self-invisibility is the specifically modern, European, masculine, scientific form of the virtue of modesty....This kind of modesty is one of the founding virtues of what we call modernity. This is the virtue that guarantees that the modest witness is the legitimate and authorized ventriloquist for the object world, adding nothing from his mere opinions, from his biasing embodiment. And so he is endowed with the remarkable power to establish the facts" (Haraway 1997, 23–24).

If we think about the work of archaeoacoustic researchers, we can observe that their approach inherently challenges the archetypal laboratory scientist; the embodied nature of their fieldwork actively incorporates sounding, listening, and acoustic tests. The risk of the "god-trick" of a disembodied observer appears to be, at least concerning the fieldwork activities, allayed. This could indicate the beginnings of an investigation which fundamentally challenges such a dominant trope and re-questions what subjectivity and objectivity mean. *Situated Knowledges* argues against a charade of a supposedly neutral objectivity in favour of an acknowledgement of subjectivity rooted in an "always a complex, contradictory, structuring, and structured body" (Haraway 1988, 589). By understanding the processes of knowledge production in archaeoacoustics as situated, and by acknowledging the specificities of the field in terms of its partiality, the embodiedness and situatedness of sonic knowledge production must be taken seriously.

Haraway wrote that much of Western scientific and philosophical discourse is organized around positioning and the imagery of vision; this should always be understood as a part of the "politics and epistemologies of location, positioning, and situating," where partiality and not universality make up the conditions of "being heard." (ibid: 587, 589). However, Haraway reverts back to the metaphor of vision to reassert the importance of embodied vision, the actively seeing eye, and the concept of diffraction as a "double-vision" which unlike reflections, do not displace the same elsewhere, "Rather, diffraction can be a metaphor for another kind of critical consciousness[...] one committed to making a difference" (Haraway 1997, 273). Concurrent to this reappraisal of the technologies and epistemologies of vision in Haraway' *Situated Knowledges*, a complementary approach can turn to auditory epistemology and sonic knowledge – in fact, there are even clues in Haraway's essay gesturing towards this, "Feminist accountability requires a knowledge tuned to *reasonance*, not to dichotomy. Gender is a field of structured and structuring difference, in which the *tones* of extreme localization, of the intimately personal and individualized body, *vibrate* in the same field with global high-tension emissions" (my emphasis) (Haraway 1988, 588). The metaphors for reflection and diffraction work similarly well for the auditory as they do for the visual (Goh 2017, 296). The auditory activities of sounding and listening suggest a groundedness of sonic knowledge production in archaeoacoustics which can harness the critical potential of feminist epistemologies.

There are two interconnected levels which are pertinent to address in applying feminist epistemologies to archaeoacoustics, namely: firstly, the level of identity, representation, or subjectivity, and secondly, the epistemological level, which is aimed at challenging dualistic Western models of thought. The first is directly applicable to archaeological methods and interpretations, whilst the second belongs moreover to the philosophy of archaeology, yet also has applications for contemporary archaeoacoustic research (Wylie 2002). Gender archaeology, which emerged as a field since the 1980s has argued that the subjectivities of researchers has palpably and demonstrably affected how archaeologists have hitherto undertaken research and interpretation. Conkey and Spector state: "Most anthropologists have been western, white, and middle- or upper-class men, and their own position within a race, class, and gender system has shaped their perspective on research," whilst commenting that a similar statement can be applied to archaeology (Conkey and Spector 1984, 4). In particular, they criticize the pervasive "man-the-hunter" model as an example of androcentrism: by presuming specific "male activities" such as hunting, and specifically "female activities" such as child-rearing, historically and culturally specific ideas about gendered arrangements are being reflected onto human ancestors in ways which imply such roles are ahistorical. Such implicit assumptions present an idea of continuity in gender arrangements and thereby imply a kind of inevitability and immutability of these in these spheres of social life (ibid.: 7). Gendered biases influence and affect our interpretations of the past (Gero and Conkey 1991; Wylie 2001; Díaz-Andreu 2005), just as historical and contemporary notions and prejudices around race and ethnicity similarly affect archaeological interpretations (Orser 2001; Lydon and Rizvi 2016). Aside from practising self-reflexivity in research methodologies to minimize the encroachment of such identity-based biases into archaeological interpretations, one can hope that greater attention to political and ethical questions around research will enable such practices to become increasingly engrained.

If the identity or subjectivity-based issues are, seen as such, fairly easily identifiable, then it is worth pointing to the second level of sonic knowledge production in archaeoacoustics, the irrevocably entwined epistemological questions. Drawing on Haraway's work, Karen Barad develops a notion of *ethico-onto-epistemology* to denote the inextricable nature of ontology, epistemology, and ethics from one another in processes of knowledge production, "the separation of epistemology from ontology is a reverberation of a metaphysics that assumes an inherent difference between human and nonhuman, subject and object, mind and body, matter and discourse...what we need is...an appreciation of the intertwining of ethics, knowing, and being....because the becoming of the world is a deeply ethical matter" (Barad 2007, 185).

Similar to the ocularcentric models of knowledge production in Western culture more broadly, the contribution of critical analyses of knowledge production, which includes feminist epistemologies as just one strand of scholarly research, enables reflection (and perhaps a Harawayan diffraction) upon the many, deeply-embedded notions which significantly shape archaeological interpretation. Other stratifications of contemporary society, less easily named and described by terms such as gender, race, sexuality, or class will similarly have repercussions. The sensory hierarchy in knowledge production is merely one of many issues. If we regard the processes of knowledge production in archaeoacoustics ethico-onto-epistemologically, we may grasp that the philosophical and identity-based considerations are difficult to separate from one another.

On the level of identity or subjectivity, careful consideration of pre-existing gendered biases in archaeological interpretations may help to balance statements, for example, around the uses and roles of male and female voices in archaeoacoustics, or more generally around the societal activities of people of all genders and ages in archaeoacoustic case studies. On the ethico-onto-epistemological level, there are several interconnected philosophical bodies of work which could be productively examined using archaeoacoustics. A significant one of these might be the notion of mind-body dualism and the corresponding effects of ocularcentrism for knowledge production, where sonic knowledge production is but one amongst many possible alternative openings to challenge traditionally conceived, disembodied, Western, masculinist notions of knowledge.

Conclusion: Sounding Situated Knowledges

In conclusion, greater attention to all aspects of the questions being asked in archaeoacoustics can only be desirable at a time when the field is in a crucial moment of formation, as it currently is. Although the large majority of archaeoacoustics work will remain aimed at contributing to archaeological knowledge, rather than the epistemological questions surrounding sonic knowledge, as I have hoped to demonstrate in this article, the two issues are intimately related. Whether intentional or not, archaeoacoustics work is directed towards explicating forms of sensory (sonic) knowledges not commonly found in the modern, post-Enlightenment West in gathering evidence of the significance of sound and listening in past human behaviours at a site. As is clear in both archaeoacoustics and sound studies, ocularcentric modes of knowledge production have inhibited understandings of other cultures, as well as our own. By tracing through what we commonly know about Western traditions, such as its Cartesian inheritance of mind-body dualisms, it is possible to begin to theorize how sound and listening might have been used as ways of knowing which related differently to embodied experiences: these are the stakes of sounding situated knowledges.

Existing theorisations of sonic knowledges in sound studies so far, namely Steven Feld's "acoustemology" and Julian Henriques' "sonic logos" enable us to imagine how sound and listening-led practices can be considered examples of sonic knowledge production. Crucially, leaning on the key insights provided by feminist critiques of traditional science, such as Donna Haraway's *Situated Knowledges*, the centrality of the body in knowledge production should be treated carefully to foreground the importance of embodiedness and situatedness in knowledge production. As I have outlined here, archaeoacoustics already possesses the advantage of being rooted in the embodied practice of sounding and listening in its fieldwork activities; it should palpably be better positioned to avoid the mind-body dualisms responsible for previous historical ocularcentrism. However, as the body as "always... complex, contradictory, structuring, and structured" (Haraway 1988, 589) indicates, the inevitable partiality of knowledge so produced means that the ethico-onto-epistemological approach must be heeded to account for the situatedness of such processes: both embodiedness and situatedness are imperative in ensuring the accountability and responsibility of sonic knowledge production in archaeoacoustics. This includes, on one level, a reflexivity of identity or subjectivity-based biases in archaeoacoustics work to ensure, as far as possible, that deeply engrained cultural ideas taken for granted in a contemporary setting do not infringe upon archaeological interpretations without due acknowledgement. Yet, furthermore, on an interrelated level, the very sensory modes which are used in processes of knowledge production deserve greater interrogation to help attend to the manner in which these processes are operating, or being presumed to operate. It would be naïve to assume that merely paying attention to sound and listening will answer the many archaeological riddles of our time. However, archaeoacoustics's contributions can be enriched by attention to its processes of knowledge production as sounding situated knowledges as discussed here, which may enable us to better understand – even incrementally – human behaviours and cultures at times and places very different to our own.

BIBLIOGRAPHY:

Barad, Karen. 2007. *Meeting the Universe Halfway: Quantum Physics and the Entanglement of Matter and Meaning*. Durham: Duke University Press.

Blesser, Barry, and Linda-Ruth Salter. 2006. *Spaces Speak, Are You Listening? Experiencing Aural Architecture*. Cambridge, MA: MIT Press.

Butler, Judith. 2011. *Bodies That Matter: On the Discursive Limits of 'Sex'*. Routledge Classics. New York: Routledge.

Classen, Constance. 1993. *Worlds of Sense: Exploring the Senses in History and across Cultures*. London: Routledge.

Classen, Constance, David Howes, and Anthony Synnott. 1994. *Aroma: The Cultural History of Smell*. London: Routledge.

Conkey, Margaret W., and Janet D Spector. 1984. 'Archaeology and the Study of Gender'. *Advances in Archaeological Method and Theory* 7:1–38.

Day, Jo, ed. 2013. *Making Senses of the Past : Toward a Sensory Archaeology*. 27th Annual Visiting Scholar Conference (2010 Carbondale, Illinois). Carbondale: Southern Illinois University Press.

Díaz-Andreu, Margarita. 2005. *Archaeology of Identity: Approaches to Gender, Age, Status, Ethnicity and Religion*. London: Routledge.

Feld, Steven. 1996. 'Waterfalls of Song: An Acoustemology of Place Resounding in Bosavi, Papua New Guinea.' In *Senses of Place*, by Steven Feld and Keith H. Basso, 91–135. Santa Fe: School of American Research Press.

Feld, Steven. 2015. 'Acoustemology'. In *Keywords in Sound*, edited by David Novak and Matt Sakakeeny, 12–21. Durham: Duke University Press.

Gero, Joan M, and Margaret W Conkey, eds. 1991. *Engendering Archaeology: Women and Prehistory*. Social Archaeology. Oxford: Basil Blackwell.

Goh, Annie. 2017. 'Sounding Situated Knowledges: Echo in Archaeoacoustics'. *Parallax* 23:283–304.

Grosz, Elizabeth. 1994. *Volatile Bodies: Toward a Corporeal Feminism*. Theories of Representation and Difference. Bloomington, Ind: Indiana University Press.

Hamilakis, Yannis, Mark Pluciennik, and Sarah Tarlow, eds. 2002. *Thinking through the Body: Archaeologies of Corporeality*. New York: Kluwer Academic/Plenum.

Haraway, Donna J. 1988. 'Situated Knowledges: The Science Question in Feminism and the Privilege of Partial Perspective'. *Feminist Studies* 14 (3):575–599.

Haraway, Donna J. 1997. *Modest_Witness@Second_Millennium.FemaleMan©_Meets_OncoMouseTM: Feminism and Technoscience*. New York: Routledge.

Henriques, Julian. 2011. *Sonic Bodies: Reggae Sound Systems, Performance Techniques, and Ways of Knowing*. New York: Continuum.

Howes, David. 1991. *The Varieties of Sensory Experience: A Sourcebook in the Anthropology of the Senses*. Toronto, Canada: University of Toronto Press.

Howes, David. 2003. *Sensual Relations: Engaging the Senses in Culture and Social Theory*. Ann Arbor: University of Michigan Press.

Howes, David, ed. 2005. *Empire of the Senses: The Sensual Culture Reader*. Sensory Formations Series. Oxford: Berg.

Howes, David, and Constance Classen. 2014. *Ways of Sensing: Understanding the Senses in Society*. London: Routledge.

Ihde, Don. 1976. *Listening and Voice: A Phenomenology of Sound*. Athens: University of Ohio Press.

Ihde, Don. 2007. *Listening and Voice: Phenomenologies of Sound*. 2nd ed. Albany: State University of New York Press.

Irigaray, Luce. 1985. *Speculum of the Other Woman*. Ithaca: Cornell University Press.

Jay, Martin. 1993. *Downcast Eyes: The Denigration of Vision in Twentieth-Century French Thought*. Berkeley ; London: University of California Press.

Lydon, Jane, and Uzma Z. Rizvi. 2016. *Handbook of Postcolonial Archaeology*. London: Routledge.

Mignolo, Walter. 2000. *Local Histories/Global Designs: Coloniality, Subaltern Knowledges, and Border Thinking*. Princeton, N.J.: Princeton University Press.

Orser, Charles E, ed. 2001. *Race and the Archaeology of Identity*. Salt Lake City: University of Utah Press.

Pinch, Trevor, and Karin Bijsterveld. 2012. *The Oxford Handbook of Sound Studies*. Oxford Handbooks. Oxford: Oxford University Press.

Polanyi, Michael. 1962. *Personal Knowledge: Towards a Post-Critical Philosophy*. Repr. with corrections. London: Routledge & Kegan Paul.

Reznikoff, Iégor. 1987. 'Sur La Dimension Sonore Des Grottes à Peintures Du Paléolithique'. *Comptes Rendus de Académie Des Sciences* 304 & 305.

Reznikoff, Iégor, and Michel Dauvois. 1988. 'La dimension sonore des grottes ornées'. *Bulletin de la Société préhistorique française* 85 (8):238–46. https://doi.org/10.3406/bspf.1988.9349.

Said, Edward W. 1985. *Orientalism*. Harmondsworth: Penguin.

Scarre, Chris, and Graeme Lawson, eds. 2006. *Archaeoacoustics*. McDonald Institute Monographs. Cambridge: McDonald Institute for Archaeological Research.

Serres, Michel. 2008. *The Five Senses: A Philosophy of Mingled Bodies*. London: Continuum.

Spivak, Gayatri Chakravorty. 1988. 'Can the Subaltern Speak?' In *Marxism and the Interpretation of Culture*, edited by Cary Nelson and Lawrence Grossberg, 271–313. Urbana: University of Illinois Press.

Sterne, Jonathan. 2003. *The Audible Past: Cultural Origins of Sound Reproduction*. Durham: Duke University Press.

Tilley, Christopher Y. 1994. *A Phenomenology of Landscape: Places, Paths, and Monuments*. Oxford: Berg.

Wylie, Alison. 2001. 'Doing Social Science as a Feminist: The Engendering of Archaeology'. In *Feminism in Twentieth-Century Science, Technology, and Medicine*, edited by Angela N. H. Creager, Elizabeth Lunbeck, and Londa L. Schiebinger, 23–45. Chicago : University of Chicago Press.

Wylie, Alison. 2002. *Thinking from Things: Essays in the Philosophy of Archaeology*. Berkeley: University of California Press.

Zubrow, Ezra B. W. 2014. 'The Silence of Sound: A Prologue'. In *Archaeoacoustics: The Archaeology of Sound. Publication of the 2014 Conference in Malta*, edited by Linda C. Eneix, 7–9. Myakka City, Florida: OTS Foundation.

Wemyss Bay Caves Auralisation Project

Nicholas Green

NICHOLAS GREEN: MSc. Sector Manager and Senior Lecturer: Department of Audio Engineering, Perth College UHI, University of the Highlands and Islands, Scotland.

ABSTRACT: Wemyss Caves on the North Eastern firth of the River Forth are a multi-use site with significance in Pictish culture. The site shows evidence of a complex of eight coastal caves with six still accessible. Of significance are two caves containing the largest single site examples of early class 1 Pictish rock carvings. The site also represents the most Southerly frontier of Pictish culture, the Forth Valley creating a natural boundary between Southern and Northern Scotland. The site has been the subject of a recent 3D mapping project; in order to complete the multi-sensory mapping an auralisation mapping of the site has taken place between the summer of 2016 and winter 2017. Was the site a significant acoustic site for the early Picts? The site certainly presents some clues despite the challenging environment for carrying out controlled audio recording and analysis.

Introduction

Archaeoacoustic research has come of age in recent years; with exponential growth and development in digital technologies, we are now able to recreate sites in 3D with laser mapping and fully immersive soundscapes. Audio gaming engines and immersive 3D audio technologies will aid our understanding and help to deliver a meaningful experience of our shared past to the public in virtual reality environments. Digital technology is becoming the new gateway to our ancestors and to their sense of their world in a multi-sensory medium for the twenty first century. Since Professor Angelo Farina of the University of Parma first presented several papers on the practice and processing of IR's; impulse responses (Pcfarina, 2017), acoustic analysis has accelerated and with it archaeoacoustics. These papers first published in 2000 and presented at the Paris AES (Audio Engineering Society, 2000) have become the standard for field recording. Dr. Damian Murphy of the University of York developed the Open Air Library, an online repository for upload and download of various heritage sites. Using adapted and updated techniques similar to Farina's practices as the reference point for field recording IR's (Openairlib, 2017). These techniques will be discussed further in the field recording methodologies section of this paper.

Farina's work pre-dates the first archaeoacoustics conference held at Cambridge University in 2003 and chaired by Chris Scarre and Graeme Lawson, the book of proceedings becoming the first reference textbook that started many new researchers in the field of archaeoacoustics. However there has been analysis of archaeological sites and their acoustic properties in practice for years. In particular the work of Paul

Devereux and colleagues must not go unrecognised. In fact one of Paul's papers is presented in the first Archaeoacoustics publication along with main researchers of the current field, Iegor Reznikoff and Ezra Zubrow (Scarre, C., & Lawson, G. 2006, *Archaeoacoustics*).

The Site

The site of Wemyss Caves located on the North shore of the river Forth estuary is significant, featuring the largest concentration of early 'class 1' (Foster, S. 1996, *Picts, Gaels and Scots*) Pictish rock art known. The site is also significant as the river Forth represents the most Southerly boundary of the Picts as pointed out during first visits to the site in summer 2016 by archaeologist Dr. Joanna Hambly of the University of St. Andrews (Hambly, J. 2016). This natural geographical border is even more significant when a map of Scotland is considered. The rivers Forth and Clyde create a natural bottleneck between the North of Scotland most of which was inhabited by the late Bronze Age Picts and most intensively the North East where many of their carved standing stones still exist in large numbers (Wainwright, F.T. 1955).

As a culture spanning the late Bronze Age to the early Medieval, the Picts were given their now accepted name by the Romans. Roman General Agricola and his chroniclers named them for the markings on their skin (Pictavia, 2017), Picti meaning picked or tattooed in Roman Latin. There is however no archaeological evidence that this was the case, the Picts were a Celtic culture with much in common with Celtic peoples across Europe. There is evidence that they were seafaring and traded as far as the Mediterranean. They remain an enigma due in part to having no first hand records of their culture and partially due to the lack of understanding and speculation that persists around the meanings in the symbolic standing stones, their most famous legacy (Wainwright, F.T. 1955).

Image. 1 The current shoreline

Image. 2 The shoreline as the Picts would have found it

The site of Wemyss Caves, as well as being the boundary of Pictland, has also been a multi-use site over the centuries. Despite the caves' relative geological infancy, being formed by tidal erosion between 8 and 5,000 years ago, it is hard not to believe that our Neolithic/Mesolithic ancestors did not use them (wemysscaves.org/the-caves, 2017). There is however no evidence to support this. There are the scars of twentieth century industrial use still in evidence and the remains of the Medieval McDuff Castle (wemysscaves.org/the-caves, 2017). The Scottish coal and gas board both used the site, creating an access road along the shoreline and a pipeline that are both still visible today (wemysscaves.org, 2017). This, despite ironically tidal erosion, is probably the biggest threat to the site nowadays. Court Cave has two large brick-built pillars built into the entrance of what would have been an impressive opening facing West along the shoreline. These were courtesy of the Coal Board in the late nineteen thirties. McDuff Castle sits on the cliff top directly above Well Cave.

One can assume a potent site due to excavations that discovered a naturally occurring spring inside the cave, no longer active due to mining activity in the area. This was formerly known as St. Margaret's Well, a spring of holy value and no doubt valued by the pre-Christian Picts alike (wemysscaves.org, 2017).

For this study four caves were 'sampled', from East to West: Court Cave, Well Cave, Jonathan's Cave and Sliding or Sloping Cave, so named for a relatively recent landslide that has occluded much of the original entrance. Of these caves the two most impressive acoustically are Court and Jonathan's Caves, both of which contain the highest number of Pictish rock carvings. Court Cave features mainly Pictish symbol carvings with Jonathan's Cave mainly depictions of animals. If there were early Pictish rock carvings in Well Cave they are no longer visible; every available surface up to shoulder height has been over inscribed through the centuries with much evidence of Victorian and Twentieth century 'graffiti'. It may be due to the lack of natural daylight filtering into Well Cave that it actually never contained any Pictish rock carvings. This would have rendered the Pictish artists blind unless working by flickering firelight. We shall never know.

Image. 3, 4 Pictish symbol panel Court Cave and animal carving Jonathan's Cave

In addition to animal symbols found mostly on the South Westerly wall near the cave entrance, Jonathan's Cave contains the earliest known carving of a boat with one figure in it. This single depiction is of value and significance on its own. The Picts, like other Celtic peoples, were seafaring people and the shoreline location may have been used as a launch site. The cave also features later Pictish crosses interestingly carved on surfaces much deeper into the cave. It is no accident that much of the early Pictish symbol carving is found near the entrances of both Jonathan's and Court Cave due, one can assume, to the natural daylight available for their creation.

The location of each cave poses recording challenges. Court Cave has an entrance that faces West parallel to the shoreline. A partial landslide over the Southerly portion of the sizable entrance also helps to mask the sound of tidal action. This makes Court Cave particularly good for acoustic analysis where background noise is minimal comparatively speaking. Doo Cave was ruled out of this study; containing few Pictish rock carvings, it is also far too shallow to yield useful IR results and directly faces the shoreline.

Well Cave is a noise-free space upon entering, helping to exterminate external noise contamination by virtue of the small tunnel like entrance itself located at the rear of a smaller cave entrance creating an anti-chamber and acting much like the air lock entrances to professional recording studios. The entrance also has the effect of rendering the main space of Well Cave almost completely pitch dark.

Jonathan's Cave is a little more challenging, facing directly South to the shoreline and the accompanying tidal action. It is helped by being raised up the cliff face with a lip along the entrance floor, helping to eliminate tidal noise. There is however trace effect visible in spectrogram analysis and faintly audible in the IR recordings, adding background distortion to these renderings.

Finally, Sliding or Sloping Cave although containing very few carvings has a partially collapsed entrance and again benefits by being accessed via a small climb up the cliff face to enter. Only a metre or so into the cave this renders external noise almost inaudible, certainly to the human hearing mechanism.

The Picts would have experienced a different soundscape (Schafer, 1993) than the modern site. With a rise in sea level in the last few millennia, the shoreline is considerably higher than our ancestors would have found it. In the late Bronze Age, early Pictish era an extended grassy raised beach would have made the tidal noise more distant.

Field Recording Methodology

Several audio recording techniques were utilised during multiple visits to the site between summer 2016 and summer 2017. This required transporting battery operated equipment onto the site along with video recording equipment to document the field recording practice. The main field recordings were conducted on 4th of November 2016.
Access to the site is quite straightforward, although 2 shoulder bags and 1 backpack containing the field recording equipment as well as tripods and microphone stands had to be transported along a mile or so stretch of coastline. The caves were visited in an East to West fashion. To help negate the possibility of tidal noise, the timing of the field trip coincided with low tide. The weather was still, with little hint of a breeze although a chilly 5°C which can be detrimental to battery longevity. However all equipment operated without problems and lasted the duration of the field trip.

Each of the caves' acoustic signatures was recorded using two techniques: a high fidelity and a lower fidelity 'guerrilla' IR recording technique. Having visited the caves previously with members of the Save Wemyss Ancient Caves Society (SWACS) and Joanna Hambly of St. Andrews University, a good idea of microphone and source playback placement was gained. For the high–

fidelity recordings the following equipment was selected:

- Apple Macintosh Macbook Pro running Impulse Utility software
- Focusrite Scarlett USB bus powered audio interface
- Beyer Dynamic MM1 reference class 2 omni-directional microphones
- B&O Beoplay A2 Omni-directional loudspeaker system (although a consumer device it has very linear playback qualities)
- Mogami low noise XLR cables, one adapted for loudspeaker playback

The recording method employed for this field recording was to take three mono recordings from three different locations within each cave. This is a W, X, Y technique that is B-Format compatible. In other words, it is compatible with mono, stereo and through to fully immersive ambisonic playback of up to 16 directional loudspeakers with LF extension filter for sub bass frequencies (ambisonic.xyz). By taking recordings in three separate positions a surround sound audio image can be recreated in laboratory simulations.

The recording format in IR recording methods and acoustic analysis needs to be considered. A sample rate of 96KHz was chosen as the highest audio resolution available on the system hardware in use with the high fidelity recording technique. Of equal if not more importance, is the choice of Bit depth. 24 Bits is the minimum standard acceptable in IR recording methods. This retains the highest dynamic range available within a digital recording system and is vital to capturing the lowest levels of decay possible when using an IR method, particularly if this is to be convolved into a reverb algorithm. The point of reverberation decay where the reflected sound ceases to be audible is how we accurately measure reverberation time (Murphy, D. 2013)

There are two approaches that can be employed in this field recording system set up. The first is to position the microphones in a central location within the space, rotating the position three times in a 360degree arc, the source playback being repositioned accordingly at each rotation. This works in the context of standard architecture. A flexible approach to positioning needs to be considered when presented with the more complex space of a cave. The approach employed for this fieldwork is actually a better means of gathering the acoustic characteristics of the space by reversing the microphone and source playback positions. The playback source regarded as the best means of fully agitating the full audio spectrum from 20Hz to 20,000HZ was a sine sweep generated over a 10 second playback time, rising from the lowest frequency to the highest. Using a digital process of Fourier based convolution these recordings can be effectively squashed to create an IR, an IR is by its very nature a broadband, short, loud burst of acoustic energy and is well described by Tae Hong Park as "Agitating a System" (Park, T. 2010, Introduction to Digital Signal Processing). The agitation is the impulse, the system and how it responds is in this case a series of caves.

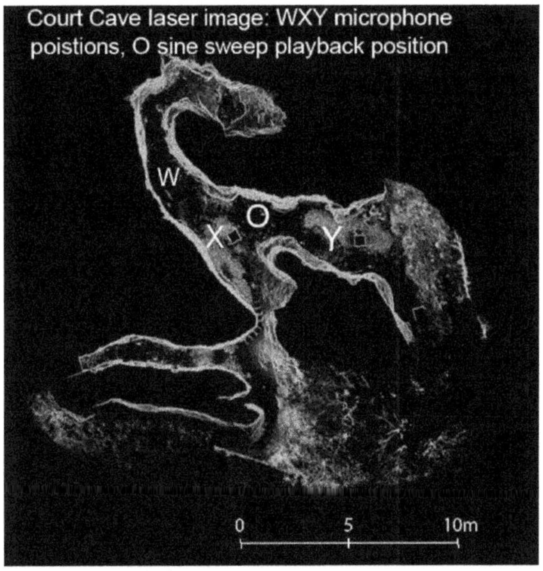

Image. 5 WXY - microphone positions,
O - sine sweep playback position: Court Cave

These recordings are of interest in an auralisation context as they can be used to recreate the reverberation and frequency response characteristics of the caves very accurately. To represent the IR visually, spectrogram recordings were also taken using various software; some integrated into the software of Impulse Response Utility on the laptop, others using a smart phone application, a freely available download. The smart phone's own microphone was not selected in favour for a relatively new reference class miniature microphone developed for use with smart phones and tablets, this also adheres to reference Class 2 standard (*IEC 61672 - A Standard for Sound Level Meters Explained*, 2017).

The Guerrilla Impulse

A second means of acoustic measurement was also employed to augment the initial high fidelity approach as detailed above. A small portable PCM audio recorder was used to record the sound of a balloon being burst within the space of each cave. This is a comparatively crude approach to IR recording but does yield impressionist results that can be very easily used in convolution processes. There are some advantages to this technique over the sine sweep playback method. In a challenging environment which exhibits background noise, such as the shoreline location of Wemyss Caves a balloon being burst gives a relatively instantaneous impulse, this can be timed to reduce the inclusion of background noise which can cause unwanted acoustic shadowing at best, distortion at worst. The equipment is exceptionally portable and inexpensive; the portable audio recorder of choice has a four capsule microphone array that can be set in a pseudo omni-directional, surround pickup pattern giving a 360degree impression of the space.

As a means of recording and analysis, as long as the shortcomings are taken into consideration, there has been an argument that some sort of digital acoustic impression of a potentially endangered heritage site is better than none (Green, N. 2015). This technique produces useful data and can be used to recreate convolution-reverbs of the space very accurately if practiced. Balloons exhibit their own resonant frequencies but do produce enough mid-range energy in the main focus of the human hearing spectrum. The balloons should be inflated to the maximum size possible with care being taken when setting the pre-amplifier gain level so as not to record a distorted or clipped signal. The recordings in the Wemyss Caves location were timed to coinside with the tidal wave sweeping out, the lull, and a point where the least background noise contamination is audible. Various field-recording approaches to IR capture are detailed in this journal article (Kemler, A. 2017, Acting on Impulse).

Archaeologist and anthropologist Riitta Rainio of the University of Helsinki used a hybrid recording method for acoustic analysis at a site of ancient painted cliffs in Northern Finland. In that instance she and her team used a handheld portable PCM audio recording devise coupled to a high-end reference class pair of microphones recording in a close stereo A, B configuration. The impulse was produced by firing a starters pistol which, similar to the balloon being burst, produces enough energy in the midrange spectrum with enough low and high frequency energy being produced to make the waveform analysis useful (Eneix, L. 2014, *Archaeoacoustics*).

All guerrilla recordings were taken using a minimum format standard in IR methods, 48KHz and 24Bit; this is B-wav or broadcast standard file (Murphy, D. 2013).

Discussion and Conclusion

SWACS (Save Wemyss Ancient Caves Society) a local community group, have been responsible for raising money from the Scottish government to help with the preservation of the site and the 3D laser mapping and website creation. The Wemyss 4D web project set out to map the site digitally for the purposes of heritage archiving and to create a virtual online tour of the caves with high-resolution images of the Pictish rock carvings. The site also features a visualisation time-line of the shoreline site from modern times through industrialisation of the site at the turn of the twentieth century, the Medieval and the Iron Age Pictish.

From these visualisations and environmental evidence, the greatest future threat to the site is the rising sea level and encroachment of tidal action, the very means by which the caves were created to begin with. The shoreline site has suffered from multi-use over the years with gas and coal industrial use along the site dominating the landscape in the mid-twentieth century. Jonathan's cave was also subject to vandalism by bored local youths in the 80's (Hambly, J. 2016). Thanks to a road built to access a gas works at the far West end of the bay and still partially in evidence today, a stolen car was driven along to Jonathan's cave where it was abandoned and set light to. The heat generated damaged the surfaces of the cave immediately surrounding it, effecting mainly the roof of the cave. The action did also damage and cause some of the carvings to be lost. Projects such as the Wemyss 4D project become vital in archiving sensitive heritage sites for current and further generations and for continued research.

All of the above field recordings were augmented with recordings of performances using a frame or shamanic drum, a shallow drum, 60cms in diameter. This was conducted to assess any further frequency analysis of the caves using an instrument. However there are no records of shallow frame drum having been used by the Picts. Rather there is evidence of a deep tom-like drum depicted on a later Christian Pictish stone relief, known as the Lethendy Tower panel (Lethendy Tower. Canmore, 2017). There is a recreation of this drum constructed using the image from the stone as the only reference point. This reconstruction artefact was on display at the now closed Pictavia Museum, Brechin, Scotland, and has recently been traced to a store in Angus museums archive. Access to record the drum has been recently granted (archive.angus.gov.uk/historyaa/museums/meffan/, 2017). To complete this study, further field recording of the drum, which cannot be removed from the Angus museum collection, can however, thanks to IR recordings and the use of convolution reverb processing, be effectively translated to the site.

Image 6: The Lethendy Pictish Drum reconstruction

The philosophy behind augmenting acoustic analysis recordings with imagined performance pieces is that our ancestors did not have access to digital and electroacoustic equipment. If there was an awareness of the acoustic properties of the site it is wholly subjective and survives only in our modern context of imagining. Chanting or some type of vocalisation is also favoured alongside IR and sine sweep recording (Scarre, C., & Lawson, G. 2006, *Archaeoacoustics*). Observations from these additional recordings can offer further insight into the spaces inhabited by our ancestors and help to answer questions such as how and why sites of acoustic significance were selected. It is the hope that the field recording and auditory mapping of the site will help to complete the picture of this site. With emerging digital technologies we can preserve sites such as this in a virtual environment. SWACS, as a local community initiative aid in the stewardship of the site as well as curating a collection of papers and artefacts at the local school associated with the caves, in turn providing education and awareness of the site's cultural importance to the local community and beyond.

ACKNOWLEDGEMENTS

Dr. Joanna Hambly, University of St. Andrews
Mike Arrowsmith, SWACS and Wemyss4D
John Johnston, Meffan Art Gallery and Museum
With thanks to research assistant Keith Harvey and reader in music UHI, Dr. Mark Sheridan

REFERENCES

A Brief History of the Picts. (2017). *Pictavia.org*. Retrieved 26 October 2017, from http://www.pictavia.org/history/history.html

Ambisonic.XYZ. (2017). *Ambisonic.xyz*. Retrieved 30 December 2017, from http://ambisonic.xyz

Angelo Farina's Home Page. (2017). *Pcfarina.eng.unipr.it*. Retrieved 26 October 2017, from http://pcfarina.eng.unipr.it

Arrowsmith, M. (2017). *Wemyss4D*. *4dwemysscaves.org*. Retrieved 26 October 2017, from http://4dwemysscaves.org

Eneix, L. (2014). *Archaeoacoustics* (p. 141). Myakka City, FL: The OTS Foundation.

Foster, S. (1996). *Picts, Gaels and Scots* (p. 71). Historic Scotland.

General view of Pictish cross-slab carving re-used as stair lintel, Lethendy Tower. | Canmore. (2017). *Canmore.org.uk*. Retrieved 15 December 2017, from https://canmore.org.uk/collection/449889

Green, N. (2017). *The Guerilla Impulse*. *Archaeoacoustics Scotland*. Retrieved 15 December 2017, from http://www.archaeoacousticsscotland.com/the-guerilla-impulse.html

IEC 61672 - A Standard for Sound Level Meters Explained. (2017). *NoiseNews*. Retrieved 26 October 2017, from http://www.cirrusresearch.co.uk/blog/2012/07/iec-61672-a-standard-for-sound-level-meters-in-three-parts/

Interview with Dr. Joanna Humbly. (2016). Wemyss Caves, Fife, Scotland.

Kemmler, A. (2017). *Acting on Impulse*. *Emusician.com*. Retrieved 26 October 2017, from http://www.emusician.com/gear/1332/acting-on-impulse/36638

Local History | Museums | Meffan Museum and Art Gallery | Angus Council. (2017). *Archive.angus.gov.uk*. Retrieved 15 December 2017, from http://archive.angus.gov.uk/historyaa/museums/meffan/default.htm

Murphy, D. (2017). Interview with Dr. Damian Murphy. University of York.

Murphy, D. (2017). *OpenAIR | The Open Acoustic Impulse Response Library*. *Openairlib.net*. Retrieved 26 October 2017, from http://www.openairlib.net

Park, T. (2010). *Introduction to digital signal processing* (p. 132). Hackensack, N.J.: World Scientific.

Scare, C., & Lawson, G. (2006). *Archaeoacoustics*. Oxford: McDonald Institute for Archaeological Research.

Schafer, R. (1993). *The soundscape*. Rochester, Vt.: Destiny Books.

The Caves. (2017). *Save Wemyss Ancient Caves Society*. Retrieved 15 December 2017, from http://www.wemysscaves.org/the-caves.html

Wainwright, F. (1955). *The Problem of the Picts. Edited by F.T. Wainwright. [The Picts and the problem, by F.T. Wainwright, ... The Archaeological background, by Stuart Piggott, ... Fortifications, by R.W. Feachem, ... Houses and graves, by F.T. Wainwright. Pictish art, by Robert B.K. Stevenson, ... The Pictish language, appendix, by K.H. Jackson, ...]*. Edinburgh: Nelson.

Images 1 – 5 courtesy of SCAPE, image 6 N. Green

Archaeoacoustics and the Fall of Jericho

Anne Habermehl

ANNE HABERMEHL, B.Sc., is an independent American/Canadian scholar who publishes on ancient history subjects from a biblical point of view. Contact: anneh@twcny.rr.com.

ABSTRACT: The conquest of the Canaanite city of Jericho by Joshua's forces, as famously narrated in the Hebrew Bible, was a fascinating archaeoacoustic event. There were horns blowing, feet marching, shouts, the sounds of the wall falling, a fierce battle, and a great fire. Bible believers do not doubt that the conquest of Jericho took place as described in the historical account, and that the wall fell by Divine action. Skeptics say that archaeology proves there was no destruction of Jericho at the time in history when it should have occurred. Discussion of these issues shows that one's worldview will ultimately determine acceptance or rejection of the Jericho event as real history.

KEYWORDS: Jericho, ancient warfare, worldview

Introduction

Ancient warfare was tumultuous, and attacking forces were ruthless in their destruction. In this, the conquest of the Canaanite city of Jericho by Joshua's forces, as narrated in the Hebrew Bible in Joshua 6, was not unusual. A parallel example of this event is found in the annals of the Assyrian king Tiglath-Pileser I, who describes (we might say with relish) how he burned, razed and destroyed enemies' cities, made them into ruin hills and heaps, cut off the inhabitants' heads and made the blood flow into the hollows and plains of the mountains (Younger 1990, pp. 80–81).

However, Jericho's defeat by Joshua's forces was distinguished by some unusual features that included a Divine miracle in making the city's wall fall. This latter aspect has naturally raised skepticism from those who do not accept the biblical narrative literally.

Many scholars deny that the Exodus of the Children of Israel from Egypt and their conquest of Canaan really happened, and claim that the stories are merely traditions and myths (see, for example, Rogerson 2014, pp. 21–22). On the other side are those who believe that these events happened exactly as the Bible narrates. Nobody denies that there is a ruined tell (Tell es-Sultan) on the Jericho plain where successive walls fell and fierce fires destroyed the city repeatedly during the third and second millennia BC (Nigro 2006).

Background of the Story

The Bible relates in the book of Exodus that in Egypt the Children of Israel, Jacob's descendants, had been under heavy bondage to a tyrannical pharaoh who forced them to do extremely hard labor under cruel taskmasters. God raised up Moses to deliver his people from the Egyptians and lead them to their promised land of Canaan. The pharaoh finally allowed them to leave after God sent ten plagues to destroy Egypt: all surface water turned into blood, frogs, biting insects, wild animals, livestock disease, boils, fiery hail, locusts, darkness and finally, deaths of all the firstborn of people and animals in the land. (The Israelites, who were living in Goshen, were exempt from all these afflictions.) This miraculous deliverance out of Egypt is commonly called the Exodus.

With Moses as their leader, the people headed en masse for their promised land, Canaan, to conquer and inhabit it. God forced them to wander in the wilderness between Egypt and Canaan for 40 years as punishment, because they had made Him angry. Finally they arrived at the Jordan River, the eastern border of Canaan, with Joshua now their leader. The first city they had to conquer was Jericho, located strategically on the eastern edge of Canaan, near the Jordan River (see Figure 1). The fame of the Exodus and related events had preceded them, and fear of the Children of Israel had spread far and wide (Joshua 2: 9–11).

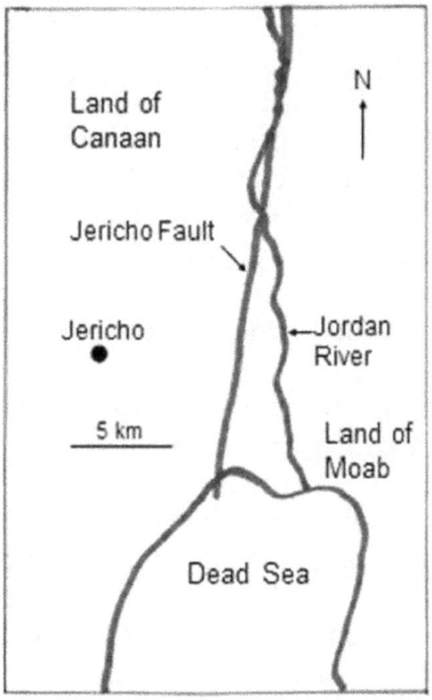

Figure 1. Map of the north end of the Dead Sea, showing the position of Jericho relative to the Jordan River and the Jericho Fault. The Children of Israel crossed the Jordan River from the east to the west and camped between Jericho and the river before conquering the city. (Habermehl 2017, after Al-Zoubi et al. 2007)

About Jericho

Situated in a key location north of the Dead Sea at 258 m (846 ft) below sea level, the ancient city of Jericho was founded about 11,000 years ago on the standard timeline of history (Kenyon 1998), making it nearly as old as Göbekli Tepe in southern Turkey. Its name most likely comes from the ancient Canaanite moon god Yarikh (also spelled Yerach); Jericho was an early center of worship of this god (Noll 2013, p. 337). It was called the city of palms because a copious year-round spring watered the area and many palms grew there. Over the millennia it had been destroyed and rebuilt many times, and had built up into a tell that is today called es-Sultan (Nigro 2006). (The modern city of Jericho is situated nearby.)

In Joshua's day, the wall probably enclosed about 3 ½ hectares (8.5 acres) (Gabriel 2003, p. 121). Clearly some of those ancient cities were quite small by today's standards. The Jericho that Joshua and his forces encountered was well fortified by a wall with towers at intervals. Rahab, the harlot, helped Joshua's spies escape by secretly letting them down outside the city wall on a cord through a window (Joshua 2:15) to avoid the pursuing soldiers of the king of Jericho.

As shown in Figure 1, the ancient tell sits almost directly on a geological fault (called the Jericho Fault) that runs north from the Dead Sea (see Al-Zoubi et al. 2007, p. 57). Because of this fault, and the geological structure of the area, Jericho has seen many earthquakes throughout its history.

Acoustics of the Event

According to the Hebrew Bible, God told Joshua how he was to conquer Jericho (Joshua 6). A procession of warriors, followed by seven priests blowing rams' horns, priests carrying the sacred Ark, and a rear guard, was to march around the city once a day for six days, and seven times on the seventh day. The mass of people was to surround the city (outside the circuit of the procession) as onlookers each day. For six days the people were to remain quiet. After the seven circuits on the seventh day, the priests were to blow a long blast on the horns and all the people were to shout loudly. God promised that the wall would immediately fall flat—which it did, no doubt with loud sounds of tumbling masonry. This was followed by the sound of the battle on top of the tell (mound), in which all human and animal occupants of the city were killed by sword (except Rahab and her family). Last of all was the roaring sound of the fire that destroyed the city. We will look at these elements individually.

The Shofar is a hollow animal horn, the traditional trumpet of the Jewish people from early times to this day. The shofarot (plural of shofar) in the Bible were rams' horns (Montagu 2015, pp. 3,4), and therefore those blown at Jericho would have been from rams. (The shofar had to be specified because there were also silver trumpets used in the tabernacle rituals (Numbers 19:1,2)). The shofar could be compared to a bugle because its pitch could be changed by the embouchure of the player. Although it was prescribed for use at different times in Jewish rituals, it was an important instrument for war; see Adler (1894, pp. 444–446) for a list of 19 biblical instances of blowing the shofar for war. It appears from these biblical citations that it was a loud instrument, and seven of them were blowing simultaneously at Jericho. In Egypt in the time of King Tutankhamun, trumpeters blew short notes to time the steps of the marching soldiers (Williams 2003); we might expect that the Children of Israel, who had lived in Egypt for 215 years, would have followed a similar military tradition. According to Goldman (1999, pp. 220–221), the blowing of these horns would not have contributed to making the wall fall because the horns' vibration frequency range was too high.

Marching. The march around Jericho was possibly not a strange custom to the inhabitants of the city. As Waltke (2007, p. 519) says: "The royal march around the city is based on widespread custom in the ancient Near East of laying claim to territory by tracing out its bounds." The number of soldiers that Joshua would have commanded, and who would have marched in the procession around Jericho, is estimated by Gabriel (2003, pp. 113–114) to have been about 8,000 men. How loud a sound their marching feet would have made on the ground around Jericho with the footwear of that time is difficult to say. They would most likely have worn leather sandals as sandals are mentioned in scripture as footwear of the Children of Israel during their trek from Egypt (e.g., Deuteronomy 29:5).

But audible sound is not the only result of the marching. Experts include infrasounds (cycles per second in the range below what the human ear can hear) in the science of acoustics. Goldman (1999, pp. 220–221), a mechanical engineer who specializes in vibration analysis and pulse theory, says that it was likely the marching feet of the soldiers in the procession that would have caused the low-frequency vibrations needed to make the wall fall. His reason was that the wall would have had a natural frequency close to that produced by the marching feet. Jones (2001, p. 2), an expert in viscoelastic vibration damping, also mentions that the vibrations of Joshua's marching soldiers supposedly are what made the wall fall. He cites the failure of the Tacoma Narrows bridge in 1940 in the State of Washington as an example of vibrations that became excessive; in that case the wind (64 km/hr) caused the damaging vibrations. However, if marching could cause city walls to fall, we might wonder why we do not have reports of the walls of other cities falling down from marching as well.

Shouting. After the 7th circuit on the 7th day, all the people were to shout at the same time as the trumpets blew. How many people were there, and how much noise did they make? According to Gabriel (2003, pp. 113–114), the number of the Children of Israel of all ages would have been about 35,000. Goldman (1999, pp. 220–221) says that the shouting would have served mainly to terrify the inhabitants of Jericho, but would not have contributed to making the wall fall because the frequency of the vibrations produced from the voices was too high.

Earthquake: Because Jericho sits almost on top of a geological fault (as noted earlier), an earthquake could have caused its wall to topple. This quake would have been accompanied by low rumblings or various other noises from the ground's movement (Hill 2011). If an earthquake caused Jericho's wall to fall, believers consider that its timing just at the moment of the 7th-day shofar blowing and the people's shouting had to have been a Divine action. Doubters, who do not accept the historicity of the biblical account, prefer to say that earthquakes may have destroyed the city at some not-specified time or other. Rucker and Niemi (2010, p. 103) warn about circular reasoning in claiming that an earthquake happened at a certain date in ancient history, and then tying specific earthquake damage to that date. (We will not consider the possibility that aliens made the wall fall, as promoted in the Ancient Aliens 2017 TV episode.)

Falling of the wall: How much sound this would have made cannot be estimated with any certainty. We do not know what the wall was built of (whether only mudbrick, or mudbrick on a stone foundation), and what its height and thickness were. Which fortifications from what era are also in question, as it is possible that archaeologists have been looking at the wrong set of walls in their discussions. They believe that there was a set of two walls in place, one around the top of the tell and one around the bottom, although the biblical account consistently says "wall," singular. (For a description of this set of walls, see Wood 1999.) Further discussion of this is in the section below on dating the walls.

Battle Sounds: After the wall fell, the soldiers were instructed to go straight up the mound into the city and kill all the inhabitants (except Rahab and her family) and all animals (Joshua 6:5). We would expect that this part of the conquest of the city would have entailed considerable sounds of terror from the occupants.

Fire. The Jericho city structures would have been made of mudbrick, as most buildings in the Near East have been since Neolithic times right up to the present day (Forget and Shahack-Gross 2016). These authors did research on heating mudbricks, and determined that ancient cities may have burned a lot faster than had previously been believed, in as little as 2–3 hours. As anyone knows who has burned wood in a fireplace, the faster the fire burns, the louder it roars. The fire that burned Jericho may have been a loud conclusion to the city's conquest.

Did This Event Really Happen? If So, When?

Today many theologians deny that the biblical stories of the Old Testament are real history. These scholars do not care whether there ever was a Joshua, let alone a conquest of Canaan. According to them, the narrative of the fall of Jericho should not be taken literally (e.g., Thompson 1999, pp.

34–44) and therefore this event did not happen.

Archaeologists, however, look to the results of their excavations for verification or denial of biblical events. One of the reasons most often given by archaeologists for disbelieving the literal conquest of Jericho as related in the Bible is that there was no walled city of Jericho in place at the time of the claimed biblical conquest; therefore they claim that the biblical account is not historical. Kathleen Kenyon, whose excavations at Jericho earned her great renown, discusses some aspects of this problem (Kenyon 1970, pp. 208–212).

This raises the important subject of dating the ruins, and it is a hotly debated topic. According to the Hebrew Bible, the destruction of Jericho occurred 40 years after the Exodus from Egypt; this was the length of time that the Children of Israel wandered in the wilderness between Egypt and Canaan as a punishment from God because they had displeased Him (Numbers 32:13). The Exodus is generally considered by bible-believing scholars to have taken place at about 1450 BC (Habermehl 2013), making the conquest of Jericho about 1410 BC.[1] Secular scholars say that Jericho was in ruins at that time because carbon 14 dating of the top ash layer shows the destruction earlier, around 1550 BC (Howard 1993, p. 97).

But the biblical and secular timelines do not mesh, as shown by Habermehl (2015) at the Archaeoacoustics II conference in Istanbul. The divergence of these two timelines is about 350 years (counting back to the end of 12th Dynasty) or 750 years (counting back to the end of 6th Dynasty) at the time of the Exodus (Habermehl 2013); this is based on these two dynasties running concurrently and ending at the same time when Egypt collapsed because of the ten plagues. This means that the conquest of Jericho should show up on the secular timeline as early as 1750 BC or 2150 BC, or even earlier, using standard dating methods. This throws confusion into the dating because Kenyon's date is not early enough (as noted above).

I believe that it is highly possible that the dating argument is over the wrong set of walls. Like Garstang, who excavated earlier from 1930–1936 (John Garstang 1998), everyone believes that the fourth set of destroyed walls is the one that fell in Joshua's conquest (the destructions are numbered from the bottom up at Jericho, and the fourth destruction is the top one). This double set of walls looks as if it could have been destroyed by an earthquake, and there is a thick layer of ash from a fierce fire (Kenyon 1970, pp. 197–198). It seems to fit the biblical story enough to satisfy uncritical believers. However, it is overlooked that these walls do not actually fit the biblical account (Joshua 6:20–25) because Joshua's wall was singular in the Bible (not "walls"). There are also difficulties in working out how the spies were let down on the rope by Rahab, if there was a wall around the crown of the mound as well as another at the foot of the mound. (The explanation commonly given is that Rahab's house was located inside and against or on top of the lower wall, and that she did not live in the city proper).

According to the Bible, after Joshua's conquest Jericho's fortifications were rebuilt some hundreds of years later by a man called Hiel in the time of King Ahab of Israel (I Kings 16:34), who ruled for 22 years from 918–896 BC (Jones 2007, p. 279). It is most likely Hiel's double wall that everyone mistakenly believes is the wall that fell in the biblical account, while the wall that actually fell in Joshua's time would have been an earlier one. This would solve the dating problem all around. We cannot go into further detail on this here.[2]

Conclusion

Believers will consider this story to have happened as narrated in the Bible, and will be certain that God made the walls fall at the moment of the shouting and shofar blasts. For them, questions of dating will be secondary. On the other side, skeptics will look for natural reasons like an earthquake or possibly even marching feet to account for the tumbling of the walls that have been excavated at the tell es-Sultan. For them, the question of whether the capture of Jericho happened as the Bible claims is secondary. It is doubtful that Jericho's conquest was an incidence of sonic warfare.

Which walls were the ones of Joshua's time, and whether the biblical and secular timelines diverge, are controversial matters. The question of whether or not there was a fortified city of Jericho in place at the right time remains. Ultimately it is one's worldview that will determine whether to accept or reject the literal biblical Jericho event as real history.

NOTES

1. Some scholars believe that the Exodus and conquest took place as late as the 13th century BC. For an excellent discussion on this, see Heater (2014, pp. 77–84).

2. A paper on this subject will be published in the future, in which I will propose that Joshua's wall was actually an earlier one, based on timeline and other considerations. (website www.creationsixdays.net)

REFERENCES

Adler, C. 1894. *The Shofar: Its Use and Origin*. Smithsonian Institution. Washington, D.C.: Government Printing Office.

Al-Zoubi, A.S., T. Heinrichs, I. Qabbani, I., and U.S. ten-Brink. 2007. The northern end of the Dead Sea basin: Geometry from reflection seismic evidence. *Tectonophysics* 434:55–59.

Ancient Aliens. 2017. Soundwaves that can shatter walls. Television series episode. *Prometheus Entertainment.*

Forget, M.C.L., and R. Shahack-Gross. 2016. How long does it take to burn down an ancient Near Eastern city? The study of experimentally heated mud-bricks. *Antiquity; Cambridge* 90 (353):1213–1225.

Gabriel, R.A. 2003. *The Military History of Ancient Israel.* Westport, Connecticut: Praeger Publishers.

Goldman, S. 1999. *Vibration Spectrum Analysis, 2nd edition.* New York, New York: Industrial Press, Inc.

Habermehl, A. 2013. Revising the Egyptian chronology: Joseph as Imhotep, and Amenemhat IV as pharaoh of the Exodus, in *The Proceedings of the Seventh International Conference on Creationism*, M.F. Horstemeyer, ed. Pittsburgh, Pennsylvania: Creation Science Fellowship, Inc.

Habermehl, A. 2015. Dating prehistoric musical instruments: The two timelines. In *Archaeoacoustics II: The Archaeology of Sound,* L.C. Eneix, ed., pp. 61–68. Myakka City, Florida: The OTSF Foundation.

Heater, H. 2014. *Bible History and Archaeology: An Outline*. Seattle, Washington: CreateSpace Independent Publishing Platform.

Hill, D.P. 2011. What is that mysterious booming sound? *Seismological Research Letters* September/October 2011. Accessed Dec. 12, 2017 @ http://www.seismosoc.org/Publications/SRL/SRL_82/srl_82-5_op/hill_op.html.

Howard, D.M. 1993. *An Introduction to the Old Testament Historical Books*. Chicago, Illinois: Moody Publishers.

John Garstang. 1998, revised 1912. John Garstang, British archaeologist. *Encyclopaedia Britannica, Inc*. Accessed Dec. 12, 2017 @https://www.britannica.com/biography/John-Garstang.

Jones, D.I.G. 2001. *Handbook of Viscoelastic Vibration Damping*. Chichester, England: John Wiley & Sons, Ltd.

Jones, F.N. 2007. *The Chronology of the Old Testament*. Green Forest, Arizona: Master Books.

Kenyon, K. 1970. *Archaeology in the Holy Land, 3rd edition.* New York, New York: Praeger Publishers.

Kenyon, K.M. 1998, revised 2017. Jericho, town, West Bank. *Encyclopaedia Britannica, Inc*. Accessed Dec. 12, 2017 @https://www.britannica.com/place/Jericho-West-Bank.

Montagu, J. 2015. *The Shofar: Its History and Use*. Lanham, Maryland: Rowman & Littlefield.

Nigro, L. 2006. Results of the Italian—Palestinian expedition to Tell es-Sultan: At the dawn of urbanization in Palestine. In *Tell es-Sultan/Jericho in the Context of the Jordan Valley: Site Management, Conservation and Sustainable Development. Vol. 2. Proceedings of the International Workshop Held in Ariha 7–11 February 2005,* L. Nigro, and H. Taha, eds. , pp. 1–40. Rome, Italy: Dep't of Antiquities and Cultural Heritage—Ministry of Tourism and Antiquities UNESCO Office— Ramallah Rome "La Sapienza University."

Noll, K.L. 2013. *Canaan and Israel in Antiquity: A Textbook on History and Religion*. London and New York: Bloomsbury.

Rogerson, J.W. 2014. "Myth" in the Old Testament. In *Myth and Scripture: Contemporary Perspectives on Religion, Language, and Imagination*, D.E. Callender, Jr., ed., pp. 15–26. Atlanta, Georgia: Society of Biblical Literature.

Rucker, J.D., and T.M. Niemi. 2010. Historical earthquake catalogues and archaeological date: Achieving synthesis without circular reasoning. In *Ancient Earthquakes*, M. Sintubin, I.S. Stewart, T.M. Niemi, and E. Altunel, eds., pp. 97–106. Boulder, Colorado: Geological Society of America.

Thompson, T.L. 1999. *The Mythic Past: Biblical Archaeology And The Myth Of Israel*. London: Random House.

Waltke, B.K. 2007. *An Old Testament Theology*. Grand Rapids, Michigan: Zondervan.

Williams, B. 2003. *Ancient Egyptian War and Weapons*. Chicago, Illinois: Heinemann Library.

Wood, B.G. 1999. The walls of Jericho. *Bible and Spade* 12 (2): 35–42.

Younger, K.L. 1990. *Ancient Conquest Accounts: A Study in Ancient Near Eastern and Biblical History Writing*. Sheffield, England: Sheffield Academic Press.

How Do the Acoustics in Cathedrals Affect the Human Subconscious?

Keith Harvey

KEITH HARVEY is a Scottish Audio Engineer Bsc Honours Graduate and currently studying a Mlitt in Archaeology.

ABSTRACT: The significant effect cathedrals have had on numerous parts of modern society is easy to see. They have affected the way cities have expanded; they were integral to not only the religious aspects of medieval life but also severely influenced the political landscape throughout Europe. The acoustics of these structures is rarely discussed, other than to comment on how well they carry the sound of the choir and organ during services, and how it can be difficult to have a conversation when the structure is full of other people. This project looks into the possibility that perhaps the acoustics of these structures had a deeper connection with the worshippers of the time and the tourists visiting in the present day.

Research Background

The original idea for this paper came from a desire to continue the research that one started the previous year for a research paper, "How has Sound Affected Human Cultural Development?" but with a more focused topic. Whilst researching this topic, a few cathedrals where visited during the research phase and the acoustics in them where so fascinating that the idea of studying them more in-depth led to the decision that it would be the perfect topic for this paper.

Once deciding upon the broad topic of acoustics in cathedrals, if became more focused when considered from the subconscious point of view. This combination came from a personal interest in standing waves and their effect on people. Standing waves may occur in some rooms when a large amount of pressure is generated, particularly in the frequencies below 300 hz. (Foley, 2015) We sometimes hear these waves in horror films when they are used to generate tension and unease. This is due to the feelings of discomfort that sounds such as these can create within people. (Derbyshire, 2010)

Understanding that standing waves can effect humans subconsciously by inducing natural, instinctive emotional responses when connecting with our subconscious, led to the development of the final question for this paper and beginning of researching which methodologies to use for this paper. The main body of first-hand research gathered for this paper was in a month long field trip around the UK to visit as many cathedrals as possible to gather both Quantitative and Qualitative data. Due to the multi-disciplinary nature of this paper, background was done on as many of the related disciplines as possible to come to a rounded conclusion.

Research Techniques

This research is based on a combination of previous research and fieldwork into the different subjects surrounding the question presented. It primarily focuses on how the human subconscious is affected by different frequencies, taking into account research done in psychoacoustics and published papers from archaeoacoustics.

Published sources on cathedrals were consulted to see if any acknowledgment has been made on the affect from the acoustics in Cathedrals has on the human subconscious if any.

To gather first-hand information to compare acoustic findings with the published sources, three main techniques where used, the first being binaural microphones where the primary tool used, in conjunction with Logic Pro X, a DAW (Digital Audio Workstation). The second main research technique was observation (Greetham, 2009) which includes observing the general public in the spaces, focusing on how they are reacting and interacting, particularly if they become louder or quieter, relaxed or anxious.

The third technique employed is Interviews and Questionnaires. (Preece, 1994) The interviews will be with academics who have studied the field, those who understand architecture and acoustics, as well as with people who have visited the buildings. This will allow a wide range of opinions from those who have a deep understanding of the subject, as well as people who have experienced them.

To make sure that the possibility of people's individual unique human perception was being taken into account whilst in the same building, a group of nine were all asked to do a smaller questionnaire with similar questions to the main interviews when in the same cathedral at the same time, the chosen Cathedral being St Mary's in Aberdeen. Chosen for its central location in Aberdeen, its neutral design in terms of age, shape and materials, whereas some cathedrals have very unique designs that could have significantly affected the results. This allows comparable data to let us see if there is a trend in the responses from the group questionaries' and the one-on-one interviews. (Greetham, 2009)

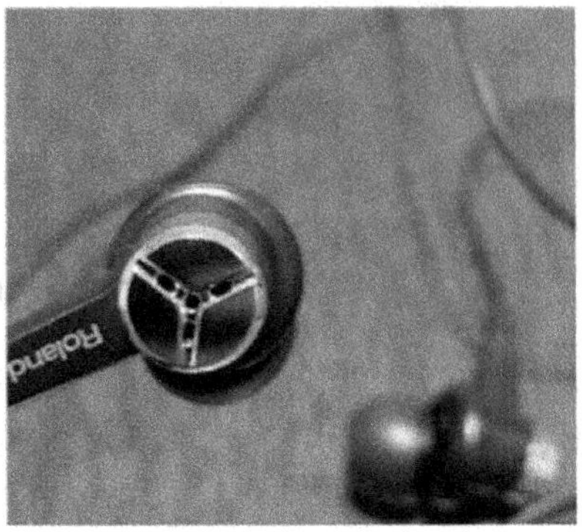

Image 1- Roland Binaural Microphones

Binaural Microphones

Binaural microphones (Image 1) are a simple piece of audio gathering technology that is easy to use. It is simply a pair of headphones that have microphones connected to the outside of the ear buds so that they can record what the wearer would be hearing which is then recorded onto a portable recording device. The audio can then in turn be put into a DAW for acoustic analysis. This technique was chosen for three reasons; firstly it is very flexible and mobile. There is very little equipment that is needed to analyse the locations when using this technique and it allows for the user to get around all of the location very quickly and listen back to an accurate sound track of the building.

The second reason being to experiment with a technique that goes against most modern conventions, which, as shown in the Wemyss Bay Cave Project and Rosslyn Chapel Project, required more equipment and, for the former, two people to carry all of the equipment, with approximately 10-20 minutes to set everything up.

The third reason was the ability to record ambient sounds in public buildings such as Cathedrals without the need to cordon off an area to set up sound analysis equipment. It puts the researcher in the room with the device, therefore recording what can be argued to be a more accurate way of gathering audio, if the goal is to see how people interact with their acoustic environment.

By using the Binaural Microphone Technique, the aim to is to find a way to gather high quality data more efficiently, by reducing the amount of time it takes to set up equipment, the quantity of data that can be gathered and to allow the recordings to be taken from locations that would not otherwise be viable due to the limitations of the established techniques.

The audio that was gathered using this technique can be heard on the portfolio that was created to showcase the work done in Archaeoacoustics by oneself as well as a documentary going into more detail on the benefits and problems with this technique. https://keithharvey95.wixsite.com/archaeo-acoustics

The Binaural Microphone Technique gave good results in the form of quality, quantity and time efficiency as was predicted. However certain problems were found with the practice at the beginning of the field trip. It proved ineffective at capturing audio for techniques like the Nick Green' *Guerrilla Impulse,* as it still required two people to be there to capture audio of good quality, one to record and one to pop the balloon. The other main problem with this technique is how accurate it is for comparing frequencies of the cathedrals, as one would need to be completely accurate in the route that one took around the cathedral, or record from the same approximate location in the cathedral such as centre of the nave. However for a primary research technique and more importantly a preliminary research method it proved very effective.

Aside from these small problems, the technique was incredibly useful for gathering the audio for this paper. It provided a simple, easy and assured way to gather audio from the different cathedrals for analysis and is the main reason that so many sites were visited on the research trip as there was no time taken up by setting up equipment. As can be seen from the photos of the audio taken from the cathedrals, it would have been almost impossible, certainly by oneself, to gather such a large amount of data in the allocated time due to the budget constraints.

This methodology has proven an extremely useful tool in the research methodologies for Archaeoacoustics and will be used in future projects of similar natures carried out by this researcher. It allows the user to take equipment into areas that would normally have to have a power outlet to plug in equipment, not such a problem in cathedrals that have had electrical outputs installed, but if used in locations such as ancient ruins like castles normal equipment might struggle to be used, at least without carrying a portable battery.

The main benefit of this technique is putting a human at the centre of the experiment. Something that one believes is vital in order to gather as accurate data as possible of how humans perceive sound due to our bodies absorbing and reflecting sound. The current methodologies that have been mentioned and used by the author before using this one all gathered high quality data, however they all had the same weakness of not focusing on the human element. This technique aimed to offer another point of view when analysing the structures to get audio that has humans present in the structure and to give accurate data with this element added.

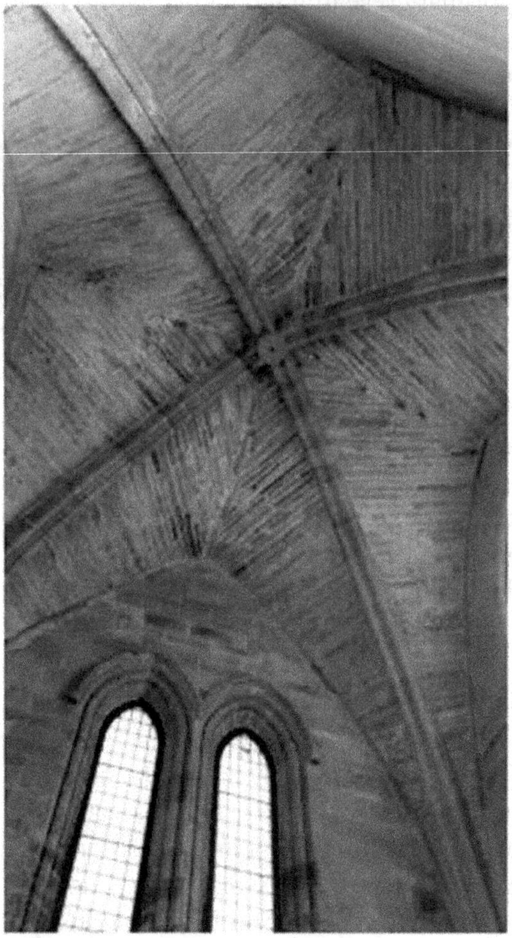

Image 2 – Arbroath Abbey Sacristy

First Field Test

Arbroath Abbey was chosen to be the test location for this technique because of the sacristy' acoustic properties. The room has a long reverb time of four to six seconds making it an ideal location to try out the different recording techniques that can be used for this project. The first to be tested was recording myself stamping my foot in all the different corners of the room to see if there were any slight differences in sound, and then repeated again with balloons to get a more accurate soundscape. The problem with using my foot was it was almost impossible to keep the volume, but it is a quick way to detect any significant reverbs, as is clapping. The balloons are an easier and more accurate system for comparing reverbs and the frequencies responses from different locations.

Then some singers where asked to sing different songs and chants to see if there were any particular frequencies that achieved a longer reverb. The use of three different singers whose voices all sat in different ranges was incredibly useful for comparing the frequencies. The results from the singers can be heard in a research video that was made discussing Binaural Microphones. (https://www.youtube.com/watch?v=i45IRC7OymY)

The Human Element

One of the most important aspects of this paper was the acknowledgment of the importance of the human interaction with the sites. Cathedrals are designed for human interaction, as the acoustic analysis supports. This was also heavily supported from the questionaries' carried out.

The interviews offered excellent insight and raised many extra questions for this paper, but still remained focused on the possible answers to the main question presented by this paper. In particular they tied back into a point made by Richard England's paper on *Neolithic Architecture – Space and Sound,* when he talks about the possible power of silence. (Eneix, 2014: 33) Many of the interviewees agreed with the view that the silence of these buildings is part of the reason they inspire awe and make them feel more relaxed and calm, with many directly discussing the power of silence without the topic being raised by the interviewer.

What's interesting about this comparison is that England's paper discusses how the power of silence for ancient societies when compared with the echo's and reverberation that can be found inside in ancient structures such as the Hypogeum in Malta, comparing the effect for ancient man to hearing "the voice of gods" when someone spoke.

(Eneix, 2014: 35) What makes this statement so interesting and this view of our ancestors, is how people are reacting in these spaces today. From the interviews and the published texts we can see that modern man is still having an emotional reaction to these buildings, but in a very different way. Whereas previous generations would have been awed by these spaces where sound became so important in their quiet world, the modern generations seems to also be awed by these buildings due to the isolation from the constant noise of modern living. Many of the interviewees thought of the Cathedrals as a space that provides calm, quiet places to think and believe that the power of these spaces was due to their silence and the way you can hear every individual sound echo.

What is interesting to note is modern research on psychoacoustics, summarised particularly well when we read an article published by Joshua Leeds *The Power of Sound* and the effects music can have on us. Leeds mentions how music can make us tap our feet in rhythm with the beat and hum along almost subconsciously and he also goes on to discuss how sound affects us emotionally. (Leeds, n.d.) Now that we consider the implications of sound affecting us emotionally and then comparing it back to the England paper, (Eneix, 2014: 33) the difference in how our society now interacts with sound and how humans used to interact with sound allows us to make some interesting conclusions and comparisons. Namely in that buildings such as cathedrals with their very distinctive sound would come to be associated with a subconscious response from the people using it as a place of worship.

First Impressions of Cathedrals

An interesting point that was observed throughout the project is the shape of cathedrals. Looking at an individual section of a cathedral it can be seen that if isolated it would have good acoustic properties, for example the shape of the sacristy at Arbroath Abbey. It can be seen they are very similar in design. What is fascinating is that this design is throughout most of the cathedrals main areas, and yet according to Wilson the architects from before the 13th century would not have had to-scale drawings available and considered it a "novelty" (Wilson, 1990: 140) and yet they quickly developed standardized templates so architects could be in numerous places at once and delegate their work to others. (Wilson, 1990: 140) The implication is that they discovered a style that worked.

What makes this so intriguing is that the shape of the cathedrals was already in practice before the 13th century. Wilson goes on to discuss that by this time building the cathedrals had become a matter of "routine" (Wilson, 1990: 140) and thus shows us that a style had already been agreed upon, but regrettably there is not much information on the architects from before the 13th century so we have little information on the reasons for this development other than it was done for structural reasons, strength and to support the arches and vaults at the top. (Wilson, 1990: 19) Iegor Reznikoff supported this by saying that if the evidence of these structures were only based of the evidence in literature then it would be assumed no churches had ever been built. (Scarre and Lawson, 2006: 81)

The other interesting thing to note is the symmetry present in these buildings. You could cut the main chamber down the middle and basically be left with two mirror images of each other. This symmetry is present in all of the sources consulted on cathedrals and confirmed in the field trip. This leads us to the question, are the acoustic properties that a cathedral has an accident of architectural design? Whilst possible, it seems unlikely that the first gothic cathedrals' acoustics were accidents. It is more likely they were inspired from the older Norman Romanesque style, like parts of their design. (Wilson, 1990: 13) With the Romanesque buildings being inspired by previous structures built by advanced cultures in the Mediterranean.

Otto Von Simson's *The Gothic Cathedral* discusses the introduction of more light into the Gothic design than what was previously available from the Romanesque design.

Southwell cathedral was an excellent example of this, as the cathedral exhibited both architectural styles with the west half being traditional Norman/Romanesque (Image 4) and the eastern half (Image 5), including the choir, being gothic as the cathedral has been expanded from a simple chapter house over hundreds of years. (Jenkins, 2016: 250) These different styles of architecture do allow different amounts of light into the buildings with the Romanesque feeling gloomier than its Gothic counterpart. However from the results gathered from the research trip using the binaural microphones and the results from the interviewees, it is clear that although some slight acoustical difference, overall the acoustics in the two different styles of cathedrals were very similar. The acoustic resonance was in the same range of under 250Hz and the interviewees offered similar opinions regardless of the cathedral they had recently been in.

The last point to be made about the structure of cathedrals is their use of vaulted ceilings, and in some cases domes. There is plenty of evidence that these structures are excellent resonators for sound, as is discussed by numerous authors throughout many of the texts that were consulted such as *Archaeoacoustics 2014 Malta Conference*, Scarre and Lawson's *Archaeoacoustics,* Templeton's *Acoustics in the Built Environment* and Ball's *The Music Instinct*. When combined with the importance of resonance on the human mind made by Leeds in *The Power of Sound* we can come to the conclusion that these buildings are excellent resonators and this is one of the keys to the acoustics in cathedrals and their possible effect on the human subconscious. (Leeds, n.d.)

Image 3 – Southwell Cathedral

Image 4 – Southwell interior, west wing

Image 5 – Southwell interior east wing

By using the research techniques and analysing the data gathered from the cathedrals we can look at any similarities that can be found between the different locations as well as the key differences and why these might exist.

Upon comparing the data gathered on the research trip in January 2017, we can immediately see the similarities between the cathedrals. The majority of the cathedrals visited are built west to east with only one of the cathedrals, Coventry Cathedral, being north to south. They all share similar shapes in the form of a cross with the nave, transepts and the choir all meeting at what is known as the crossing. The only time the cathedrals do not match this floor plan is if they have been built after WWII or have been expanded but the landscape around them has forced them to grow in unusual ways. Although with different dimensions depending on the size of the cathedrals, all the cathedrals built before the 20th century make use of the two rows of pillars running west to east through the nave and the choir, forming the aisles. The arched ceilings are present in most of the cathedrals, only being excluded in some of the smaller or older ones in favour of vaulted ceilings, being more triangular in shape. This development of cathedrals can clearly be seen in Wilson's *The Gothic Cathedral*, the website HistoricUK.com, Jenkins' *England's Cathedrals* and from photos taken during the research trip.

The recordings show that the prominent frequencies present in the cathedrals are all very similar, likely due to the similar design of the structures which can be linked back to Wilson' comment about the architects having detailed plans that they can edit to suit the needs of the design. (Wilson, 1990:140) With the time spent on the research trip it is quite clear that although there are different styles and inspirations for the architecture, such as Romanesque, Gothic, Perpendicular etc. it is clear that they all have similar features and designs.

Acoustic Analysis

Frequencies found in the recordings show that the cathedrals all shared resonate frequencies of below 250Hz, in particular they stand out around the 100 to 150HZ. (Examples provided in Images 6 and 7 and on the website https://keithharvey95.wixsite.com/archaeo-acoustics)

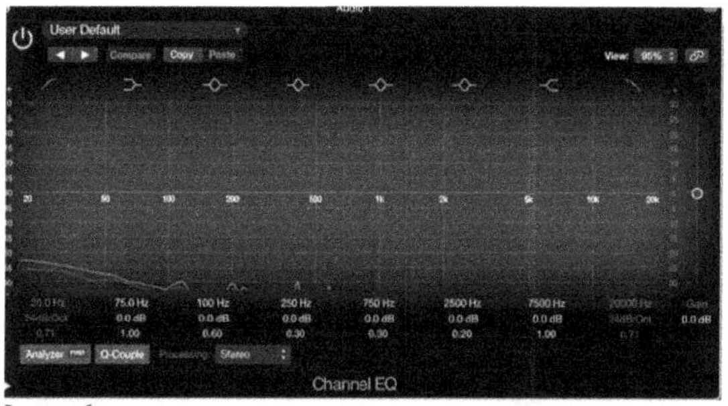

Image 6 –

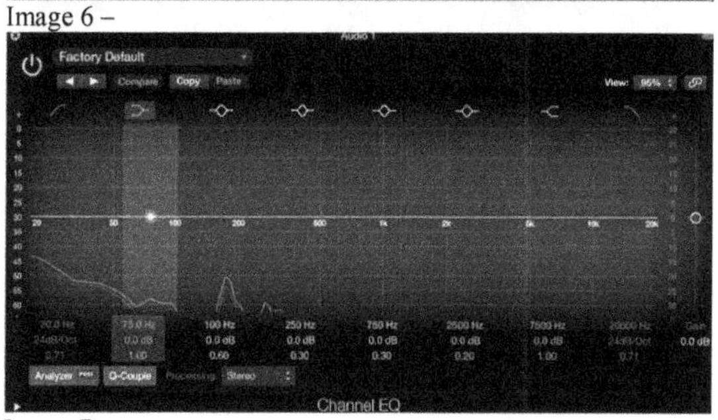

Image 7 -

This falls in line with the arguments put forward by Renzikoff, Debertolis, Tirelli and Monti' in *Archaeoacoustics 2014 Malta Conference.* This similarity that is present in such a wide range of cathedrals cannot be a coincidence when coupled with the knowledge that is discussed in Wilson's *The Gothic Cathedral*, which discusses the architects of these structures and how there is evidence of them using to-scale drawings to allow them to delegate the more day-to-day running of building these structures. (Wilson, 1990:140) If we consider that cathedrals were built by architects who worked from a Cathedral Binaural similar template, it is not so great a leap to expect similar acoustics results.

The evidence that Renzikoff points out in the similarities of acoustic pots and vases present in religious medieval architecture, being present not only in western European architecture, but eastern as well (Scarre and Lawson, 2006: 80) supports the idea that the designs of the cathedrals were based upon a template, once again reinforced not only by Wilson but Simson as well.

Another supporting source for this theory is the knowledge contained in both the *Quadrivium* and *Siencia,* which contain the information gathered by previous generations of leading scholars and scientists, providing evidence that some people of the time did understand physics and biology enough to possibly know how it would affect people. This is supported the evidence presented that St Augustine's, viewed sound and music as important and dedicated time to its study. (Simson, 1988: 21)

Rosslyn Chapel

Rosslyn Chapel was the final location on the month long research trip. Nick Green was joined here to carry out more acoustic analysis, but this time combining the research methodologies used at the Wemyss Bay Caves project as well as the Binaural Microphone technique that has been used throughout the project to offer a final comparison between the two.

The results proved to be an excellent addition to the database that had been gathered, as it allowed close examination of an isolated section of a cathedral, the choir, left unfinished due to a shortage of funds. The architecture in the structure can be seen as exceptional offering even more opportunities for resonating frequencies due to the detailed carvings that stick out from the walls.

Image 8 - Rosslyn Chapel, Exterior

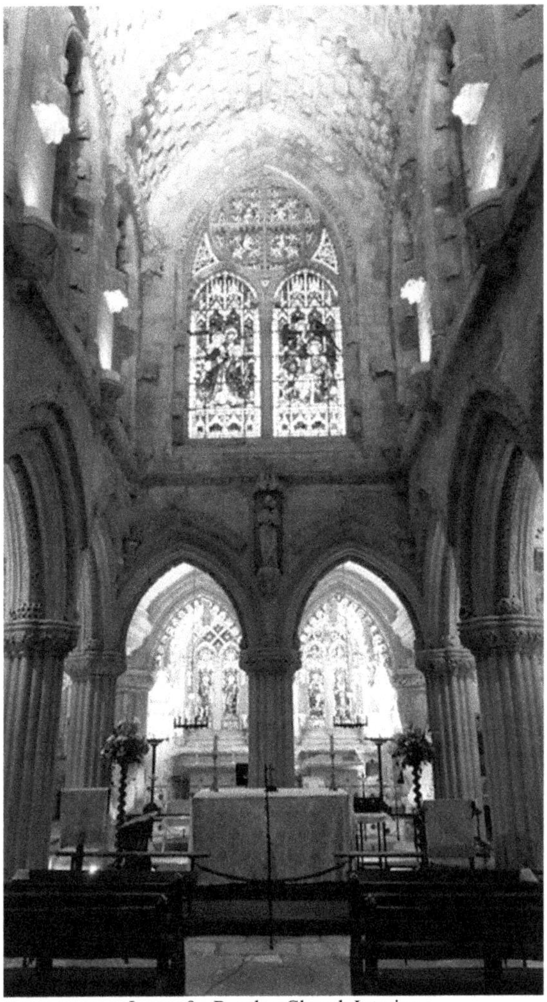

Image 9 - Rosslyn Chapel, Interior

It was very beneficial for showing the strengths of the Binaural Microphone Technique when another person is present and we can capture techniques like Nick Green's *Guerrilla Impulse*. A small Documentary on Rosslyn Chapel can be found on Nick Green's Website. http://www.archaeoacousticsscotland.com/about.html

Conclusion

The main flaw found in the research technique was the difficulty of doing more varied techniques for acoustic analysis when by oneself. From test results, such as the one at Truro cathedral, it can be heard that the binaural microphones did not get used to their full potential because they were too close to the source and the gain had to be turned down so as not to clip the tracks. A better solution would be to use them in a similar style to the Wemyss Bay Project, with another person creating the source sound, in this case a balloon.

This was tested at Rosslyn chapel where we can see the results of having someone wearing the microphones and another popping the balloon, allowing it to capture the acoustic tone of the room. This can all be heard in the previously mentioned online documentary.
(https://www.youtube.com/watch?v=i45IRC7OymY&t=288s)

Overall this technique was very successful in providing large amounts of quality data and will be used again in this field now that is strengths and weaknesses have been documented and can be worked around. It is an excellent research methodology for public spaces and as a preliminary research technique for larger projects.

In summary, it can be seen that the main limitation was hard scientific evidence that would pinpoint the exact cause or frequency that affects people when they are inside cathedrals. However to get this information more equipment and time would be needed to gather such data to then be analysed. Other main limitation include time and budget constraints.

Do the acoustics in cathedrals affect the human subconscious?

From the data gathered from the field research, the published texts consulted and the interviewee's response, we can come to a solid and rounded conclusion in response to the question presented by this paper.

The common frequencies from the field trip show that in the majority of cathedrals there are resonant frequencies below 250Hz. When compared with the reading from the research papers from *Archaeoacoustics*

2014 Malta Conference and the discussions about the 110Hz, even if it is unlikely that it is that exact frequency, there are lower frequencies around that one that affect people. We can conclude that there are frequencies present in many cathedrals that will have some form of effect on the human subconscious.

This is supported by the results and opinions of the interviewees and questionaries' that have been gathered during this project. The overall consensus was that cathedrals have a calming and relaxing effect on peoples' psyche and that many people find that being in them allows them to calm their minds, focus and "gather their thoughts." https://keithharvey95.wixsite.com/archaeo-acoustics

We can conclude that the Acoustics in Cathedrals do have an effect on the human subconscious, however this affect has likely changed over the generations as our world and our culture have changed. When they were originally built, cathedrals were a place of worship, music and noise in a silent world. Whilst still a place of worship they have grown to provide a place of peace, quiet and calmness in our modern world where we are constantly under a barrage of sounds and music. Cathedrals allow us to escape to a quieter time in human history, allowing us to return to the silence that we have lost in our modern world.

In response to the main questions asked by this paper we can summarise that, sound does have an effect on us as supported by and agreed upon by the previously published texts. That our modern lifestyle and the way we interact with music has changed how we perceive these buildings and in doing so has changed the effect cathedrals have upon us. Finally we can see from all the evidence gathered that cathedrals do have an effect on the human subconscious. The most obvious discussed in this paper is the calming and relaxing affect and how it is likely that the architects had some knowledge of acoustics and how it would affect us. The human response to these buildings has evolved as our culture has developed, however, and therefore the subconscious affect has most likely changed from the one originally intended.

BIBLIOGRAPHY

Ball, P. (2010). *The music instinct.* 1st ed. New York: Oxford University Press.

Biggam, J. (2008). *Succeeding with your master's dissertation.* 1st ed. Maidenhead: McGraw Hill/Open University Press.

Briggs, M. and Briggs, M. (1974). *Cathedral architecture.* 1st ed. London: Pitkin Pictorials.

Byrne, D. (2012). *How music works.* 1st ed. San Francisco [Calif.]: McSweeney's.

Chion, M., Gorbman, C. and Murch, W. (1994). *Audio-vision.* 1st ed. New York: Columbia University Press.

Corfield, P. (2017). *Why History matters - Articles - Making History.* [online] History.ac.uk. Available at: http://www.history.ac.uk/makinghistory/resources/articles/why_history_matters.html [Accessed 14 Apr. 2017].

Cox, T. (2014). *The sound book.* 1st ed. W. W. Norton & Company, Inc

Derbyshire, D. (2010). *Ever wondered why the music in horror films scares us? The harsh sounds tap into instinctive fears.* [online] Mail Online. Available at: http://www.dailymail.co.uk/sciencetech/article-1281385/Ever-wondered-music-horror-films-scares-Theharsh-sounds-tap-instinctive-fears.html [Accessed 14 Apr. 2017].

Duncan, B., Sinclair, I., Brice, R., Hood, J., Singmin, A., Davis, D., Patronis, E. and Watkinson, J. (2009). *Audio Engineering.* 1st ed. Oxford: Newnes.

Eargle, J. and Foreman, C. (2002). *Audio engineering for sound reinforcement.* 1st ed. Milwaukee, Wisc.: Hal Leonard.

Eneix, L. (2014). *Archaeoacoustics.* 1st ed. Myaka City, Florida: The OTS Foundation.

Foley, D. (2015). *What Are Standing Waves and How Do They Affect Your Listening Experience?.* [online] Audiophile Review. Available at: http://audiophilereview.com/room-acoustics/what-are-standingwaves-and-how-do-they-affect-your-listening-experience.html [Accessed 14 Apr. 2017].

Greetham, B. (2009). *How to write your undergraduate dissertation.*

Historic-uk.com. (n.d.). *Cathedrals in the UK | Interactive Map.* [online] Available at: http://www.historic-uk.com/HistoryMagazine/DestinationsUK/Cathedrals/ [Accessed 14 Apr. 2017].

Huber, D. and Runstein, R. (2010). *Modern recording techniques.* 7th ed. Boston: Focal Press/Elsevier.

Ingold, T. (2013). *Making.* 1st ed. Hoboken: Taylor and Francis.

Jenkins, S. (2016). *England's cathedrals.* 1st ed. London, UK : Little, Brown.

Leeds, J. (n.d.). *Psychoacoustics, defined « The Power of Sound.* [online] Thepowerofsound.net.
Available at: http://thepowerofsound.net/psychoacoustics-defined/ [Accessed 19 Jan. 2017].

Martineau, J. (2010). *Quadrivium.* 1st ed. Glastonbury, UK: Wooden Books.

Martineau, J. (2011). *Sciencia.* 1st ed. Glastonbury, UK: Wooden Books.

McAdams, S. and Bigand, E. (2001). *Thinking in sound.* 1st ed. Oxford: Oxford University Press.

Preece, R. (1994). *Starting research.* 1st ed. London: Pinter Publishers

Salford.ac.uk. (n.d.). *Psychoacoustics | Acoustics Research Centre | School of Computing, Science & Engineering | University of Salford, Manchester.* [online] Available at: http://www.salford.ac.uk/computing-science-engineering/research/acoustics/psychoacoustics [Accessed 27 Jan. 2017].

Scarre, C. and Lawson, G. (2006). *Archaeoacoustics.* 1st ed. Cambridge: McDonald Institute for Archaeological Research.

Simson, O. (1988). *The Gothic cathedral.* 1st ed. Princeton: Princeton University Press.

Templeton, D. (1999). *Acoustics in the built environment.* 1st ed. Oxford [etc.]: Architectural Press.

Wilson, C. (1990). *The Gothic cathedral.* 1st ed. New York, N.Y. (500 5th Ave., New York): Thames and Hudson.

2011-2017, (. (2017). *Writing your Dissertation: Methodology | SkillsYouNeed.* [online] Skillsyouneed.com. Available at: http://www.skillsyouneed.com/learn/dissertation-methodology.html [Accessed 3 Feb. 2017].

https://keithharvey95.wixsite.com/archaeoacoustics

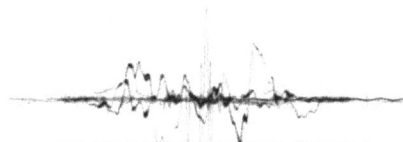

ARCHAEOACOUSTICS
III

TOMAR AND MAÇÃO, PORTUGAL

Getting in Tune with the Earth and the Sky
Healing in the Temple Colonnade

Alvin Holm

ALVIN HOLM, AIA, is an architect who has lectured and taught widely, having initiated a course in Design with the Classical Orders in the National Academy of Design in 1981 and subsequently at many other institutions.

ABSTRACT: Although way below the level of audibility, the extremely low frequency (ELF) vibrations emanating from Greek Temples should be of interest at this conference. As an architect practicing in the classical design tradition today, and as perennial student and long term teacher of classical architecture of the past, I have developed a novel hypothesis that explains both 1) the proliferation of wonderful columns on typical temples that far exceed any structural justification, and 2) the particular subtleties incorporated invariably although almost invisible to the eye. Those refinements in the shaping of the stone are termed "entasis," and although they are described by Vitruvius in the first century they have never been adequately understood. By analogy to the design of musical instruments and the subtle curvatures that accommodate the Pythagorean Comma I believe we can explain how the entasis of columns has been developed to enhance their harmonic efficiency in transmitting the Schumann resonance.

Because the Schumann frequency of 7.83 Hertz corresponds directly to our brainwaves in the Alpha State, a person in the immediate vicinity of the resonating column will be affected sympathetically. As the effect of a single column is modest, when dozens of columns surround the temple a great battery of these resonators is formed. I believe these Temples were built as might generators of electromagnetic energy in the ELF range. And as such they were designed as healing centers for both the body and the spirit. I will describe how they were used in Greece, and I intended to make the case that their early ancestors were the stone circles of Britain, like Stonehenge, also designed for healing.

KEYWORDS: Healing, Entasis, Schumann Frequency

The first conference on Archeoacoustics held in 2014 in Malta was a wonderful inspiration for me. It is strange for me to say that because I did not attend it, and I only learned about it from a report in the NEARA journal many months later. But it rang a bell within me; so to speak, the idea that a well-attended international conference could take place on the subject of the acoustics of ancient sacred architecture was exciting!

As a practicing architect passionate about classical Greek temples it reminded me of the notion that there are "buildings that sing." In a collection of essays, I read many years ago by the French poet Paul Valerie there is a pseudo-Socratic dialogue called "Eupalinos the Architect." Eupalinos explains that most buildings we see these days stand there on the street mute. They are often simply bland and featureless, sullen and silent. Sometimes, they are strident in their juxtaposition to the visual context, but they don't really speak. Then there are buildings that we encounter from time to time that can be said to speak. They may speak of the dignity of our government or perhaps the importance of the school system, or the majesty of the church, etc. Many of them will, of course, also speak of their functional aspect i.e.: where you enter in and how you are sheltered, but that's generally just "talk." Occasionally you run into a building that really "sings." Eupalinos the architect, this mystical architect, just sort of leaves it at that, and I've been worrying about it ever since. I have a "poetic" sense as to what it means for a building to "sing," but I think there is both a simpler and deeper meaning that I am now beginning to understand.

Another venerable axiom that has enchanted me for many years is that "beautiful architecture is frozen music!" This was one of those things we learned early on. Whether we're architects or not, almost everybody must have heard the expression! I don't know whether it was Goethe who said it first, or Schiller perhaps, or whoever. Everybody's got a nomination for who said it first.

"What pleases the eye, pleases the ear," is another axiom that has vexed many, many architects over the years... Palladio writes about this great mystery. Why is it that what pleases the eye pleases the ear? What's the relationship? But it seems to be the case, that if you design it harmonically it will be beautiful, or vice versa. As most of the acoustical research has involved chambers, cavernous vaults and more recently more modern works like Gothic cathedrals, all of which are enclosed spaces, my interest here is in external areas like the Temenos of a Greek temple where it is the sub-acoustic vibrations that create the, "Buildings that Sing."

With all due respect to the achievement of the Greeks, they only did two great kinds of building: temples and theatres. It was pretty much one temple and one theatre, each with hundreds of variations. It's a little bit like the people who say Vivaldi composed a concerto, and he did it really well, but then he did dozens of variations on that one concerto for the rest of his career. And we don't fault him for that, but it's more or less true that most Greek temples are simply temples. Some are big and some are small but they are all pretty much the same. The theatres are all pretty much the same, too. And it strikes me that they're both acoustic constructions.

Acoustics in the Greek theatres are said to be just astonishing. I have experienced a few! We know some of the tricks that the Romans took from the Greeks about what they put on the floors of the theatre and the resonating chambers under the orchestra, etc., but most interesting to me is that the one, the theatre, is carved out of the earth while the other is set upon a promontory, usually a rock. They are both resonating devices for sacred purposes. We know that the theatre was originally for sacred drama, not entertainment, and to that degree it is sacred space, as the temple is.

Let me sum this up before going further; My title is this: *Getting in Tune with the Earth and Sky; Healing in the Temple Colonnade*

Although way below the level of audibility, the extremely low frequency (ELF) vibrations emanating from Greek Temples should be of interest here today. As an architect practicing in the classical design tradition these days, and as a perennial student and long-term teacher of the classical architecture of the past, I have developed a novel hypothesis that explains both 1) the proliferation of wonderful columns on typical temples that far exceed any structural justification, and 2) the particular subtleties incorporated invariably although almost invisible to the eye. That refinement in the shaping of the stones is termed "entasis," and although it is described by Vitruvius in the first century, "entasis" has never been adequately understood. By analogy to the design of musical instruments and the subtle curvatures that accommodate the Pythagorean Comma I believe we can explain how the entasis of columns has been developed to enhance their harmonic efficiency in transmitting the fundamental rhythm of the earth that is known as the Schumann frequency.

Because the Schumann frequency of 7.83 Hertz, corresponds directly to our brainwaves in the Alpha State, a person in the immediate vicinity of the resonating column will be affected sympathetically. As the effect of a single column is modest, when dozens of columns surround the temple a great battery of these resonators is formed. I believe these Temples were built as mighty generators of electromagnetic energy in the ELF range. And as such they were designed as healing centers for both the body and the spirit. I will describe how they were used in Greece, and I intend to make the case that their early ancestors were the stone circles of Britain, like Stonehenge, also designed for healing.

One of the fundamental things about Greek ruins is that they are just as beautiful in a ruinous state as they are intact. One of the reasons for this is our romantic imagination, of course, but there is more to it than the imagination. It is that every temple is designed in such a way that every part is in a harmonious relationship with the adjacent part and the next larger part, and that next larger part is a part of the whole, so that you can take the whole thing apart and each of the elements will also be beautiful! There is a peculiar integrity about a pile of stones like this. Doric columns are said to be Apollonian. Apollo you know is god of the sun, and of reason, cool rationality, and of music.

Above is a scene from the wedding of the Lapiths and the Centaurs. The Centaurs are half human and half beast and they were invited to the wedding party of the Lapithians. Everything was pleasant until the wine flowed too long; at a point the centaurs got unruly (really beastly so to speak) and started running off with the Lapithian women. A big riot ensued. Apollo then came down with his bright blue eyes and with this magisterial gesture brought harmony back to the earth. That's what the Doric does.

In the next image you will see Robert Fludd's version of the Harmonic kind of Apollo type as we all are - Man as the microcosm. And here he is being tuned like a fiddle. There is the fundamental, the midpoint makes the octave and there's the diatessaron. That's the musical metaphor of the cosmos and of the micro-cosmos which is man.

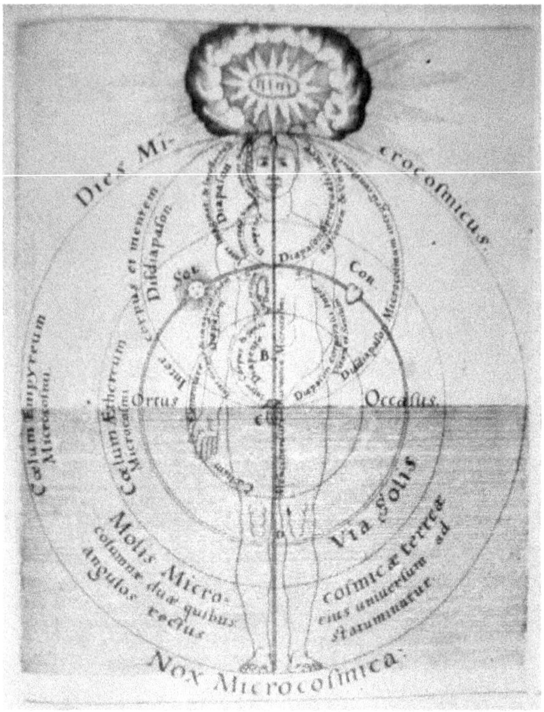

In another old illustration there is the hand of God tuning the monochord, the Pythagorean instrument, showing the same thing. Here it shows the same parts, diapason is octave, diapente is the fifth, diatessaron is the fourth etc.

And this is this glorious instrument that just cannot be improved upon. There are many beautiful predecessors of this instrument as you probably know, but once they got this in the 17th century with its perfect ratios and its perfect form, all the beautiful balanced curves and proportions, it could not be improved upon.

Now I am certainly not the first to have noticed that there is a resemblance between the figure of a woman and the figure of the fiddle, or the cello, or the viola. Manray and other artists have really worked this over.

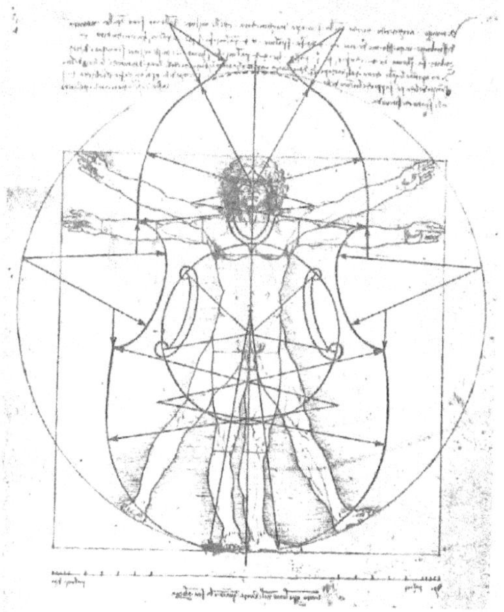

Now here is the Da Vinci man pretending to resemble a violin with his own perfect ratios. One to two is the octave, two to three is the fifth, three to four is the fourth; four to five is the major third. And then there's the minor third, and there's a couple of notes you don't want to play, and there's the leading tone into the next octave. It is all here and embodied in the violin as it is in our bodies.

Now, let's talks about chakras. Most of you know something about your chakras and there are seven of them. All of you, no matter how tall or small you are, have seven chakras, and they're all in the same ratio to one another. So does every stone you raise upright. If you jam a stick, or perhaps an iron rod into the ground, it will become an antenna. It will establish a circuit between the heaven and the earth. There is always an electromagnetic differential between the earth and the sky, and when you put a rod, a stone, or a broom stick upright into the earth a current will flow in both directions from the earth to the heaven and the heaven to the earth. And when you drive a current through a rod or a wire, as you know, a field is generated that you can feel outside the rod. The current flows in both directions in a spiral form much as the serpents twine upon the caduceus, and more significantly, as our own DNA is described. This is directly analogous to the illustration of our spiral energies, and where the spirals cross are known as chakras.

These are the same chakras that we all possess, and it's always seven. And they always start at the base and end at the crown. Even stones that aren't shaped like us have these chakras, it doesn't have to be suggestively shaped. Any stone will do. Consider the Menhir of Champ de Remps in Brittany. It is an experience just to stand in the presence of the one of these mighty menhirs.

There are five basic orders of classical architecture as defined by Vitruvius in the first century B.C. From left to right they are the Tuscan, the Doric, the Ionic, the Corinthian, and the composite.

The Paragon of temples to my mind is the temple of Apollo at Bassae in the Doric manner which is regarded as male. It is solar, it is rational, maybe kind of militaristic, protective. The Ionic is feminine, lunar, intuitive and it's organic. Then there is the Corinthian which is something else altogether. If the Doric and the Ionic are in any sense evolutionary forms deriving from earlier, sort of proto-Doric or proto-Ionic, they reach a point of refinement and development, and they stop. You cannot say the same thing about the Corinthian column, it appears that it was born, it wasn't developed. It evidently was created by somebody, and the fellow has a name. It is Callimachus, and it happened one day, and ever since we've been imitating that. He did it in remembrance of a dead maiden who we associate with being born and coming to life again (Persephone). One marvelous thing about this temple is that it is male and protective on the outside (in the Doric order), female on the inside, (the interior columns are in the Ionic mode,)

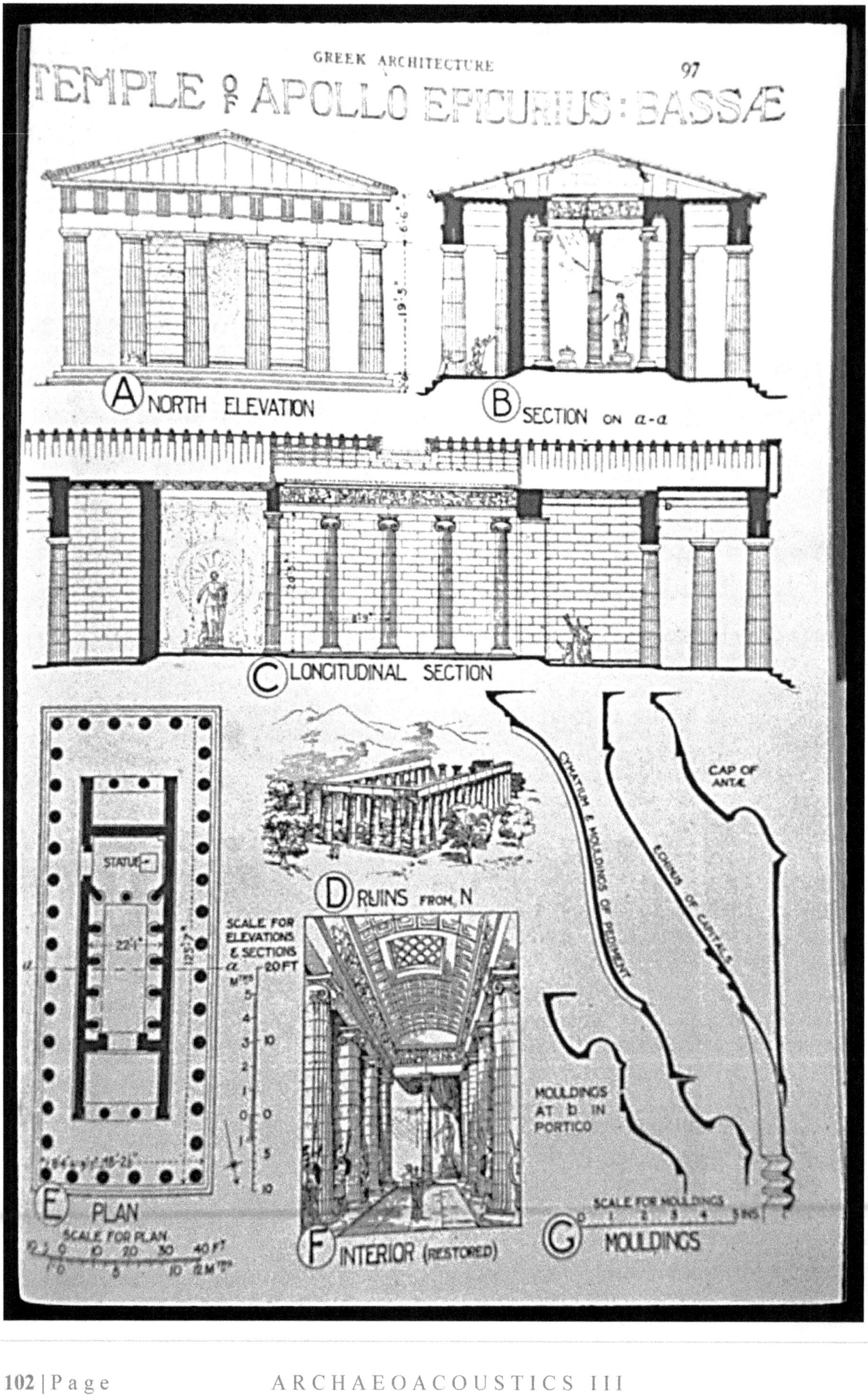
GREEK ARCHITECTURE
97
TEMPLE OF APOLLO EPICURIUS : BASSÆ
A NORTH ELEVATION
B SECTION ON a-a
C LONGITUDINAL SECTION
D RUINS FROM N
CAP OF ANTÆ
CYMATIUM & MOULDINGS OF PEDIMENT
ECHINUS OF CAPITALS
STATUE
SCALE FOR ELEVATIONS & SECTIONS
MOULDINGS AT b IN PORTICO
SCALE FOR MOULDINGS
E PLAN
SCALE FOR PLAN
F INTERIOR (RESTORED)
G MOULDINGS

and right at the end of the central axis, in a very unusual place, there is a single Corinthian column. You never see columns in the very middle except under very unusual circumstances and here she is, as if she is the cult figure itself, (the Corinthian maiden). This is universally acknowledged to be the very earliest Corinthian column ever found.

It doesn't look a whole lot like a girl either although it's supposed to represent one. It could hardly be improved upon and for thousands of years it has remained very much the same.

Unlike the Corinthian order which was born one day and remained perfect forever, the violin evolved over thousands of years until it reached a point of perfection in Cremona in the 16th century. This day the subtle curvatures and exquisite proportion achieved by the Stradivarius, Guarneri, and Amadi families have never been improved upon. This is much like the Corinthian order. Here is a drawing of mine happily examining the contours of the violin which I compare to the Entasis of Greek Columns.

Surely they don't need that many columns for practical or structural reasons; I think that they are there because they are resonators. This is like a great battery, and when you walk through there and when music is played in there all these columns resonate, and you are processed as you move from the great court through this hypostyle hall into the sanctuary,

Persepolis with a hundred columns. Looking at this, you would know at once, that you wouldn't need all these columns because wood trusses were doing the job of holding the roof. This is a spectacular exaggeration without any structural sense at all. So the question arises again why so many columns? And the answer is because all together they create an enormous battery resonating to the heart beat of the earth , the ELF frequency of 7.8 Hz, the Schumann frequency, and that is exactly the frequency of our own brain wave in the alpha state of complete relaxation + mental clarity

Now then, the classical temples descend from megalithic monuments. The grand ancestors, predecessors of the Classical column, are rude "menhirs". Everywhere in the world we have these standing stones. Everywhere mankind has set a foot, there seems to be the impulse to take a recumbent stone and pull it up vertically, then dance around it, or mark a boundary, or pray or sing or something like that. So these Menhirs as they are called, are the predecessors of the Classical column and the Greek temples.

It has been taught a long time that the predecessor of the Greek temple was the primitive hut." At some point clearly we did build huts, but long before we had huts, I feel certain we had temples. I believe that the colonnaded temple derives from the megalithic alignment, the standing stones, and the stone circles and not the primitive hut. And I've got a thousand slides to demonstrate how that works (for another time.)

To review:

1. The temple derives from Megalithic Alignments

2. The Greek Temples were designed as musical instruments were designed for many of the same reasons, incorporating the same geometric subtleties that we find in the violins of the 17th century

3. Enlightenment and healing took place in the presence of the temple energies which were electromagnetic at extremely low frequencies known as ELF levels. so it was not within the temple that enlightenment and healing tool place, it was outside in the temenos

From what we know of the Greek temple the action took place outside. The altar was outside the temple on axis with the entrance, where public and private sacrifices could be made. Healing, from what we know, took place outside between the columns. Here's the protocol: you go to the temple with your problem, it may be a psychic issue or a broken leg, and you tell your troubles to a temple attendant and they take your case, or not, then they do a little intake interview, and ask critical questions. And then you are given at the end of the day some kind of instruction to return. Perhaps it was a mantra or the name of the deity to invoke or something you must do when you returned on a certain date in the evening.

Then you are given a bed roll and a position between two columns in the outer aisle outside the temple, and you lie down between two enormous columns. You then begin your mantra, and you go to sleep, having prepared yourself for an appropriate dream. Hopefully, you have a vision, a visitation, or something like that. Then in the morning you wake up, you throw away your crutches and walk away. You are healed!

To Review:

When a stone or a rod or a column is thrust vertically into the earth an electromagnetic current is induced because the surface of the earth has a permanently negative charge and the inner surface of the ionosphere is positively charged. An electromagnetic differential always exists between the heaven and the earth. Periodically throughout the day and the night lightning strikes somewhere on earth as that differential is discharged. It has been ascertained by a physicist named Schumann in the 1950's that this occurs on average about 8 times every second somewhere around the world. This is now known as the Schumann Frequency, precisely 7.8 Hertz. We could call this "the heartbeat of the earth," and the fundamental rhythm of earth and sky. Also as we know this is equivalent to the alpha state of our brain, relaxed and clear at the point we are about fall into sleep, perfect for inducing deep harmony into our bodies and mind and hence healing.

The amazing Healing power of the Greek temples, as well as the great stone circles like Stonehenge, is a result of these stones and those great batteries of columns vibrating with the heartbeat of the earth, and generating fields of energy that induce in us the harmonies of the heaven and earth.

Ecstatic Sounds,
- a Hypothesis on Relationships between Sounds and Various States of Consciousness

Torill Christine Lindstrøm

TORILL CHRISTINE LINDSTRØM, Dr. philos., professor, Psych. Faculty & SapienCE, Centre for early Sapiens Behaviour, CoE, Faculty of Humanities, University of Bergen, Norway. Fields: Roman archaeology, paleolithic era, theory of science. Torill.Lindstrom@uib.no

ABSTRACT: This paper starts with claiming that there are elements of human universals in music: That certain music, rhythms and sounds tend to evoke certain rather predictable reactions and interpretations. Examples are given. Then, how ecstatic states can be achieved through two opposite routes: ergo-tropic activation/arousal and tropho-tropic activation/arousal, with their respective correlates to various states of consciousness, is explained. Based on this, a hypothesis is proposed: That there is a relationship between particular characteristics of sounds in music and the state of consciousness called ecstasy, but also regarding semi-ecstatic states (lighter forms) of varying intensity, but all implying a changed, and elevated, state of consciousness. Examples are given. It is concluded that music can contribute to induce ecstasy and other elevated ecstasy-resembling states of consciousness.

KEYWORDS: ecstasy, states of consciousness, human universals

Are There Elements of Human Universals in Music?

Human beings are not alone in using sounds to impress and impact the minds of other individuals. Mostly these sounds are directed towards conspecifics, but also other species may react to them, - in the same manner as conspecific individuals. I remember well how my mother's cat, "Ophelia", heard an Englishman singing a very romantic song in a television-series, - and how the cat came running through two rooms and directly to my mother's lap, purring and looking lovingly at my mother! Obviously, the cat was put in a loving mood by the human song! So apparently, certain sounds and sound-combinations can have the same effects across "the species barrier". If it can be so between species, I assume that there are also certain human universals with regard to reactions to certain sounds, - for example music.

Yet it has, from social-constructivist quarters, been argued that: the kind of sounds that affect us varies, both culturally and individually, - and music-fashion-wise. Music is vastly different from culture to culture, and personal taste varies between individuals. And indeed, this is the lesson from anthropology, ethnography, and cultural studies: Music is diverse, it varies not only across cultures but also through history. – But, can it be that there is also something in music that is common to humans? Something in human nature? Human universals? Human invariants regarding sounds?

Is there something particular about these sounds and sound-combinations in these examples of music that will create the same reactions in (more or less) everybody?

Can the Vienna-waltz "An der schönen blauen Donau" by Strauss be regarded as anything else than romantic? (https://www.youtube.com/watch?v=hBPUundF1dA).

Can "Je t'aime" sung by Jane Birkin and Serge Gainsbourg be heard as anything else than seductive? (https://youtu.be/k3Fa4lOQfbA).

Isn't the "Dovregubbens hall" ("Hall of the Mountain King" by Grieg, a bit frightening to everyone? (https://youtu.be/mlaAKQyOfwE.

Does not **"**The destiny symphony" by Beethoven signal a warning, signaling danger? (https://youtu.be/_4IRMYuE1hI.

And is it possible to perceive hard metal rock such as "Progenies of The Great Apocalypse"by Dimmu Borgir as anything but powerful and aggressive? (https://youtu.be/NiNTrKsQ8TU).

And does not "The stars and stripes forever" by Sousa express and inspire self-assertiveness? https://youtu.be/6hzXjizmX1Y?list=PLA7no0L9zTk7rLSoHp6YSRw1rJTb-JUCC

If So, Does Ecstatic Music Exist?

The term 'ecstasy' is derived from old French: estaise, which derived from late Latin: extasis, which was derived from Greek: ἔκστασις. ékstasis, meaning both enthusiasm and derangement, madness. It implies a feeling of elevation, exaltation, euphoria, well-being, entactogenetic affects (increased empathy), relaxation, reduced anxiety, inner peace, mild to strong hallucinations, enhanced perception and sensation, and an altered sense of time.

I, like so many others, and perhaps you my reader, have several times experienced emotional states resembling light ecstasy when listening to music. And this has happened in relation to pieces of different kinds of music. It has made me wonder if there still could be something that these different pieces of music have in common. Would it be possible to identify (at least some) characteristic of "ecstatic music"? Music that gives you a feeling of elevation? Music that makes you draw your breath deeply? Music that gives you chills down your spine? – Now, clearly not all individuals would necessarily experience the same music as "ecstatic" or as "slightly ecstasy-inducing" in this way. And not at all times. This would partly be because people could be in different somatic and psychological states when hearing the music, and therefore be in different receptive modes, - even in a non-receptive mode. Clearly, that would interfere with both the perception and the experience. Yet, when and if a person is receptive and perceptive to the music, is it still possible that music can induce ecstasy or ecstasy-resembling states?

Can Music Induce Ecstasy-Resembling States?

Is it possible that ecstatic and semi-ecstatic states of mind, (as altered states of consciousness), can occur in connection with music? And what are "altered states of consciousness? (And I am now referring to non drug-induced states). In my opinion, there seem to be two main characteristic "roads" to music-induced ecstatic states, and they are connected to bodily movements: One is wild, loud and intense music with rapid rhythms, for instance music that one hears when dancing and moving intensely. The other is the more restrained, controlled, often slow music of the concert hall or church when one has to sit still.

There is a theory connecting mental states (states of consciousness) with physical (bodily) activation. In 1971 Roland Fischer published a paper in "Science" called: "A Cartography of the Ecstatic and Meditative States". In this fascinating article he relates the various mental states to corresponding variations in physiological activation and arousal. (The concepts "activation" and "arousal" refer essentially to the same processes, and will here be used interchangeably).

When stimuli are perceived, signals go from peripheral nerves towards the brain, this activates the medulla oblongata (upper stem of the spinal cord nerves), and the hypothalamus, the hypophysis (the pineal gland) which produce adrenalin (epinephrine and norepinephrine), which in turn sends signals to the body: heart, cardiovascular system, lungs, adrenal cortex. It leads to energy activation, and to a general activation of the body. And of course, the cortex interprets, "makes sense of", the perceived stimuli and the bodily reactions. Fischer postulates that that there are two "roads" that can lead to ecstatic states: One goes along the perception-hallucination-continuum of increasing ergo-tropic arousal. This may lead to the hyper-aroused states of creative, psychotic, and ecstatic experiences. Then there is another that goes along the perception-meditation-continuum of increasing tropho-tropic arousal. This may lead to the hypo-aroused states of Zen, Zazen and Yoga samadhi. These two continua represent mutually exclusive states of arousal (Hess 1938, 1949, Gellhom 1970).

Put simply: from a "normal" state of activation (which is not one state, and definitely fluctuating, and varying between persons, and within persons over even short periods of time, is still experienced as "normal" for the person), - from this "normal" state, one can move in either of two directions: towards increased activation: the ergo-tropic arousal, or towards decreased activation: the tropho-tropic arousal. As I understand Fischer, states of ecstasy can be achieved at both the tropho-tropic and the ergo-tropic ends of the activation continuum. Accordingly there are two different kinds of ecstasy. Yet, at the extreme points of each of them, there may even take place a crossing-over, a rebound, and rapid shifts from one (extreme) state to the other. These two ecstatic states are impressively expressed in the statue of St. Theresa by Bernini in Santa Maria della Vittoria, Roma (tropho-tropic arousal); and in depictions of dancing maenads on ancient Greek and Roman pottery, sarcophaguses and other classical art (ergo-tropic arousal).

From here, I will talk (primarily) about the second kind of activation: the tropho-tropic arousal ecstasy. I will give examples of what I will call ecstasy-inducing music, - music that draws you towards ecstasy, and try to describe their common characteristics. This is music that is soft, melodious and repetitive (almost like a mantra), but which then lifts, goes higher than one would expect from the rest of the music, exceeds its spectrum of tones, often with a crescendo – and does so particularly at the end of the piece. So, that in the midst of the soft, melodious and repetitive, there is an element of surprise, something unexpected, something unpredictable: something that stretches beyond the expected and the "normal", sounds. Something that makes you focus your attention, draw your breath deep, and experience a pleasurable, hedonic, happy state of mind. Something "celestial", – a changed state of consciousness, - perhaps an ecstasy.

I will suggest the following music as typical examples: Händel's "The Messiah", The

Hallelujah Choir, appears primarily to induce ergo-tropic active arousal, but it has one particular point late in the piece that is "celestial". That part creates another, and different "letting-go": a peaking sensation: it raises, and raises, and raises beyond and above the rest of the tone-variations in the song. And this cannot have been coincidental, but intentional. "Hallelujah!" ("Alleluia" / "Alleluija") is an ancient Jewish & Christian exclamation, and in many Christian connections, it is an exclamation during, or just before, people experience ecstasy.

The next example is of another, and very different, "Hallelujah: A record called: "Passion und Osternacht in der Ostkirche" with Oldslavic Easter liturgy sung by a choir of monks from Monastère de Chevetogne in Belgium. Several of the songs have ecstatic components, but the "Offertoriumsgesang der Karsamstagsliturgie (offertorium-song from the Easter Eve liturgy): Da moltschit wsjakaija" ends with a "Alleluja" that is clearly ecstatic in nature. Also the "Prozession: Woskresenie Twoe, Christe Spase" with bells and voices that appear and disappear, also evokes an ecstatic sentiment.

The Norwegian Psalm, "Eg veit i himmerik ein borg" ("I know a castle in Heaven") does not have an elevation at the end of its verses, so in a sense it does not quite fulfill that criterion, yet it is so melodious with "waves" rhythmically rising and lowering, that I will characterize it as having some ecstatic inducing properties.

My final example is "O, helga natt", (French original titel «Minuit, Chrétiens»), a Christmas-song written by Placide Cappeau i 1843 and given melody by Adolphe Adam in 1847. It has been sung and recorded by many singers, but perhaps the most famous is "O, helga natt" as sung by Nils Bech in the famous series "Skam" made by Norwegian television, NRK, in 2016,
(https://youtu.be/zk0sARbPqnI?t=74).

It has the quality of being melodious and softly rhythmic like the psalm in the example above, "Eg veit i himmerik ein borg". But unlike that psalm, it raises very high at the end, and the high clear contra-tenor voice of Nils Bech takes "O, Helga natt" to quite surprising and unexpected elevated levels.

Image from audience posting to open access web; photographer not identified

Summing Up

I admit that this idea about "ecstatic music": that certain musical pieces have components and characteristics that can contribute to create an ecstatic or semi-ecstatic changed state of consciousness in the listener, is just a hypothesis. It may be farfetched, or it may be correct. Psycho-neurological investigations (EEG or fMRI), with people listening to music with these characteristics, might give an answer.

All the examples above are from religious music. That may not be incidental. Religious experiences are often of ecstatic nature, and religious ceremonies and music are to some extent "designed" to create such mental states. Yet, by no means do I think that these musical characteristics are restricted to religious music. – As a Norwegian stand-up comedian, Dagfinn Lyngbö, said in his show "Stereo": (admittedly, an approximate quote): "If you want to make music and a song that will make your audience sense a chill down the spine, you should start slowly, mention nature-phenomena, sing with a hoarse and/or low voice, perhaps sing with two voices in harmony, good idea to add something ethnic. Have refrains that mention common human situations and predicaments, life and death, and something that can give hope. It is also a good idea to have drums that start in the middle and then increase in strength and speed, and if you add light-effects: You've got it!" – Ecstatic states of consciousness by ecstatic sounds.

REFERENCES

Fischer, R. (1971). A cartography of the ecstatic and meditative states. Science, 174 (4012), 897-904.

Gellhorn, E. (1970). Psychologisches Forschung, 34, 48.

Händel, G. F. 1741. "The Hallelujah Choir" from the oratorio "The Messiah".

Hess, W. (1938, 1949). Das Zwischenhirn und die Regullerung von Kreislauf und Atmung. Thieme, Leipzig, 1938; & Das Zwischenhirn. Schwabe, Basel, 1949.

Cappeau, P. 1843. «O helga natt» (French orig. Title: «Minuit, Chrétiens») er ein kristen, melody by Adolphe Adam in 1847. https://youtu.be/zk0sARbPqnI?t=74

"Passion und Osternacht in der Ostkirche". Oldslavic Easter liturgy sung by the choir of monks in Monastère de Chevetogne in Belgium.

Støylen, B. (translator). (year unknown) "Eg veit i himmerik ein borg» Partly tranlated after the German psalm «Ich weiss ein ewiges Himmelreich», partly after H.C. Sthens Danish translation of the same psalm: «Jeg veed et evigt Himmerig». The German text is first known from Geistliches Gesangbuchlein, Hamburg, 1612. Melodie of the Norwegian psalm: folksong from Hallingdal.

Tart, C. T. (Ed.) (1969). Altered States of Consciousness: A Book of Readings. New York, NY: Wiley.

Tart, C. T. (1972). States of consciousness and state-specific sciences. Science, 176, 1203-1218.

https://www.youtube.com/watch?v=hBPUundF1dA

https://youtu.be/k3Fa4lOQfbA

https://youtu.be/mlaAKQyOfwE

https://youtu.be/_4IRMYuE1hI

https://youtu.be/NiNTrKsQ8TU

https://youtu.be/6hzXjizmX1Y?list=PLA7no0L9zTk7rLSoHp6YSRw1rJTb-JUCC

https://youtu.be/zk0sARbPqnI?t=74

Irén Lovász helps conference participants explore the acoustics of the medieval synagogue in Tomar, Portugal photo:OTSF

Archaeoacoustic Approach to the Rotunda in Bény

Irén Lovász, Paolo Debertolis

IRÉN LOVÁSZ, PhD is an associate professor in the Institute of Arts Studies and General Humanities at KRE University in Budapest, Hungary. Her research has included ethnomusicology, anthropology of religion and music, sacred communication. She is also a professional singer, applying traditional singing in voice therapy.

PAOLO DEBERTOLIS, M.D., aggregate professor at Department of Medical Sciences, University of Trieste (Italy), President of Super Brain Research Group (*)

ABSTRACT: We would like to draw the attention to a medieval sacred place in Central Europe in the Carpathian basin, where unusual sound phenomena can be experienced. We focus our research on the "12 Apostoles' Rotunda" of *Bény.* The settlement used to belong to the Hungarian Kingdom for centuries during the Middle Ages and also continuously until the first part of the 20th century. Now it is situated in the southern part of Slovakia. Recent studies suggest that the rotunda was already built by the 10-11th century, which raises more questions about who built it and for what purpose. The unique feature of the Rotunda are the 12 mysterious vaulted niches within. Each of these niches strengthens different resonances, which gives a unique sound to the human voice there, according to our hypothesis they were very probably tuned on purpose. In 2016 and 2017 fieldwork was undertaken to test the rotunda's archaeoacoustic and resonant properties. Equipment, methods, results with our conclusion are described in this paper.

KEYWORDS: acoustics, resonance, rotunda, niches, medieval, Central Europe

Introduction

We would like to draw the attention to a special early medieval sacred space of Central Europe in the Carpathian basin, where unusual sound phenomena can be experienced. We focus our research on the 12 Apostoles' Rotunda in *Bény* (Bína). The settlement used to belong to the Hungarian Kingdom for centuries during the Middle Ages and also continuously until the first part of the 20th century. Now it is situated in the southern part of Slovakia, closed to the present Hungarian border. The settlement is populated mostly by Hungarians (90%) belonging to ethnic minority nowadays.

According to the previous concepts on the origin of the building, it was purportedly built at the beginning of the 13th Century[1], together with the Premontrei Monastery and Abbey in the neighborhood.

(*) Note. SB Research Group (SBRG) is an international and interdisciplinary team of researchers, researching the archaeoacoustic properties of ancient sites and temples throughout Europe and Asia (www.sbresearch-goup.eu).

1 Cervers –Molnár Vera: A középkori Magyarország rotundái. Akadémiai Kiadó, Budapest, 1972. 39.

Fig. 1 – The location of Bény *(Bína)* in Slovakia today.

Fig. 2 – The rotunda together the abbey placed in front of it.

Recent studies however, suggest that the rotunda was built already in the 10-11th century,[2] which raises more questions about who built it and for what purpose? There is no consensus as archaeologists and art historians have different theories. The Rotunda uniquely features *12 mysterious vaulted niches*. Each of the 12 niches strengthen different resonances, which give a unique sound to the human voice there. All of them are little different in size, and according to our hypothesis they were very probably tuned on purpose.

Fig. 3 – Some of the 12 niches within the Rotunda

2 Szilágyi András: A Kárpát- medence rotundái és centrális templomai, Semmelweis kiadó, Budapest, 2008. 249.
Németh Zsolt: A Kárpát.medence legkülönösebb középkori templomai, BKL Kiadó, Szombatheely, 2017. 103-119.
Sabasosova, Elena – Havlik, Marian: Rotunda Dvanastich Apostolov v Bíni- dokumentácia odkrytych murív s vyhodnotením nálezov. 2010.
Sabasosova, Elena – Havlik, Marian: Rotunda Dvanastich Apostolov v Bíni In: Valeková, Anna (ed) :Ranostredovekásakrálna architektúra Nitrianskeho kraja: Zborník zo seminaára a katalóg ku vystave. Nitra, 2011.

There is a rumor in the village, shared also by the former local priest: „The niches are tuned as If you sit in the first one from the South and put your ear to the wall, you hear the lowest sound, and at the last one, from the Northern one you can hear the highest sound."

Our hypothesis is that there was probably a definite awareness of resonance and the conscious application of artificial niches into the architectural construction in order to improve sound quality thus serving spiritual purposes, since „niches, recesses or alcoves were used as natural resonators"[1] in medieval architectural construction.

After discovering its marvelous mysterious acoustics while giving a solo concert in the neighboring abbey, initial fieldwork was carried out in 2013. Oral history as an anthropological method of study was used and local people were interviewed to understand local legends relating to the rotunda. A priest now in service in Bratislava, who was born in Bény and spent all his childhood in the village said that sitting and singing in the niches to find which of them his feet could reach the floor, was his favorite activity as a small boy.

Unfortunately, we discovered the rotunda's acoustics had changed after some constructions, reconstructions and renovations were carried out during the centuries, as described bellow.

In 2016 and 2017 fieldwork was undertaken in partnership with Irén Lovász and Paolo Debertolis to test the rotunda's archaeo-acoustic and resonant properties.

Historical Background

Bény (Bína) is located on the right side of the river Garam, 17 kms North from Esztergom, the first medieval capital of the Hungarian Kingdom. *Esztergom* is 46 kilometres northwest of the capital Budapest. It lies on the right bank of the river Danube, which forms the border with Slovakia there today. Its cathedral, the Esztergom Basilica is the largest church in Hungary. Esztergom is one of the oldest towns in Hungary. The first people known by name were the Celts from Western Europe, who settled in the region in about 350 BC. A flourishing Celtic settlement existed there until the region was conquered by Rome. Thereafter it became an important frontier town of Pannonia, known by the name of *Salvio Mansio*, or Solva of the Roman Empire. Slavic people immigrated into the Pannonian Basin at about 500 AD. The Magyars (Hungarians) entered the Pannonian Basin/ Carpathian Basin in 896. In 960, the ruling prince of the Hungarians, Géza, chose Esztergom as his residence. His son, Vajk, who later came to become Saint Stephen, the first king of Hungary, was born here around 969-975. In 973, Esztergom served as the starting point of an important historical event: during Easter of that year, Géza, the ruling prince sent a committee to the international peace conference of Emperor Otto I in Quedlinburg. He offered peace to the Emperor and asked for missionaries. And it might be a crucial point for us!

There was a church also built by the German missionaries from around Passau between 973- 1000 in Esztergom. The center of the hill was occupied by a Basilica dedicated to St. Adalbert who, according to legend, baptized St. Stephen. Stephen's coronation took place in Esztergom on Christmas Day 1000.

---

[1] Reznikoff, Iegor: *The Evidence of the Use of Sound Resonance from Palaeolithic to Medieval times, in: Archaeoacoustics*, Lawson, G. and Scarre, C. eds. University of Cambridge, Cambridge, 2006, 80.

Bény

After the importance of the city of Esztergom in Hungarian History, *Bény* should be considered as an important small village, especially given it is located 17 kilometers North of Esztergom on the eastern side of the river Garam.

Historical records first mention the village, written as *Byn* in 1135. Earlier the Romans built a fortress here, and Emperor Marcus Aurelius wrote his famous diary in this region in 173. According to local knowledge, proved by archeological excavations, Bény used to be the Northern *limes,* border of the Roman Empire. Later, King Stephen, the first king of Hungary, gave the region to Bény, son of Hont. It was already Géza, the father of St. Stephen, who invited the Kraut knights, Hont and Pázmány to stay in Hungary on their way to the Holy Land.[1] It is also possible that the rotunda was built by Byn himself or by his descendants at the centre of genus for the leader. [2]

The great system of escarp or redoubt with 5 kilometers in its inside diameter, was the biggest and strongest that time in Europe. Finally it lost its original function during the 11-12th centuries, it was left empty. But one can still find remains of it around the village, about 1 kilometer from the rotunda. This military post might has been the gathering basis of the army of St. Stephen the first Hungarian king, before going against the tribes of the pagan leader, Koppány. Very probably it was the place where Stephen I was given his *first sword* as a symbol of his puberty-ritual or of age of discretion as part of an ancient *rite de passage* ages 21 in 998 AD. (There was a monument erected a few years ago in front of the abbey for the memory of this historical fact by the proud local citizens of Bény). During the time of early Christianity every 10 villages was ordered to build a church and several rotundas were built in this time. According to some presuppositions, one of them might has been the rotunda of Bény. But we do not share this idea. We argue in accordance with the latest literature[3] that it was already built at the end of the 10th century.

The Rotunda

In 1217 the Premontre Abbey monastery was founded in Bény in the Romanesque style. An earlier built rotunda stands in front of the abbey today. The patriarch of the rotunda is the 12 Apostles. (Though it easily might have got the name of 12 apostles afterwards).

Fig. 4 – The rotunda today

During the Turkish invasion of 16-17th Century the rotunda was damaged. The village got empty, uninhabited. The rotunda became the *filia* of the neighboring Kéménd at the beginning of the 18th Century.[4] In 1755 it was reconstructed in the baroque style, however it was given a new interior, design and ornaments. Much of the early medieval roman style was lost, for example a new baroque window was cut into the wall which would have originally been closed.
During World War I. in 1918, Czechoslovak troops occupied the area, and the region

[1] Szilágyi 249.
[2] Szilágyi, 249.

[3] Sabadosova-Havlik 2011, 141, Németh Zsolt 2017, 103-119.
[4] Haiczl, 95.

became part of Czechoslovakia. Between 1938 and 1945 Bény once more became part of Hungary. Then until 1990, it was part of Czechoslovakia. Since then it has been part of Slovakia and officially called Bína.

There was bomb damage during World War II, which revealed the medieval walls behind the baroque facade and an additional room was added to the south of the rotunda in 1945.

Fig. 5 - The Rotunda after the Second World War.

From 1961- 1978, a big archaeological excavation and reconstruction was undertaken by Slovak archaeologists.[1] All traces of the baroque design including its windows and crude electric wires were removed. More recently, Slovak archaeologists, Elena Sabadosova-Marian Havlik undertook archaeological studies and reconstructions between 2006- 2012, [2] was also supported by the Hungarian ministry of culture. After the last renovation the so called "original" structure was finally reconstructed: The ground became 40 cm deeper, with all the niches equally positioned 43cm above ground level. Was this the original case or not? If we listen to the memory of the priest who grew up playing and sitting in these niches, then we must say no.

The structure of the rotunda reminds us also of Saint Donatus in Zara, Dalmatia (9th cc). Although this one in Bény is much smaller. The size of the rotunda: inner diameter: 720 cm, 12 niches: 80-82 cm, the excedra 360 cm.

In this structure there was no separate sanctum, but the whole rotunda itself served as

[1] The first excavation was led by Alojz Habovstiak,1963, 1978.

[2] Sabasosova, Elena – Havlik, Marian: 2010, 2011.

such. It either shows that there was a different type of Christianity in the region before the 10th century or the Latin liturgy and proxemic attitudes changed so radically by the first millennium and consequently there is a lack of knowledge about. In this region during the 5-6th century, Arian Christianity was the main religion and still had some power among the *avars* after the 7th century[3]

Fig. 6 – The Rotunda's interior

As far as the structure and shape of the rotunda in Bény is concerned, it is suitable to serve ritual and ceremonial purposes of special democratic communities with equal members. All the members of the community (lets say monks) could see and hear the others while looking and turning to the middle, to the central point of the rotunda, or to the leader of them, who might have stood or set in the extedra, in the 13th and greatest niche. It might also have served as a Chapter –house, in which monks met daily to discuss business and to hear a chapter of monastic rule. But it should be noted that the acoustics of this rotunda supports chanting, singing, and not the spoken voice!

In this respect, it differs from churches, basilicas with naves, typical of the 2nd millennium. There is also a presupposition about using the special acoustic power of the space of the rotunda, for healing, initiating ritual purposes" where the human voice had an important role. Since medieval monks used magical practices and techniques to enter altered states of mind, and likewise helping others to achieve this, something very different from normal daily experiences."4

The Shape of the Rotunda

The rotunda has historical and architectural value, its shape was widespread in medieval times. A great number of parochial churches were built in this form in the 9th to 11th centuries in Central Europe and examples can be found throughout Hungary, Slovakia, Poland, Croatia, Austria, Bavaria, Dalmatia, Germany, and the Czech Republic. It was believed to be a structure descending from the Roman Pantheon, however, it is rare to find such examples in former Roman territories. They are far more common in Central Europe. (Note: Bény is located at the Northern border of the Roman Empire). Generally its size was 6–9 meters inner diameter and the apses were directed toward the east, sometimes 3 or 4 apses were glued to the central circle and this type has relatives in the Caucasus.5

3 Németh Zsolt, 118.

4 Németh, Zsolt, 121.

5 First main publication on the topic in Hungarian: Cervers –Molnár Vera: A középkori Magyarország rotundái. Akadémiai Kiadó, Budapest, 1972.

There are about one hundred rotundas in the Carpathian Basin. The greatest most recent summary of them was written by a Hungarian author who undertook a deep study of their origin along with related archetypical buildings of the type. 6 Several types of rotundas are found in the Carpathian Basin, within the former boundaries of Medieval Hungary. Many of them still stand today. In many places the ancient foundations have been excavated and conserved. Rotundas of six apses, a most interesting form, are found in Hungary, in Ukraine and several places in Armenia. There is no possibility here of mentioning all the archetypical examples of round shaped sacred architecture of the world from The Church of the Rotunda in Thessaloniki, built as the "Tomb of Galerius" in 306, to the Hall of Prayer for Good Harvests, the largest building in the Temple of Heaven, construction completed on 1420 during Yongle Emperor who also constructed Forbidden City of China, Beijing.

There is much literature on ancient buildings from all over the world that discuss the significance of the rotunda's shape. To reconstruct sonic and spatial experiences of the past, a new multidisciplinary collection of essays explores the intersection of liturgy, acoustics, and art in the churches of Constantinople, Jerusalem, Rome and Armenia. [7] Other ancient rotundas and round shaped sacred spaces are also considered, from the shape of Zuart'noc', and its relation to Syrian and Mesopotamian monuments, of the same type to its obvious prototype in the martyria of the Holy Land, above all the Anastasis Rotunda. Completed by 336 to shelter the traditional site of Christ's burial and resurrection, this structure formed the focal point of Christian Jerusalem, and indeed, of medieval Christendom more generally."[8] The book studies the structure of the rite, revealing the important role chant plays in it, and confronts both the acoustics of the physical spaces and the hermeneutic system of reception of the religious services. The result is a rich contribution to the growing discipline of sound studies and an innovative convergence of the medieval and the digital. Pentcheva made remarkable studies on the acoustics of Hagia Sofia, the cathedral of Constantinople itself.[9]

Whatever the origin of the rotunda form, the symbolic meaning of the archetypical round shape is related to the universal meaning of perfect completeness, wholeness, oneness, unity, deity. It also expresses power and nobility.[10] In rotundas and round shape sacred sites it is possible to feel the positive energy and esthetic effect on the mind and psyche, as the ancient Greek authors have contemplated.[11]

Materials and Methods

For recording two dynamic high-end microphones both extended over the audible band

6 Szilágyi András: A Kárpát- medence rotundái és centrális templomai, Semmelweis kiadó, Budapest, 2008.

7 Pentcheva V.,Bissera (ed.): Aural Architecture in Byzantium: Music, Acoustics, and Ritual, Routledge 2017.

8 Pentcheva V.,Bissera: Introduction in Pentcheva V.,Bissera(ed.), 11.

9 Pentcheva V.,Bissera : Hagia Sophia: Sound, Space, and Spirit in Byzantium, Pennsylvania State University Press, 2017, http://hagiasophia.stanford.edu 2. (2018.01.03.)

10 Szilágyi , 144

11 see: Marcus Vitruvius Pollio: de Architectura, inLacus Curtius: Vitruvius on Architecture, Book IV.Chapters 8. http://penelope.uchicago.edu/Thayer/E/Roman/Texts/Vitruvius/5*.html2017. 12.30.)

(Sennheiser MKH 8020, response frequency 10Hz - 60.000Hz) with shielded cables (Mogami Gold Edition XLR) and gold-plated connectors together with a digital portable recorder with a maximum sampling rate of 192KHz (Tascam DR-680 of TEAC Group) at 24bit were used. The microphones were placed in a number of positions in the Rotunda, but the best recordings were obtained when the mikes were located in the centre. Before recording a spectrum analyser Spectran NF-3010 from the German factory Aaronia AG, was used to detect any electromagnetic phenomena present which could influence the results. Praat program version 4.2.1 from the University of Toronto and Audacity open-source program version 2.1.2 for Windows and Linux were used to analyze the various recorded audio tracks.

The rotunda was tested by male and female voice from several locations; within every niche and within the central area of the Rotunda. No electronic or synthetic tones were used for testing, only the natural voice to recreate as far as possible the original reverberation and resonance. A drum was also used, (its frequency response and extension were identified and used in previous research) using a protocol for testing temples (Debertolis et al. 2012-2017). The music and songs used to test the acoustical properties of the structure ranged from simple tones and musical scales to Gregorian chants and romances of the XIII-XIV centuries. To establish the environmental sound characteristics (infrasound) a recording was made in silence.

Physical phenomena was measured using a Geiger counter to detect radioactivity and UV imaging to detect magnetic fields as these can influence the state of mind. The purpose was to research common characteristics found in ancient "sacred" sites as discovered in our previous research. In fact many medieval structure are built over veryancient sites with peculiar physical characteristics (Debertolis *et al.,* 2011 - 2017*)*. Every aspect of the niche was explored to search for different resonance characteristics.

Results

We observed that all the niches have a different depth, but the same height and distance from the ground. This aspect create a different resonance at different frequencies, so the 12 niches act as the tubes of an organ.

So there is not a singular frequency of resonance, but various frequencies of resonance very good for both male and female voices. The extension of resonance is wide and the result is to amplify all singing voices. But a requirement to achieve this is to have a strong volume. The spoken voice alone does not stimulate the structures resonance, there is just a little reverberation coming from the dome.

In the silence a constant frequency of 20-21 Hz can be heard and is accompanied by two peaks at 48db and 56-57Hz at 50-51db. These frequencies are not found in the surrounding area (Fig. 11). Because similar-characteristics were found at other sacred sites, we have assumed they are most likely originating from a subterranean water source deep below the surface with the vibrations concentrated in the Rotunda. Such frequencies affect the mind creating a sense of sacredness. Using various devices, the presence of magnetic fields inside the Rotunda was measured to confirm the presence of water. The main power switch which provides electricity to the building was turned off so that accurate measurements of the-magnetic fields could be recorded. We discovered a number of magnetic fields inside, but none outside or in the neighboring area (Fig. 12)

Fig. 7 – The Sennheiser microphones positioned at the centre of the Rodunda.

Fig. 8 – Every niche was tested by voice and drum.

Fig. 9 – The imaging operation with UV camera

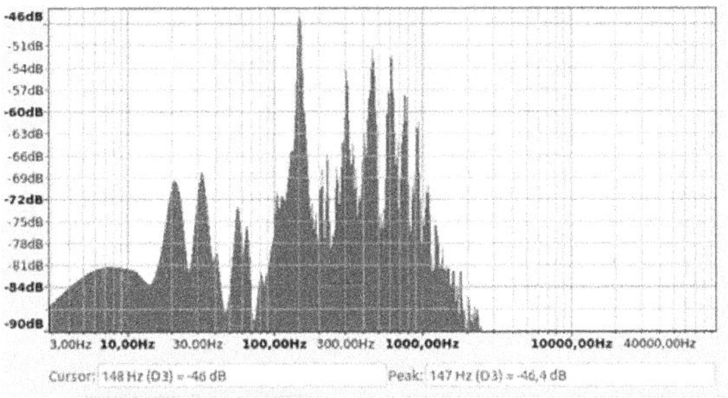

Fig. 10 –During the chanting some harmonics were produced without a single resonance frequency, but with a lot of resonance which acted like an amplifier for the voice.

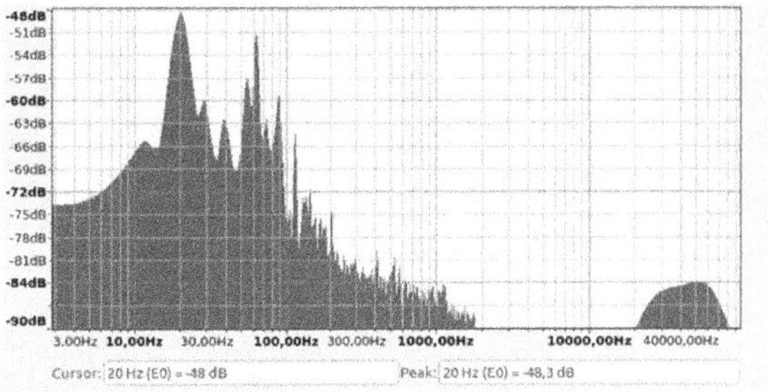

Fig. 11 – The constant peak of underground noise.

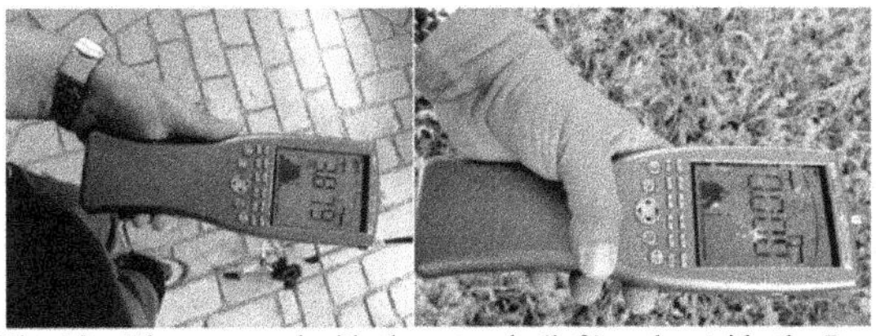

Fig. 12 – The surveys inside the rotunda (left) and outside the Rotunda (right). There were a lot of natural magnetic fields on the ground (all electric devices were switched off).

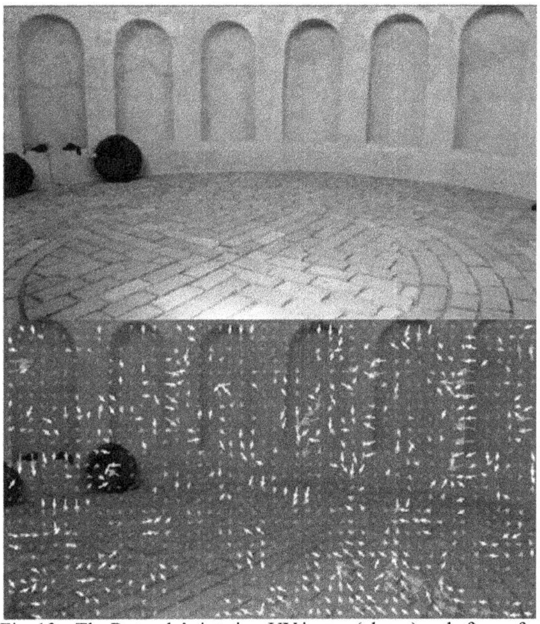

Fig. 13 – The Rotunda's interior; UV image (above) and after software analysis measuring velocimetry of air molecules (below). This shows the presence of small natural torsional magnetic fields inside the Rotunda.

Fig. 14 – Radioactivity - inside the Rotunda is double that of the outside.

By imaging the UV band it was possible to evidence the natural spiral magnetic fields originating from the movement of underground water (Fig. 13). Software analysis was made using Dante Dynamics (Canada) software.

Radioactivity looks very different inside and outside the Rotunda that is contrary to what is found in normal structures. Outside the Rotunda, natural radioactivity slowly swung in a range between 0.08 to 0,14 μSv/h (microSievert/hour) within a normal range. Inside the Rotunda radioactivity was double that of the outside, around 0,28 μSv/h (Fig. 14), but below danger level (0,4 μSv/h). There was no difference if the Geiger counter was placed close to or distant from the floor, so we can suppose it was not related to the floors material, but very likely due to the movement of underground water.

Discussion and Conclusion

Analysis was carried out on two separate occasions over two years. No single resonance frequency was found within the structure, rather there several frequencies were obtained. Through sound analysis we found that each of the 12 niches have different resonances which strengthen and enhances the both male and female voices. All niches are different in size and according to our hypothesis they were tuned on purpose. The niches function like organ tubes each with a different resonant frequency. Despite numerous renovations the Rotunda has perfect acoustics even now. The optimal acoustic effect is achieved if the singer sings in the centre of the chapel, where the reverberation from the dome and the resonance from the niches have the maximum effect. The structure of other rotundas in the region and beyond were compared to that of Bény, none featured the same design feature of 12 niches, which makes it unique. We can therefore conclude the structure was built with an acoustic scope as a sacred temple and its architect had great knowledge of acoustics. We can also assume it was built over an earlier sacred site, because it has some natural phenomena inside which are not found outside. This is typical of very ancient sacred sites we studied in Europe. This along with its acoustical properties make it worthy of broader research from experts in architecture, art history, acoustics, early music, and we hope this can be undertaken in the near future. From our side we can consider these results as preliminary, but perhaps stimulating other researchers to continue the research.

ACKNOWLEDGMENT

We are grateful to the following people for their support, information and help during and/or organising and managing our fieldwork and study in Bény: Dr. Imre Molnár, (Pozsony/Bratislava), Himler György (Párkány/Sturovo), Tamás Molnár (Pozsony/Bratislava), parish priests: Dean András Nagy, Sándor Kassa, András Szalai, representatives of the local museum and people: Gyula Koczka, Katalin Koczka, Szabolcs Kremmer (Bény), Dr. István Horváth (Esztergom). A sincere thank you to Nina Earl for her support in editing this text. We are grateful to Daniele Gullà, vice-president of Super Brain Research Group (SBRG), for analysing the imaging aspect of the Rotunda.

REFERENCES

V. Cervers –Molnár: "A középkori Magyarország rotundái. Akadémiai Kiadó", Budapest, 1972.

I. Ciulisová: Pamiatková obnova Kostola Panny Márie v Bíni. Pamiatky a múzeá 4. 1999. 32-36.

P. Debertolis, H.A. Savolainen: "The phenomenon of resonance in the Labyrinth of Ravne (Bosnia-Herzegovina). Results of testing", Proceedings of ARSA Conference (Advanced Research in Scientific Areas), Bratislava (Slovakia), December, 3-7, 2012, pp. 1133-1136.

P. Debertolis, N. Bisconti: "Archaeoacoustics in ancient sites" Proceedings of the "1st International Virtual Conference on Advanced Scientific Results" (SCIECONF 2013), Žilina (Slovakia) June, 10-14, 2013, pp. 306-310.

P. Debertolis, N. Bisconti: "Archaeoacoustics analysis and ceremonial customs in an ancient hypogeum", Sociology Study, Vol.3 no.10, October 2013, pp. 803-814.

P. Debertolis, G. Tirelli, F. Monti: "Systems of acoustic resonance in ancient sites and related brain activity". Proceedings of Conference "Archaeoacoustics: The Archaeology of Sound", Malta, February 19-22, 2014, pp. 59-65.

P. Debertolis, D. Gullà, F. Richeldi: "Archaeoacoustic analysis of an ancient hypogeum using new TRV camera (Variable Resonance Camera) technology", Proceedings of the "2nd International Virtual Conference on Advanced Scientific Results" (SCIECONF 2014), Žilina (Slovakia) June, 9 - 13, 2014, pp. 323-329.

P. Debertolis, N. Bisconti: "Archaeoacoustics analysis of an ancient hypogeum in Italy", Proceedings of the Conference "Archaeacoustics: The Archaeology of Sound", Malta, February 19-22, 2014, pp. 131-139.

P. Debertolis, A. Tentov, D. Nicolic, G. Marianovic, H. Savolainen, N. Earl: "Archaeoacoustic analysis of the ancient site of Kanda (Macedonia)", Proceedings of the 3rd Conference ARSA (Advanced Research in Scientific Areas), Žilina (Slovakia), December, 1-5, 2014, pp. 237-251.

P. Debertolis, F. Coimbra, L. Eneix: "Archaeoacoustic Analysis of the ĦalSaflieni Hypogeum in Malta", Journal of Anthropology and Archaeology, Vol. 3 (1), 2015, pp. 59-79.

P. Debertolis, D. Gullà: "Archaeoacoustic analysis of the ancient town of Alatri in Italy", British Journal of Interdisciplinary Science, September, Vol. 2, (3), 2015, pp. 1-29.

P. Debertolis, M. Zivic: "Archaeoacoustic analysis of Cybele's temple, Roman Imperial Palace of Felix Romuliana, Serbia", Journal of Anthropology and Archaeology, Vol. 3 (2), 2015, pp. 1-19.

P. Debertolis, D. Nicolić, G. Marianović, H. Savolainen, N. Earl, N. Ristevski: "Archaeoacoustic analysis of Kanda Hill in Macedonia. Study of the peculiar EM phenomena and audio frequency vibrations", Proceedings of the 4th Conference ARSA (Advanced Research in Scientific Areas), Žilina (Slovakia), November 9-13, 2015, pp.169-177.

P. Debertolis, N. Earl, M. Zivic: "Archaeoacoustic Analysis of Tarxien Temples in Malta", Journal of Anthropology and Archaeology, Vol. 4 (1), June 2016, pp. 7-27.

P. Debertolis, D. Gullà: "Preliminary Archaeoacoustic Analysis of a Temple in the Ancient Site of Sogmatar in South-East Turkey. Proceedings of Conference 'Archaeoacoustics II: The Archaeology of Sound', Istanbul (Turkey), Oct 30-31 Nov 1, 2016, pp. 137-148.

P. Debertolis, D. Gullà: "New Technologies of Analysis in Archaeoacoustics", Proceedings of Conference 'Archaeoacoustics II: The Archaeology of Sound', Istanbul (Turkey), Oct 30-31 Nov 1, 2015, pp. 33-50.

P. Debertolis, D. Gullà: "Healing aspects identified by archaeoacoustic techniques in Slovenia", Proceedings of the '3rd International Virtual Conference on Advanced Scientific Results' (SCIECONF 2016), Žilina (Slovakia), June 6-10, 2016, pp. 147-155.

P. Debertolis, D. Gullà, F. Piovesana: "Archaeoacoustic research in the ancient castle of Gropparello in Italy", Proceedings in the Congress "The 5th Virtual International Conference on Advanced Research in Scientific Areas" (ARSA-2016) Slovakia, November 9 - 11, 2016: pp. 98-104.

P. Debertolis, N. Earl, N. Tarabella: "Archaeoacoustic analysis of Xaghra Hypogeum, Gozo, Malta", Journal of Anthropology and Archaeology, vol.1 no. 5, June 30, 2017. In press.

P. Debertolis, D. Gullà: "Archaeoacoustic Exploration of Montebello Castle (Rimini, Italy)", Art Human Open Acc J 1(1): 00003, DOI: 10.15406/ahoaj.2017.01.00003.

P. Debertolis, D. Gullà, H. Savolainen: "Archaeoacoustic Analysis in Enclosure D at Göbekli Tepe in South Anatolia, Turkey", Proceedings in Scientific Conference "5th HASSACC 2017 - Human And Social Sciences at the Common Conference", Slovakia, Žilina, September 25-29, 2017: pp. 107-114.

A. Habovstiak: Archeologicky vyskum v Bíni. Vlastivedny casopis 12. 1963. 173-177.

A. Habovstiak: Bína. In: Vyznamné slovanské náleziská na slovensku. Bratislava, 1978. 23-24.

K. Haiczl: A bényi prépostság temploma. Galánta, 1937. In: Haiczl K. (ed.): Kakath, Dsigerdelen-Csekerdén, Párkány., 2nd edition:. Párkány és vidéke Kulturális Társulás, 1997. 89-99.

K. Kozák: Téglából épített körtemplomaink és centrális kápolnáink a XII.-XIII. században. Móra Ferenc Múzeum Évkönyve. 1976-77.

Z. Németh: A Kárpát-medence legkülönösebb középkori templomai II., BKL Kiadó, Szombathely, 2017. 103-119.

V. Pentcheva,Bissera(ed.): Aural Architecture in Byzantium: Music, Acoustics, and Ritual, Routledge 2017.

V. Pentcheva,Bissera : Hagia Sophia: Sound, Space, and Spirit in Byzantium, Pennsylvania State University Press, 2017, http://hagiasophia.stanford.edu 2. (2018.01.03.)

V. Pentcheva,Bissera :The Sensual Icon: Space, Ritual, and the Senses in Byzantium Pennsylvania State University Press, 2010, paper back 2013, www.thesensualicon.com

I. Reznikoff: The Evidence of the Use of Sound Resonance from Palaeolithic to Medieval times, in: Archaeoacoustics. Lawson, G. and Scarre, C. eds. University of Cambridge, Cambridge, 2006,

E. Sabasosova, M. Havlik: Rotunda Dvanastich Apostolov v Bínidokumentácia odkrytych murív s vyhodnotením nálezov. 2010.

E. Sabadosova, M. Havlik: Rotunda Dvanastich Apostolov v Bíni In: Valeková, Anna (ed): Ranostredovekásakrálna architektúra Nitrianskeho kraja: Zborník zo seminaára a katalóg ku vystave. Nitra, 2011.

A. Szénássy: Felvidéki Árpád-kori templomok lexikona. I. A Nyitrai kerület- Szlovákia, Komárom, 2005.

A. Szilágyi: A Kárpát-medence rotundái és centrális templomai, Semmelweis kiadó, Budapest, 2008. 2nd edition 2009.

N. Tarabella, P. Debertolis: "Archaeoacoustics in Archaeology", Proceedings in19th International Conference and Assembly of the Experts of the Foundation Romualdo Del Bianco "HERITAGE FOR PLANET EARTH 2017 - Smart Travel, Smart Architecture and Heritage Conservation and its Enjoyment for Dialogue", Florence, Italy, 11-12 March 2017: 240-246

Experiencing an Ancient Performance in a Roman Theatre

Maria Cristina Manzetti

MARIA CRISTINA MANZETTI, Archaeologist, Ph.D. candidate in Digital Media and Cultural Heritage at the Technical University of Crete (Greece). Specialized in Roman theatres and 3D modeling. Email: cristina@ims.forth.gr Laboratory of Geophysical - Satellite Remote Sensing & Archaeo-environment, Foundation for Research & Technology, Hellas (F.O.R.T.H.), Institute for Mediterranean Studies (I.M.S.), Rethymno (Greece)

ABSTRACT: The first aim of this research is to identify what kind of performance was played in the theatre of Aptera, one of the several Roman theatre in Crete (Greece).The theatre is not well preserved but we have many information about its structure so that it is possible to reconstruct its 3D model and therefore investigate it through virtual acoustics analysis.

The second aim is to create some auralised sounds, in different positions of the seating area of the theatre, in order to simulate the experience an ancient Greek performance.

KEYWORDS: Auralisation, Virtual Archaeology, Roman theatres

Introduction

Until now, archaeology has often been a "visual" discipline. The first thing an archaeologist does when he/she arrives on an ancient site, is observing what is around him/her. The accurate visual examination of the terrain and the landscape provides some suggestions regarding what kind of settlement might be found there. It is the same for the study of the artefacts: specialists observe the material, the quality and the decorations of the object they have in their hands, in order to formulate some preliminary hypotheses about their date, provenance and function. Probably it is also because nowadays humans are much more accustomed to be focused on what they see, more than what they hear, touch or smell.

May we affirm that the sight is the easiest and fastest way to understand and also to memorize something? The answer of the archaeologists would be probably "yes", if we consider that in the world of archaeology, the visualization of ancient sites, monuments and artefacts, also through digital media and 3D models, has become so popular and widely used. Even if it might be true that our perception of things around us is mainly due to the sight, we cannot forget that other senses are fundamental too, and in the past, they may have been even more used than the vision, in particular in some contexts, as ritual spaces or places reserved to communication, and performances, as it may be a theatre.

Finally, the hearing is gaining importance too in the study of the ancient sites, as it is demonstrated by the new term "archaeo-acoustics"(Scarre and Lawson, 2006). Many acoustics analysis are conducted in archaeological sites (in situ or virtually) in order to understand something more about their past, their function, their use. The auralisation of the ancient buildings is spreading as well (more by the acoustic sector than by the archaeological one), offering the opportunity to experience feelings and emotions in the same places our ancestors were frequenting.

Auralisation in Archaeology

Even if acoustics analysis is recently more frequently applied to cultural heritage context, it is still rare to find works involving the auralisation of ancient monuments or archaeological sites.

The auralisation gives us the perception to be present in a specific space, whose acoustics characteristics produce a unique sound and effect. Today, the possibility to reconstruct virtual buildings and rooms, allows the auralisation of buildings and monuments that are not existing anymore, or that are only partially standing. The auralisation is not only useful to experience the sound of the past and consequently to be virtually immersed in a different reality, but also to trigger a series of new feelings and letting us identify ourselves as actual participants/ inhabitants of that place. The auralisation can also help the interpretation of the results of the acoustics analysis, because it is actually the subjective perception of the objective values of the acoustics parameters measured: can we well understand the content of the sound? Is the sound enjoyable or annoying? Is there any echo?

The AudioLab of University of York is one of the few that are involved in the study of acoustics characteristics of cultural heritage sites, also through the use of virtual reconstructions and virtual acoustics analysis. The researchers employed the auralisation to validate the results obtained by in situ measurements (in the medieval Church of St. Margaret at York, UK) and by virtual calculations: the subjective perception of the auralised files obtained though the different methodologies, show a different result in the length of the reverberation (Foteinou and Murphy, 2011). Furthermore, they created the OpenAIR Library that consists in a collection of acoustic data of historical sites around UK, which are available to produce auralisation from the different sites (available at: http://www.openair-lib.net/about).

A multi-disciplinary team from the North Carolina State University elaborated an online visual and acoustic model of St. Paul's Cathedral in London. The purpose is the reproduction of a sermon pronounced by John Donne in 1622 in the courtyard of the sacred place, together with the reproduction of the soundscape, composed by the clamour of the people, the dogs, the birds and the bells. It is possible to select eight different locations from where someone listened to the preacher, and also to compare each one of them at different hours, that means in a more or less crowded area (500, 1200, 2500, 5000 people) (available at: https://vpcp.chass.ncsu.edu/).

A larger application of acoustic analysis and auralisation in archaeological sites, in particular ancient theatres, has been done during the ERATO project (audio visual conservation of the architectural spaces in virtual environment). Again, the auralised files of the theatre of Aspendos (Turkey), were obtained through several virtual reconstructions and virtual acoustics measurements, to notice the differences in the sound, caused by the presence (or the absence) of some architectural elements in the 3D models of the theatre (Lisa et al., 2006).

Case Study and Methodology

Aptera was an ancient city located in the north-west area of the Greek island of Crete. Its theatre was first built in the III century BC but then the Romans partially modified its shape, in two different phases. The theatre has been largely investigated and the archaeological excavations lasted about ten years, conducted by the Ephorate of Prehistoric and Classical Antiquities of Chania. The archaeologist and the architect who worked on the site, produced a detailed report about the second Roman phase of the theatre, accompanied by plans and sections of it (Niniou-Kindeli and Chatzidakis, 2016). Therefore, even if the theatre is not fully preserved, it has been possible to recreate a 3D model representing its original aspect during the last phase, based on the archaeological information acquired.

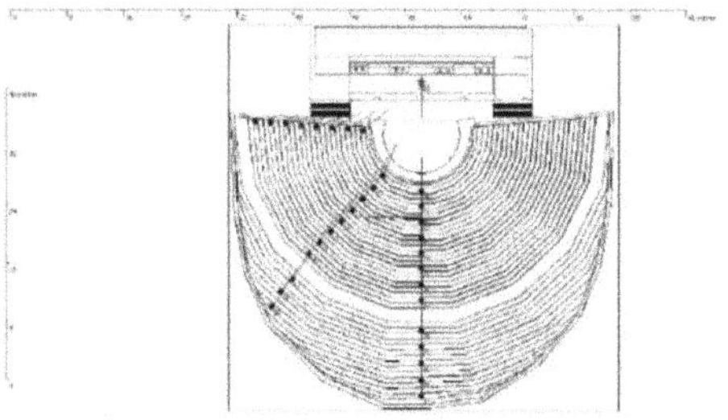

Figure 4. Plan of the 3D model of the theatre of Aptera in Odeon Room Acoustics, with receivers (blue dots [in seating area]) and source (red dot [on stage]).

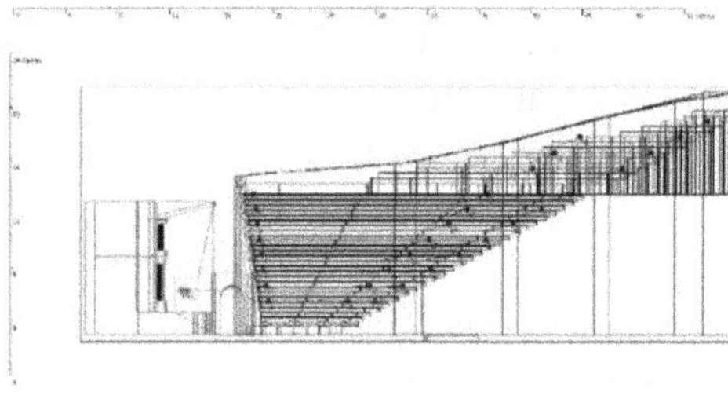

Figure 5. Section of the 3D model of the theatre of Aptera in Odeon Room Acoustics, with receivers (blue dots [in seating area]) and source (red dot [on stage]).

The plan has been first imported in AutoCad, it has been scaled and then it has been traced. The digitalized plan has been then imported in 3D Studio Max in order to extrude it, according to the measures given in the section of the theatre. Once the basic 3D model of the second Roman phase of the theatre has been completed, it has been imported in Odeon Room Acoustics in order to proceed with the virtual acoustics analysis. First of all, a grid of receivers has been built: 32 receivers have been placed in one half of the cavea, one every three seats, forming three lines respectively of 8, 10, and 12 receivers; they are located 75 centimetres higher than the correspondent seat in order to simulate the average height of the eye of a seated man/woman. An omni-directional source has been placed at the centre of the stage, 160 centimetres higher than its floor to simulate the average height of a standing man/woman, and its overall gain has been set to 60 dB (figs.1 and 2).

The second fundamental step has been the assignation of the right materials to each surface of the 3D model, in order to attribute the correct absorption coefficient, which greatly influence the acoustics of a room. The general settings for the measurement and for the auralisation have been modified to adapt them to this specific case.

The acoustics parameters that have been considered in order to evaluate the quality of the sound in the theatre of Aptera are: Reverberation Time (T60), Early Decay Time (EDT), Clarity (C80), Definition (D50) and Speech Transmission Index (STI). The suitable values for T60, C80 and D50, depending if the place is reserved for speech or for music performance, are indicated in table 1; about the STI, values above 0.60 are good for the intelligibility of the talk, above 0.75 are excellent, under 0.60 are fair and bad(Spagnolo, 2014). The EDT should be between 0.2 and 0.4 seconds lower than the T60, as it has been demonstrated during the ERATO project (Gade and Angelakis, 2006).

	T60	C80	D50
Speech performance	~1s	>3dB	>0.50
Music performance	~2s	<3dB	<0.50

Table1. Ideal values of Reverberation Time, Clarity and Definition, for speech and for music performance.

Results

The partial results obtained by the virtual acoustics analysis are listed in the table 2.

To summarize, only the average values of all the receivers, for the range of frequencies 125 Hz-2000 Hz, are shown. As it is visible from the table 2, the average value of the reverberation time is between 0.66 and 0.86, the average value of the early decay time is between 0.2 and 0.3 seconds lower than the respective values of reverberation time, the clarity is between 7.58 and 13.59 dB, and the definition is between 0.7 and 0.85. The average value of speech transmission index is 0.79.

	125 Hz	250 Hz	500 Hz	1000 Hz	2000 Hz
T30	0.86	0.75	0.7	0.66	0.7
EDT	0.64	0.4	0.34	0.35	0.35
C80	7.58	11.38	13.29	13.4	13.59
D50	0.7	0.8	0.84	0.84	0.85

Table 2. Average values of 32 receivers, between 125 and 2000 Hz, of Reverberation Time, Early Decay Time, Clarity and Definition.

Comparing the results obtained from the virtual acoustics analysis with the ideal values for each of the parameters aforementioned, it is evident that the Roman theatre of Aptera was suitable to have speech performance.

At the Laboratory of Acoustics Mario Corbino at CNR of Rome-Tor Vergata (with the assistance of Dr. Paola Calicchia), an anechoic file has been recorded in order to be convolved with the various impulse response obtained by the virtual acoustics analysis. Martina Giovanetti, a student in philology from the University of Rome Tor Vergata, played a monologue from "The Trojan Women" by Euripides, lines 634-648. The file has been auralised for all the receivers and all of them have been listened to in order to have a subjective perception of the sound within the ancient theatre. The hearing of the auralised file has confirmed the results obtained by the virtual acoustics analysis because, from all the positions, the monologue is clear and understandable, without perceiving any long reverberation.

A very short video has been created using the auralised files and the rendered images of the textured 3D model of the theatre of Aptera. The images have been rendered placing the camera in the same positions of the receivers (fig. 3). In the video, each image is linked to the corresponding sound so that the visualization and the hearing are combined in the same work.

Figure 6. Render of the texturixed 3D model of Aptera in 3D Studio Max

Conclusion

This paper has briefly illustrated the applications of auralisation in archaeology. Auralisation can be a subjective instrument to enhance the interpretation of objective results obtained through acoustics analysis. Furthermore, it can be used in digital applications (for smartphones, tablets, and PC), together with 3D models of the ancient monuments or buildings, in order to create a virtual archaeology project that helps the general public to empathize with our ancestors and with the places of our past.

It is interesting to note that large part of the studies about acoustics and auralisation in ancient space have been conducted by acousticians, rather than by archaeologists. It is desirable that, in the future, there will a deeper collaboration between the two fields, in order to produce more appropriate interpretations of the acoustics data, according to the peculiarity of each archaeological site analysed, and also to take advantage of all the possibilities that the acoustics analysis offers.

ACKNOWLEDGMENTS

This research has been carried out during my Ph.D. at Technical University of Crete (Greece), thanks to the funding from IKY-State Scholarship Foundation of Greece.

Thanks to Dr. Paola Calicchia who has granted me to use the facility of the Laboratory of Acoustics Mario Corbino at CNR in Rome.

Thanks to Prof. Dr. Nadia Magnenat-Thalmann, Director of MIRALab at University of Geneve to allow me to use their 3D character representing Cassandra in my 3D model of the Roman theatre of Aptera.

Thanks to Dr. Apostolos Sarris for his continuous help and his useful suggestions.

REFERENCES

Foteinou, A., Murphy, D.T., 2011. Perceptual validation in the acoustics modelling and auralization of heritage sites: the acoustic measurement and modelling of St Margaret's Church, York, UK, in: The Acoustic of the Ancient Theatres. Presented at the The Acoustics of the Ancient Theatres Conference, Patras.

Gade, Angelakis, 2006. Acoustics of ancient Greek and Roman theatres in use today. The Journal of Acoustical Society of America 120, 3148–3156.

Lisa, M., Rindel, J.H., Gade, A.C., Christensen, C.L., 2006. Acoustical Computer Simulations of the Ancient Roman Theatres, in: ERATO Project Symposium. Istanbul, Turkey, pp. 20–26.

Niniou-Kindeli, V., Chatzidakis, N., 2016. The Roman theatre at Aptera: a preliminary report, in: Roman Crete. New Perspectives. Oxbow Books, Oxford & Philadelphia, pp. 127–153.

Scarre, C., Lawson, G., 2006. Archaeoacoustics, McDonald Institute Monographs. ed. McDonald Instituire for Archaeological Research, Cambridge.

Spagnolo, R., 2014. Manuale di acustica applicata. CittàStudi, Torino.

Exploring Aural Architecture: Experience, Resonance, Attunement

Cláudia Martinho

CLÁUDIA MARTINHO, PhD in Music - Sonic Arts (Goldsmiths, University of London, UK), is an architect (FAUP, University of Porto, Portugal), with a Master of Science and Technology in Architectural Acoustics (UPMC, Paris, France). c.martinho@gold.ac.uk

ABSTRACT: This research project aimed to established connections between aural architecture and archaeoacoustics. It explored a methodology to create ancient acoustic effects nowadays, and to study the human affective experience of it. The project experimented the design of aural architecture as a communication channel through the language of vibration, to open up a gaping moment in spacetime and a resonant dialogue between inside and outside worlds, or what has been described as mind-body experiences. The methodology included techniques of sensory variation and resonant design based in energetic geometry. It addressed a qualitative study of the affective experience of acoustic phenomena, which is less developed than quantitative study and would benefit from further experimentation. The experiment created a public experience of resonance and attunement, to observe its results with a survey.

KEYWORDS: aural architecture, mind-body experiences, resonance frequencies

Introduction

This paper reports on a practice-lead research project, entitled *Passage* (2017). It explored aural architecture design methods for the transformation of the acoustic space experience through resonance and attunement, a technique known to our ancestors. The term aural refers to the human experience of a sonic process, and aural architecture refers to the properties of space that can be experienced through listening (Blesser and Salter 2007, 2-5). This project aimed to establish connexions between aural architecture and archaeoacoustics, by the experimentation of possible manifestations of ancient acoustic effects nowadays, with the purpose of studying the human affective experience of it. Acoustic effects are largely studied with the aid of mathematical and computational models (acoustic ray tracing prediction, electroencephalography), providing important data for the advancements of archaeoacoustic studies. However, the qualitative study of the affective experience of acoustic phenomena is less developed and would benefit from further experimentation. In this sense, the results of this experiment were observed through public's experience with a survey. This approach may provide a line of relevant research on human affect of acoustic phenomena, which could complement and enrich others, constituting an additional tool.

An affective phenomenon has been described as something that is felt without registering consciously, registering only its effects (Massumi 2008, 4). Therefore cultural and social aspects were not relevant in this study. Affect was approached as an instinctive human reaction, as primary, non-conscious and intensive (Massumi 2002, 27). Affect is linked to the experience of attunement. Attunement has been described by various disciplines (music, cognitive science, philosophy of science, physics and psychology) and is linked to sympathetic vibration, resonance and entrainment. This project approached attunement as a phenomenon based in sympathetic resonance that affects the quality of that experience. The project aimed to experiment ways to trigger an innate capacity of attunement (Morton 2014) to self, other beings, and the environment. Creative design was engaged to discover applications of specific acoustic qualities, to understand how its physical and psychic experience could influence behaviour. The emphasis was on the exploration of the aural experience as an ecological and/or spiritual relationship with the surrounding environment. The experiment focused on the physicality of resonance and vibrational force, linked to the corporeality of environmental sound, enhanced by space's acoustics. The purpose was to explore how an aural architecture experience may act as a communication channel, towards what has been described in the field of archaeoacoustics as mind-body experiences.

Aural Architecture and Archaeoacoustics

Physical acoustics and aural architecture, while directly related, have profoundly different emphases. The former uses a scientific language to describe the way in which spatial acoustics changes attributes of sound waves, while the latter considers the experiences and behaviour of inhabitants in a space. One emphasizes discrete measurement and modelling, while the other explores a complex interactive phenomenon. (Blesser and Salter 2007)

According to acoustician Barry Blesser and environmental psychologist Ruth-Linda Salter, aural architecture focus on the way that listeners experience space (Blesser and Salter 2007, 2-5). Any environment, natural or built, generates an aural architecture. Every space has an aural architecture. It is the attributes of a space, such as surfaces, objects, materials and geometries that will determine its specific acoustic aspects. And it is the human experience of that space that determinates its aural qualities. The acoustic cues orientate our navigation but provide also sensory stimulus which define the space's aural specificity and influence our associations and moods. The fields of archaeoacoustics and aural architecture share similar aspects, and perhaps common methodological points. Both are multidisciplinary fields, and may require knowledge from archaeology, architecture, acoustics, neuroscience, cognitive science, philosophy of science, sound design, musicology, anthropology. The difference is that one pertains with the past and the other with present and future. This aural architecture research project was based in a knowledge of the past, to act in present with ecological forms of experience and contribute for a sustainable planning of the future.

The aural qualities of a space are recognized by the human being since pre-history. Several ancient sites of different cultures provide evidence on the importance of sound and acoustics for ancient civilizations. Musical instruments or sound-producing devices (Lund 1981, 246) were found in ancient sites which indicate sound timbres and frequency spectres. But we do not know exactly what the ancient cultural forms of experience were like, its rituals, music, dances, rhythms and melodies. Nevertheless, we do know that particular resonance frequencies are found in ancient sites, with similar acoustic effects and affects. There are also evidences that natural phenomena (rain, thunder) and the elements (water, air, earth, fire) were important for ancient cultures, and most of the time associated with supernatural forces of creation. In fact, a large number of acoustic effects has been reported. These include: particular resonance frequencies at prehistoric sites such as *Lascaux* cave (c. 15000 BC), France, at megalithic chambers such as *Newgrange* (c. 3200 BC), Ireland, and at excavated temples such as the *Oracle Room* at *Ħal Saflieni Hypogeum* (c. 3300-3000 BC) in Malta, Europe; reverberant chambers inside or beneath pyramids such as the *Pyramid of Djoser, Saqqara* (c.2667-2648 BC) or the *Great Pyramid of Giza* (c. 2580-2560 BC) in Egypt, the *Pyramid of the Sun* in *Teotihuacan* (c. 200 AD), Mexico; reverberant ceremonial *kivas* in Native America such as *Anasazi kiva* at *Aztec* (c. 1100AD), New Mexico; infrastructures of reverberant water channels such as *Chavin De Huantar* (c.900-200 BC) in Peru; chirped echo of *El Castillo* pyramid of *Chichén Itzá* in Mexico; acoustic design for sonic amplification in *Palenque* (c. 675) in Mexico (Zalaquett et al., 2008); to more recent temples such as the *Charola* of *Convento de Cristo, Knights' Temple* (1520) in Tomar, Portugal, the *Whispering Gallery* of *St. Paul's Cathedral* (1697) in London, England. Archaeoacoustics research of sound behaviour in ancient places of different eras and cultures has revealed that our ancestors used spaces with very particular resonant qualities, most likely to enhance ritual-based experiences.

While the primordials of aural architecture might have resulted from unplanned acoustic accidents, it certainly resulted as the origin of inspiration for intentional aural architecture constructions, from which has emerged knowledge and cognitive frameworks on universal geometry. Universal geometry became an important tool in architecture to integrate in its design relationship between sound, light, shape, frequencies, vibrational patterns, waves' propagation, acoustic effects. Universal geometry was implied in architectural design until the 19th century. It devised very particular acoustic qualities to the experience of space and involved a deep knowledge on its consequences. However, this universal knowledge has been lost in most of current architectural practice. Since the advent of the industrialisation in the 19th century, architecture became standardised. Until that time, architecture integrated the knowledge of acoustics. Architects designed based in the relations between geometry, materials density, spatial proportions, acoustic effects and the human multi-sensory experience. But since the 19th century, an education system separated architecture and acoustical engineering into different disciplines and specialisations. As a consequence, in most of current architectural practice, acoustic features such as geometry, proportions and materials are not taken into account in early design stages (except in concert halls). It is mostly considered as an acoustic correction, mainly concerning noise control and acoustic insulation techniques. Besides, the science of acoustics is constrained to investigate sensation and perception in a laboratory context.

In the theory of modern architecture we find very little about the relationship between sound, space and body, and this is reflected in architectural practice. While the acoustic features of the built environment are shaping our experience of the world, architecture practice usually neglects the auditory experience and acoustic space in its design process. The lack of quality and diversity in acoustic space experiences in the everyday produced by a generic architecture is a problematic issue, as it contributes to turn our sensorial interplay into a dormant state. Urban space is becoming a saturated amount of poor quality acoustic experiences; and because modern architecture has underestimated if not completely ignored these phenomena, it certainly has caused substantial damage (Leitner 1998, 293). The sound quality of our everyday experiences has serious consequences in our well-being, as it directly affects our nervous system, heart, breathing and blood pressure which are largely beyond conscious control.

In this context, as most of architecture practice usually neglects acoustic space in its design process, there is a need to re-integrate ancient forgotten knowledge in innovative ways. The challenge today lies in re-integrating the knowledge of acoustics, geometry and the human experience of sound in architecture's design process. This practice-based research project addressed this gap in knowledge, exploring connexions between archaeoacoustics and aural architecture. Furthermore it experimented the application of acoustic effects found in ancient places into a contemporary aural architecture practice. Nevertheless, while recognising the value of ancient knowledge, this project explored new forms of experience, resonance and attunement adapted to our culture.

Experiment

The research project, entitled *Passage* (2017), was commissioned for the event *Lisboa Soa,* an annual event on sound art, urbanism and auditory culture in Lisbon, Portugal. The aim of *Lisboa Soa* was to take citizens on a sonic journey to re-discover their cities' green spaces, and value them. This edition took place in *Estufa Fria*, a human-made ecosystem of hundreds of species of plants, running water, water falls, stones, steps, caves and tunnels, which produced particular acoustic effects. The project *Passage* proposed an aural architecture installation to explore ways to listen to this ecology through acoustic resonance. The experiment engaged an active and immersive listening, in a mode of attunement with self, the place and the surroundings.

Aural Architecture as an Acoustic Communication Channel

It has been discussed that ancient architecture, besides providing shelter, was created based in the language of vibration and universal geometry, with the intention to facilitate the human experience of connection between Earth and Cosmos. The interconnection between body, mind, space and time, achieved through sound, affects even our innermost being, most of the time unconsciously, and architecture determinates the quality of this relation. Architecture might had the function of an acoustic communication channel through resonance, for attunement to specific frequencies and dimensions.

[It] is like a kind of dialogue which is determined by the acoustics premises. This dialogue enables us to experience ourselves in the sound of a room. (Leitner 1998, 294)

Architecture's physical hearing engages the entire body, as being in a constant dialogue between the inner-self with the outside and surroundings. In much the same way architecture also resembles such a dialogue connecting inside and outside in a specific way (Leitner 1998, 299). Was it this kind of dialogue what ancient civilisations were experiencing in their resonant temples?

The intention of the project *Passage* was to experiment the design of aural architecture as a communication channel through the language of vibration, with creative forces, to open up a gaping moment in spacetime and a resonant dialogue between inside and outside worlds, or mind-body experiences. Therefore, this project aimed to study an affective response to ancient acoustic effects nowadays.

Through the experience of specific frequencies and sounds in acoustic space, it experimented the enhancement of mind-body experiences, in order to study the human affective experience of it, here named affective attunement.

In this sense, the experiment developed two methods of aural architecture design, as follows:

- Sensory variation - the experience of sound in space as powerful vibrational forces, with the use of natural phenomena and elemental (primordial) sounds;

- Resonant design - architectural space as an acoustic resonator, by the appropriation of existing places with particular resonance qualities or building new ones.

Sensory Variation

Sensory phenomena varies in general in terms of frequency, patterns, rhythm and tone; it can be sensed as movement, temperature, light or sound. The purpose for this project's sensory variation was to draw the public's attention to subtile forces at work in the place, usually gone unnoticed. In other words, it engaged a translation of languages, as a transduction into an hearable spectrum. So the aural architecture of *Passage* was designed as a channel for acoustic communication through the language of vibration. This technique explored the language of vibration by the modulation of natural phenomena, to achieve a sensorial transformation. Ancient civilisations used natural spaces and built architecture for specific purposes with sensory variations of sound, light, smell and temperature. In this place, the presence of water was subtle yet its experience could become intense and overwhelming. This intervention was inspired by ancient places such as *Chavin de Huantar*, or *Tihuanaco*, in South America, where amazing acoustic infrastructures of water channels are found. It is possible that when water was passing in, a deep roaring sound was produced, probably associated with an entity of rain or thunder, or as an acoustic matrix, where oracular pronouncements could be deciphered (Devereux 2001, 143). For this experiment, similar acoustic effects were sought. Here, the water presence (water falls, water drops, water channels) was amplified through acoustics and became a powerful vibratory force that modulated the whole ambiance. The aural architecture installation explored forms of experience of environmental sound through acoustic phenomena and embodiment. The notion of embodiment considers the human body as the natural mediator between mind and physical environment, and may engage the experience of space and time as a unified field of resonance. Here, acoustic embodiment was achieved through the space's resonance. The magnification of environmental sounds produced a sense of physicality and entered into sympathetic vibration with the audience's body and mind.

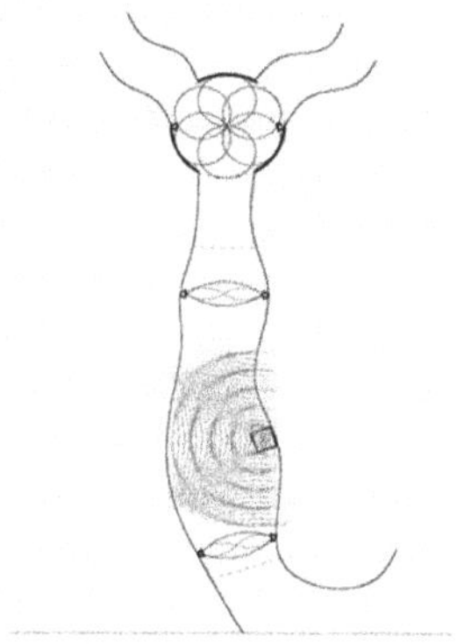

Figure 1 - plan and section of the tunnel and the zome

The experiment *Passage* created a walking path with two different interrelated forms of experience, and engaged the public in two different kinds of sensory variation (fig. 1):

- The tunnel - an intervention in an existing architectural space - a soundscape installation in resonance with the acoustic space of a passage way.

- The zome[1] - the creation of a new architectural space - an architecture installation as a relatively quiet moment for attunement.

This intervention enhanced the water presence, and amplified it in the acoustic space of the tunnel, which was a cold, dark, humid passage. The tunnel turned into a water channel, a passageway between different time-spaces, a communication channel to the whole network of the thousands of species of plants living there. The vibratory forces of water turned into a deep roaring sound and modulated specific resonance frequencies. An immersive passage was created, as an in-between zone of low frequency sound and sympathetic resonance, a void, with portals of standing waves (figures 2, 3, 4).

As an extension, the zome was built in a strategic place, where the path coming out of the tunnel divided in two paths. It embraced the path coming out of the tunnel and transform it into another form of experience. The zome offered a cosy, dry, warm shelter, to turn the experience to the interior of self (figures 5, 6, 7).

Resonant Design

Through this method of resonant design, based in acoustics and geometry, architectural space became an acoustic resonator. It may be applied by the appropriation of existing places with particular resonance qualities or by creating new ones. Here, three techniques were experimented: space as resonator, resonant soundscape and space as frequency.

Figures 2, 3, 4 - the tunnel experience

Figures 5, 6, 7 – the zome experience

1 The term zome was coined in 1968 by thinker Nooruddeen Durkee, combining the words dome and zonohedron.

Space as Resonator

The acoustic space of the tunnel was activated with its resonance frequencies. As this tunnel was not an enclosed space, an accurate acoustic space measurement could not be done, neither a reverberation time calculation. Anyway, these studies were not needed to conduct the experiment. The aim was to reach out for particular resonance frequencies, to create standing waves, which was simple to calculate based in the distance between two parallel walls. To reach out the acoustic effect of a standing wave pattern, the primary resonance frequency between two parallel walls was calculated. The results in two points were of 108 Hz and 110 Hz. Coincidently these are resonance frequencies recurrent in ancient sites, as it has been investigated in the field of archaeoacoustics.

Resonance Frequencies and Mind Body Experiences

Experiments conducted in prehistoric caves with megalithic art on their walls and chambers provided evidence of particular resonance qualities (Reznikoff and Dauvois 1988; Scarre 1989; Dayton 1992; Devereux and Jahn 1996; Watson and Keating 1999). Prehistoric megalithic chambered sites in England and Ireland, such as Newgrange, Ireland (c. 3200 BC), were tested with acoustic measurement for their natural resonance frequencies (Jahn *et al.* 1996). A strong resonance at a frequency between 95 Hz and 120 Hertz was identified, with most at 110–112 Hz, despite major differences in chambers shapes and sizes (Jahn *et al.* 1996). Some of them had rock drawings that resembled the chamber's resonant modal patterns (Jahn *et al.* 1996). An acoustic experiment with drumming took place at the Orkney chambered mound of *Maes Howe* (Watson and Keating 1999), and an Helmholtz resonance of 2 Hz was found (Devereux 2001, 101). A state of relaxed body and alerted mind was reported. Other bodily sensations were felt, such as infrasound and the illusion that the sound was being generated inside the participant's head. It was suggested that some of these pre-historic acoustic spaces were ideal environments for producing dramatic sound effects (Watson and Keating 1999, 335). It has been discussed that particular resonant qualities of the ancient structures could have been evoked by ritual chanting and might have facilitated mind-body experiences (Devereux 2001, 89; Jahn *et al.* 1996). To further an understanding on the apparent tuning of ancient structures to particular resonance patterns, other experiments tested the effects of these frequencies on human adults. In a pilot project conducted by Ian A. Cook *et al.* (2008), 30 adults were exposed to tones at these frequencies, monitored with electroencephalography (EEG). These studies showed that frequencies between 108 and 112 Hz specifically affected regional brain activity. Listening to a tone of 110 Hz, EEG revealed that the brain activity was significantly lower corresponding to "a shift in prefrontal activity that may be related to emotional processing" (Cook *et al.* 2008, 96). In another study assessing brain activity conducted in experienced practitioners of meditation while meditating, an increased activity in the prefrontal cortex was reported (Xu *et al.* 2014, 5). These authors argued that this activity relates to a relaxed focus of attention, which allows spontaneous thoughts, images, sensations, memories and emotions. It has been discussed that future experiments might assess whether these frequencies could lead to shifts in emotional state, in the content of the listener's thoughts, or in a sense of "disorientation" (cf. Cross and Watson 2006) (Cook *et al.* 2008, 101).

This research project *Passage* addressed this kind of qualitative experiment, to further an understanding on the role of specific frequencies and sounds in human behaviour, as it will be later discussed.

Resonant Soundscape

In the experiment *Passage*, the tunnel underwent a process of acoustic transformation, engaged as an overwhelming force, an immediacy, like an immersion that takes experience into a void. A resonant soundscape composition and acoustic spatialisation sublimated the presence of the vibratory force of water in its different dynamics, and modulated distinct layers of frequencies, pitch, rhythms, intensities.

The presence of infrasound was amplified, to enact the physicality of vital force of environmental sound, enhanced by space's acoustics. Primary resonance frequencies of 108 Hz and 110 Hz as tones and its harmonics activated standing waves to draw mind-body experiences into levels of relaxation and meditation. After going through this passage, the audience would go inside the zome, for an inner-outer form of experience.

Space as Frequency

Inspired by the surrounding geometry of plants, the geometry of the flower of life was experimented into the construction of a zome. Zomes are geometric volumes composed of lozenges arranged in a double spiral. The geometry of this zome was based in a frequency of six, which is a diagram of equilibrium and balance of forces (a triangle up and a triangle down, same distance between the points) (figure 8). This is also the diagram at the base of the geometry of the flower of life, which is known since ancient cultures as a geometry that includes all existing geometric patterns (figure 9). This geometry expanded then vertically as a double spiral or helix, one spiral curving clock-wise, the other spiral curving anti-clock wise (figure 10).

The zome geometry resonated with the patterns of the plants and experimented a sustainable mode of building with natural materials efficiency (bendable wood, jute wire, cork). The zome was self-constructed and incorporated specific acoustic qualities, due to its geometry, material's density and spatial volume. The resulting spatial volume and energetic geometry engaged some sort of dynamic stillness, a protective environment, absorbent but at the same time allowing permeability. It created an inner intimate experience, connected to the outside environment but at the same time protected and isolated. The aim was to open up a dialogue between inside and outside.

Fig. 8, 9, 10 - hexagram, flower of life and zome geometry

Therefore an acoustic communication was engaged as a dialogue, an inner-outer listening towards a meditative state. A binaural soundscape subtly illuminated the zome's experience, highlighting the essence of the place with water drops, birds signals and drones of insects (cicadas). It resonated to what the audience had experienced in the tunnel, but with a lower intensity, higher pitch and clarity, switching from an exterior form of resonance to an interior form of resonance.

Energetic Geometry

Ancient civilisations used natural spaces' acoustics and built architecture based on an embodied knowledge of vibration, energetic geometry, cosmology, light and sound as creative design forces. Architecture might had the function of an acoustic communication channel through design with particular energetic geometry, for attunement to specific frequencies. The principles of vibration and energetic geometry were embodied in architectural design. Vibration creates form (geometry, material density, spatial volume). The nature of forms as vibrating structures or periodic systems was vastly investigated in meticulous experiments carried by physician and natural scientist Hans Jenny, to which he called Cymatics [1], a study of wave phenomena to visualise examples of patterns' formations. He studied how vibrations generate and influence patterns, shapes and moving processes. Cymatics, along with other earlier experiments, such as the Chladni figures, the Lissajous figures or harmonograph studies, allowed the visualisation of vibration and frequency patterns, clearly revealing visually that form is a vibrating structure. It has been argued that these patterns are the expression of a dialogue between the vibration of the tone and the 'answering' matter, between the motion energy contained in the vibration, and the matter which is either resonating in co-movement or paused inertia (Lauterwasser 2006, 42-46). The visualisation of these vibrational patterns also revealed similarities to patterns found in nature and universal geometry principles and symbols (vesica pisces, the flower of life, the golden ration, the five platonic solids, the star matrix).

The experiment *Passage* was an attempt to connect to the universal principle of vibration, known to our ancestors, as creative force of forms of experience, resonance and attunement.

Results

We often disregard experience, even our own, perhaps because experience cannot be seen and measured, and frequently not even communicated properly. (Leitner 1998, 302)

The experiment addressed a qualitative study of the affective experience of acoustic phenomena. Methods were sought to understand, explore and communicate the affective experience of space, which acoustical engineering cannot explain nor predict; and moreover, to transcend an anthropocentric view of experience towards an ecological understanding of space as not empty, as a field of relations, of matter-energy. This approach aimed to contribute to a line of relevant research on human affect of acoustic phenomena, which could complement and enrich others.

[1] Cymatics is a term originated from the Greek "*to kyma*, the wave; *ta kymatika*, matters pertaining to waves, wave matters" (Jenny 2001, 20).

In this sense, the experiment created a public experience of resonance and attunement with the environment, to observe its results with a survey. The event received a great amount of audience. A survey was conducted by leaving a notebook with the question: how do you feel with this experience? A notebook for comments was used instead of interviewing the audience. A few interviews were tried and it resulted as a mental feedback, with preconceived ideas, rather then a direct affective expression of the experience. Next follows a transcription of some of the notebook comments.

<u>The Tunnel - Audience's Feedback</u>

- *Impressive! I felt like a plant for the first time! Seeing with vegetable ears.*
- *No time. A dimension with a lower frequency.*
- *I feel like I'm falling into a hole. I like the strong sensation in the sounds, it seems like I feel it inside myself. I closed my eyes and it becomes much more powerful!*
- *With my body inside the wall, but with my ears out.*
- *I'm in a menthol cave, with water drops going up and going down (signed: Tiago, 9 years old).*
- *Inspired. The entrance was not very inviting, but the way out opens up a new world.*
- *A passage is normally an in-between space, but here it becomes a place in it-self, like a reality bubble.*
- *I remembered I breed!*
- *It reminded me how sensual the sound of water is... I wonder how would sound the fire, the earth, the wind...?*
- *Flying in the water*
- *Refreshed*
- *Relaxed*
- *Peaceful*
- *Water transports me in harmony and balance*
- *The sounds travel through the stones like if they are communicating between themselves*

(C. Martinho, *Passage* survey, September 15-17, 2017)

<u>The Zome - Audience's Feedback</u>

- *Sliding in life*
- *I don't want to go out*
- *Floating*
- *In another world*
- *In my world*
- *In balance*
- *Free*
- *Very calm*
- *Quiet*
- *Relaxed*
- *Is this what bees hear? I like this shape. It smells good like the earth (signed: Joana, 7 years old)*
- *Suspended*
- *Tuned with the vibrations of this space*
- *I feel the transmission of nature on an human shape, like if it was a real person.*
- *Centred*

(C. Martinho, *Passage* survey, September 15-17, 2017)

Discussion

The audience's feedback was a valuable input to understand the affective results of the experiment. The aural experience is a fundamental process in the development of the human being, that is shaped by architecture. This research project valued the direct experience, and the process of experience as a way to unify life sensations: visual, acoustic, tactile, kinetic. It has been argued that our personal experience is the only way by which we can understand acoustic space. Therefore these kinds of artistic-empirical investigations and acoustic experimentation, from field work to a laboratory-like situation, have been pointed out of outermost importance since it may engage us in the hearing of forms, materials, and perspectives (Leitner 1998, 302).

Affective attunement

The project *Passage* resulted in two aural architecture experiments with acoustic effects and resonance frequencies inspired from ancient sites. Drawing from the comments, all bodies were attuned, finding difference in unison. There was a diversification in the affective experiences. It might be interpreted that there was an enhancement of mind-body experiences, here named affective attunement, with different forms of experience of non-verbal communication between being and place. It engaged the inner and outer dynamic of the auditory experience towards deep or meditative states. On one hand, it created an affective experience of environmental sound as a presence in the now, a sense of self as part of a powerful field of energy, of water, of plants, as a driving force. On another hand, the binaural architecture of the zome acted as a point of centring in the balance of self.

Contribution

The aural architecture design of *Passage* aimed to re-integrate the art of vibration, universal geometry and resonance, known by ancient civilisations. The experiment was conducted based on the resonance phenomenon to explore how it affects the human experience of self, space, time, in its relationship to the environment. The resulting multi-sensory experiences seem to have contributed to balance the senses interplay and to an embodied understanding of self and the environment as an unified field of relationships, all interconnected through vibration.

Further Developments

It would be interesting to conduct aural architecture experiments in ancient sites to study the human affective experience of resonance and attunement. This could be a way to extend further investigations on aural body-mind experiences as an ecological and/or spiritual relationship of the human being with its environment. This would help to understand design potentials of specific ancient acoustic qualities, to experiment applications towards more ecological forms of experience. The resulting knowledge could be integrated into the design of experimental, innovative and sustainable projects of bio-architecture and acoustic ecology; with new forms of experience, resonance and attunement adapted to our societies' needs.

http://spacefrequencies.org

REFERENCES

COOK. I. A., Pajot. S. K., Leuchter. A. F. 2008. *Ancient Architectural Acoustic Resonance Patterns and Regional Brain Activity,* Time and Mind. Volume 1. Issue I, 95-104.

DEVEREUX, Paul and R.G. Jahn. 1996. *Preliminary investigations and cognitive considerations of the acoustical resonances of selected archaeological sites, Antiquity 70*(269), 665-666. doi:10.1017/S0003598X00083800

JAHN, R. G., Devereux, P., Ibisox, M. 1995. *Acoustical resonances of Assorted Ancient Structures.* Journal of the Acoustics Society of America. 99, 649-658.

LEITNER, Bernhard. 1998. *Sound:Space*, Cantz.

LUND, C. S. 1981. *The Archeomusicology of Scandinavia*, World Archaeology 12 (3): 246-265,

MASSUMI, Brian. 2002. *Paraboles for the Virtual: Movement, Affect, Sensation*, Duke University Press.

MORTON, Timothy. 2014. Accessed February 13, 2016. http://ecologywithoutnature.blogspot.pt/2014/11/attunement.html

REZNIKOFF, I. and Michel DAUVOIS.1988. *La dimension sonore des grottes ornées,* Bulletin de la Société Préhistorique Française 85: 238-46.

SCARRE, C,. 1989. *Painting by resonance*, Nature 338: 382.

WATSON, Aaron and David Keating, 1999. *Architecture and sound: an acoustic analysis of megalithic monuments in prehistoric Britain. Antiquity*, 73(280), 325-336. doi:10.1017/S0003598X00088281

XU, J.; Yik, A., Groote, I. R., Lagopoulos, J., Holen, A., Ellkgsen, O., Haberg, K.,; Anddavanger, S. 2014. *Nondirective meditation activates default mode network and areas associated with memory retrieval and emotional processing.* Frontiers in Human Neuroscience. 8. 1-10.

ZALAQUETT, Francisca, Clara Garza, Andrés Medina, Pablo Padilla, Alejandro Ramos. 2008. *Arqueoacústica maya. La necesidad del estudio sistemático de efectos acústicos en sitios arqueológicos.* Estudios de cultura maya vol.32 México. Accessed July 23, 2017. http://www.scielo.org.mx/scielo.php?script=sci_arttext&pid=S0185-25742008000200003#notas - accessed August 2017.

The Silent Space; Examining Women's Voices in Archaeoacoustics Research

Sarah McCann

SARAH McCANN, Bachelor of Education and Bachelor of Contemporary Music, is a self-taught musician with over 15 years of performance and recording experience. An Independent Researcher, having presented papers in Archaeoacoustics Research in Australia through the Australian Archaeology Association, she is currently seeking a PHD Candidature and scholarships to pursue Archaeoacoustics Research endeavouring to examine correlations of Women's experiences within sacred sites from a global context.

Archaeoacoustics is a relatively new field in Musical Archaeology which examines and records the sound and vibrations of space and place relative to music and natural sound. Previous research in Archaeoacoustics demonstrates a focus on various interconnected relationships to space, place, resonance and human behaviour. These archaeological "soundtracks" examine many relationships of sound including sine waves, electromagnetic frequencies, musical instrumentation, universal tones and the effects on the human brain and body at certain resonances. It also aims to provide a voice for ancestral remains and artefacts. The concept of our ancestors selecting significant sites for particular acoustic properties can also help to define and shape a broader contextual knowledge base in regard to Archaeology and redefine the contribution of women's voices within society.

Whilst some within the Archaeological community may view Archaeoacoustics with contention, past field work has provided further complimentary in-depth analysis of significant sites in relation to behavioural patterns and social constructs of past societies. Archaeoacoustics can also present a bridge to understanding for the Archaeological community and global Indigenous/First Nations communities by adding a sense of connectedness from past to present use of sites. The study of Archaeoacoustics investigates and sometimes reproduces the presence and relationships of ancestors to significant sites; it is an academic, scientific and analytical research but can also provide a 'vocalisation' and recreation of sound in relationship to rock art and fire, the songlines, the handprints, the etchings and spirals of our ancestors. In this context, Archaeoacoustics provides an experiential relationship, intertwining the archaeological field, ancient knowledges and contemporary culture. It may be argued that Archaeological research seems incomplete without the study and examination of sound within our historical contexts as our ancestors did not sit in spatially constructed vacuums of silence.

It has been evident within Archaeoacoustics research that certain rock formations, shelters, standing stones and natural amphitheatres were in frequent use by our ancestors to assist with amplification, projection and dilution of sound, dependant on the use of sites. Lascaux is home to a complex of caves in south-western France famous for its Palaeolithic cave paintings. Dr Iegor Reznikoff has studied the caves and found links between the positioning of rock art and the acoustic properties. In the Great Hall of the Bulls, the resonance and effect of acoustic reconstruction, reproduced stampede like noises interrelated to the positioning of the cave art. The research suggests that there is a causal relationship between the selection of sites for ritualistic purposes aligned with an emphasis on particular acoustic properties.

"Most of the caves are highly resonant, remarkable as acoustic vaulted pipes, and can produce quite astonishing echo effects. If the people using these caves chanted or used sound, it would therefore seem likely that they would have chosen for their rites, accessible locations with the best acoustics. The result appeared clearly: there is a link between the locations of the paintings (engravings and carvings) and the quality of resonance at these locations in the cave. To state the result simply: the more resonant the location, the more paintings or signs are found at this location." (Reznikoff, 2014).

The research is not limited to naturally formed environments inhabited and selected by our ancestors, but also constructed archaeological sites. The archaeoacoustic studies and reconstructions of Stone Henge by Dr Rupert Till of the University of Huddersfield and Bruno Fazenda of Salford University, have theorized "that the famous ring of stone could have sung like a crystal wine glass with a wet finger rubbing the rim, stimulated in this case by percussion played in time to the echoes of the space." (Till, 2009).

After reconstructing the Henge for the study of acoustic properties, it was noted that the positioning of the stones also produced an amplification effect, ensuring that ritual or ceremonial participants within and surrounding the henge were afforded a clear sound of the interior acoustics. Whilst there have been many studies speculating the purposes of the construction of Stone Henge, Till and Fazenda's research brings new information and possible reasoning for the design of the structure.

It is this research and the work of Devereux's ringing rocks, Waller's rock art studies, EMAP, ISGMA and others in the field of Archaeoacoustics that have started to shape and determine a new paradigm in respect to understanding our ancestral relationships with significant sites through sound. These findings have also established new ways of working, knowing and being with respect to Archaeological research and connection to culture, particularly dynamic Indigenous contemporary cultures.

In 2014, the research methodologies of archaeoacoustics and physical phenomena in ancient sites was still at a developmental stage. Researchers in Europe from various universities formed a group to create a global uniformity and standard practice for measuring Archaeoacoustics on site. This includes ultrasounds, infrasounds and audible low frequencies; materials, equipment and methods; data, editing and results.

Through the recording at archaeological sites, various strands of archaeoacoustic research have included selection of sacred space, correlations with Rock Art, altered states of consciousness, effects on the human brain and emotional intelligence and reconstruction of ancient musical artefacts which are now played in contemporary spaces.

Archaeological research has examined gender representations of women within historical frameworks however representations of women's voice (or the absence thereof) distinctively leads to further understanding of the social constructs and status of women in past and present societies. It is to yet be determined if historical representations of women within archaeological and anthropological examinations are represented purely as a *scopophilia* as opposed to engaged participants. The 'female experience' is a fragmented representation, constructed from an engendered lens and cultural and societal ideologies that produce *assumed* context and praxis for examination. For example, the Angkor Watt temple complex in Siem Reap, Cambodia, shows thousands upon thousands of representations of the female form in the shape of *Apsaras* or *Devatas*. Banteay Srei has been labelled by Khmer peoples and western archaeologists as the "women's temple". They list reasons such as the "carvings are more intricate so it must be smaller hands", the stone is a "hue of pink" (it is constructed from red sandstone) and the complex is "smaller in dimensions" than others in the temple complex. It is essential to examine these gendered assumptions based on the ideas of the 'construction' of women and what these assumptions do to further construct, compartmentalise and reinforce gendered archaeology.

There can be a contested space not only in applying 'gender' to certain artefacts and sites but also the culturally contested space as archaeology and anthropology come from a hegemonic paradigm. It is necessary to explore the constructed archaeological assumptions of artefacts, the roles and relationships that women experience to place and space and to establish underlying themes of gendered contestation, providing further understanding of the contemporary contested space.

"Focusing on 'traditional' or 'contemporary' Aboriginal women, who are represented as the different 'other', means that this universal proposition is credible in feminist and other academic discourses. Women anthropologists seek to investigate how 'traditional' and 'contemporary' Aboriginal women's role and status are in substance unlike or like those of their White sisters through the centring of their analyses on the areas of concern to White western women's discourse. That is, the sites analysed reflect or symbolise those sites under interrogation in western society by the women's movement, such as: marriage; kinship; women's economic activity; sexuality; reproduction; ritual; and socialisation." (A. Moreton-Robinson, 1998).

As Moreton-Robinson examines the notions of a *sisterhood* which creates a connectedness through gender, she also places emphasis on the importance of examining the contested space through a socio-cultural lens. The representations, research and findings of archaeology and anthropology shift dynamically whilst looking through diverse or possible lenses that create a plethora of paradigms which inevitably lead back to the individualistic lens of the observer; which, within itself, includes bias and cultural standpoints. Put simply, gendered, cultural and social capital play a role in shaping archaeological research.

Anthropologists have often been recognised and in fact lauded for their ability to disconnect from the 'subjects' or 'objects' and view societies from an 'outside' perspective. It was felt that this method kept the 'non-participant' impartial, however this kind of methodology in practice inevitably leads to a sense of 'othering' and exoticism; a scopophilia of women represented in various states of undress performing various tasks. As the field evolves in creating new methodologies of research, there is a necessity to embrace the cultural interface which places value on both sets of knowledges within the contested gendered; *Masculine/Feminine/Androgynous* and cultural; *Indigenous/Western* sense to ensure a more sensitive, authentic approach to researching and understanding collective human history.

Archaeoacoustics is a vehicle for bridging the divide between the isolation and constructed theories of the past to ideologies and representations within contemporary theoretical frameworks.

"This gap can only be dealt with adequately if we develop conceptual tools and theoretical structures with which to reinscribe the past into the present, to realize their interaction." (Shanks & Tilley, 1987)

The performance of vocalisation, musicality, ritual and lore in archaeological sites avoids a conceptual exclusion of the agents and agency of the past historical narratives and brings into the contemporary space a less speculative methodology for research. This *'re-imagining'* or *'re-constructivist'* approach to archaeology can provide a greater scope for understanding.

By examining the use of women's voices in sacred sites and sites of significance, archaeoacoustic research can demonstrate that knowledge of acoustic phenomena has played a considerable role in shaping women's historical and contemporary narratives. On a localised level, the Songlines of Australia's First Nations peoples have already aurally and orally documented a method of recording historical narratives. Many of these songlines criss-cross in the sense that they go east and west, north and south and recur according to the journeys of the ancestors. They follow topographical pathways and create a cultural network of remembering that ties all of First Nations language groups together whilst also assuring their individual group narratives. The presence of resonance and songlines is particularly evident in the construction of Rock Art and also contemporary Art for Indigenous women.

Women of Western desert communities are known for sometimes singing songlines whilst they paint, representing the stories of past ancestors, kinship groups and moieties. Close examination of Western Desert Art and songlines provides a correlation between the symbolism within art and the reproduced patterns of sand through cymatics. Cymatics is the visualisation of acoustic frequencies and it is possible to argue that the symbolism within the paintings of Western desert women is very similar (sometimes emulating) these cymatic patterns; producing a visualisation of songlines, frequency, resonance and energy of the ancestors. To suggest that First Nations art in sites of significance and sacred sites is correlated to sound, frequency, historical memory and the selection process of such sites is not implausible.

It is known that particular sites are an engendered space within which First Nations Australian women were predominantly the occupiers of such sites. Birthing Caves and Defloration Caves hold particular significance for women and these sites may have been chosen not just for the safety and location of space but also the acoustic properties. It is proposed that Archaeoacoustic research and study within these sites may assist in providing further perspective of 'space' and 'place' and a deeper understanding of women's contributions to archaeological narratives.

There is also the perspective of communal non-gendered collective spaces and women's resonance and relationship to sound at such sites. Women's voices resonate at different frequencies/hertz to those of men, creating different tonalities. Songlines in solo and collective performance by women can also alter the acoustic measuring of sites within archaeoacoustic research. This research could also assist in preserving and resurging the renaissance of Indigenous languages and provide further accessibility to languages through song.

Recent research has examined the musicality of First Nations Australian women's songs by linguists, anthropologists and musicologists. Walsh (2009), argues "one cannot come to such a song text cold and hope to glean anything approaching the richness of these song traditions. Indeed, such apparently elliptical songs receive interpretations from singers during performances so that a better understanding of the song text can be built up by collecting interpretations from many performances over time." This assertion from an archaeoacoustic methodology, reinforces *re-constructivist* approaches to research and implies it is imperative that community consultation, collaboration and connectivity to space are assured before commencing any research or recording in the field. "The challenge for western research and researchers is to engage research as an interface where conceptual, cultural and historical spaces interface or come alongside each other based on new relationships to knowledge, to research and to self." (Martin,2008)

Musicality, frequencies, resonances, language, meaning and symbolism of songlines for Kamilaroi women on country would differ from Bidjara or Darkinjung or Western Desert women and across all language groups dependant on environment. Archaeoacoustic research may assist in examining the selection of space and place for language groups and *why* spaces *are* and *become* sacred. Further examinations are required of engendered artefacts that play percussive roles within ritual and performance that are attributed primarily to the ownership of women within different language groups such as bark bundles, clapsticks and grinding stones.

The research proposed would follow guidelines set by the archaeoacoustic research global community with an emphasis on women of community. It aims to document and identify the standpoints of the designers of sites and artefacts and examine women's voices in the silent space. Without the interweaving or the intellectual input of participants within the socio-cultural, gendered space being examined, it is best methodological practice to share knowledges rather than observe from the sidelines as a voyeur with disconnectedness. If we are able to attain a sense of universality regarding our historical past, irrespective of culture or gender, we may be able to attain an interwoven partnership of learning in the present. This is not to suggest a homogenised approach to research would be preferable but rather a *'re-constructed'* platform of learning that ensures space for all voices to retain their qualities whilst producing a complimentary harmonic discourse and differing standpoints. Archaeoacoustics can help to shape these new parameters within methodology, offering musicality and sound as a universal connectedness. By examining our past voices, sonic representations and frequencies within archaeological sites we are able to connect not only with our ancestors but with each other in the field.

BIBLIOGRAPHY

Greer, S., Harrison, R., Tamwoy, S. (2002) *Community-Based Archaeology in Australia*
World Archaeology, Vol. 34, No. 2, Community Archaeology
Loose, R. (2011) Archaeoacoustics; *Adding a soundtrack to site descriptions*; Papers of the Archaeology society of New Mexico.
Martin, K. (2008) *Please Knock Before You Enter: Aboriginal Regulation of Outsiders and the Implications for Researchers*, Retrieved from http://www.econtentmanagement.com/books/283/please-knock-before-you-enter-aboriginal
Moreton-Robinson, A (1998) *When the Object Speaks, A Postcolonial Encounter: anthropological representations and Aboriginal women's self-presentations*, Discourse: Studies in the Cultural Politics of Education, 19:3, 275-289, DOI:10.1080/0159630980190302
Reznikoff, I. et al (2014) *Sounds Like Theory*. XII Nordic Theoretical Archaeology Group Meeting in Oulu. Monographs of the Archaeological Society of Finland
Shanks, M. & Tilley, C. (1987) *Social Theory and Archaeology*. Polity Press
Stig- Sorenson M. (2000) *Gender Archaeology*. Polity Press
Till, R. (2009) *Sounds of Stonehenge* Retrieved from https://soundsofstonehenge.wordpress.com
Waller, S. (2002) *Psychoacoustic Influences of the Echoing Environments of Prehistoric Art*. Paper presented at the Acoustical Society of America First Pan-American/Iberian meeting on acoustics, Cancun.
Walsh, M. (2009) *Australian Aboriginal Song Language: So Many Questions, So Little to Work With* Retrieved from https://www.questia.com/library/journal/1G1-175181481/australian-aboriginal-song-language-so-many-questions

The Singing Loom: The Importance of Textile Production in the Roman Domestic Soundscape

Magdalena Öhrman

MAGDALENA ÖHRMAN (PhD Lund) is a senior lecturer in Classics at the University of Wales Trinity Saint David. Currently (2016-2018), she is a Marie Sklodowska Curie research fellow at the Centre for Textile Research in Copenhagen.

ABSTRACT: This paper traces Roman acoustic experience of domestic soundscapes, particularly the soundscape of textile production (especially weaving) through a combination of philological analysis of Latin poetry and experimental archaeology. Based on a comprehensive survey of textile sound-mimicking in Latin poetry, the paper highlights consistent features of Roman domestic soundscapes rather than the soundscape of any one specific setting, site, or period. Spectrographic anlaysis of audio recordings of weaving experiments conducted at the Centre for Textile Research in Copenhagen provides the experimental archaeological basis for the literary analysis. Passages from Tibullus' *Elegies* 2.1 and the *Ciris* provide representative examples of literary sound-mimicking of craft processes.

Scholarly interest in Roman urban environments (and the interplay of people with them) has soared in the last decade[1*2], contributing to an increase of work on related Roman soundscapes. The range of methodologies for investigation of sound in Roman contexts is rapidly expanding. Increasingly, scholars use sound experiments, architectural and acoustic data collected from unusually well-preserved sites or objects in comparison with lexical and literary analysis to discuss the soundscapes of ancient sites.[3] This is a welcome expansion of the hitherto dominant analysis of explicit comments about sounds and noises of the ancient environment in extant literary sources.[4]

Yet in tracing generic Roman soundscapes, the increased use of acoustic modelling can and should be complemented by other tools, as also suggested by Vincent in his 2017 paper on the sounds of the *tuba*. This goes particularly for domestic soundscapes, where we have a high degree of variation depending on a home's wealth, class, location and region in the Roman world. For most domestic settings, the archaeological record does not yet provide sufficient material for acoustic modelling. Scholars of later periods have drawn on literary sources commenting on domestic sounds together with

* This project has received funding from the European Union's Horizon 2020 research and innovation programme under grant agreement No 701557. I am grateful to my colleagues Centre for Textile Research at the University of Copenhagen for adding much to my understanding of ancient textile craft.

2 E.g. Edwards 1996; Weaver 1997; Laurence and Newsome 2011; Hemelrijk and Woolf 2013; Bjørnebye, Malmberg, and Östenberg 2015; Hartnett 2016.

3 Vincent 2017; Veitch 2017. Two recent volumes on the soundscapes of Graeco-Roman antiquity demonstrate the importance of ancient sound studies, cf. Emerit, Perrot, and Vincent 2015 and Courtil and Courtray 2015.

4 Betts 2011; Grand-Clémant 2015; Francois 2015; Hartnett 2016; Laurence 2017. Cf. Gurd 2016. Vincent 2017, 149-151 highlights the methodological challenges inherent in such literary material.

other types of documentary evidence to assess the composition of their soundscapes.[5] In this paper, however, I will show that ancient literary sources may throw light on features of domestic soundscapes even when they do not make explicit comments about experiences of sound. When combined with an experimental archaeological approach, literary sources may in fact reveal additional features of common Roman domestic soundscapes.

My interest lies with literary texts that mimic the sounds of the environments they describe. Roman authors often develop sound mimetic effects to contribute to the artistic, overall impression of literary settings: Vergil's hedges buzz with bees in the *Eclogues*, cut-off repetition of phrases and syllables distinguish Ovid's description of echo in the *Metamorphoses* (3.359-401), to mention but two examples.[6] Less heavily signposted versions of such literary mimesis of sound throws additional light on the soundscapes associated with both places and activities in the Roman world and the way that they were perceived by those members of the elite by and for whom Latin poetry was primarily produced.

Here, I will explore how sound-mimicking features in Roman poetry on weaving underscore the importance of textile production and its sounds as a part of the Roman domestic soundscape from the Republic and into Late Antiquity. Using experimental reconstructions of craft working processes, I will show that poetic authors describing weaving in progress often transpose the working rhythm and the sounds arising from weaving into the new medium of the text, by means of different types of sound play.

The assumption that writers of Roman poetry and their audiences were interested in or aware of the soundscape of textile production is based on explicit comments on such sounds in their texts. Several passages in Greek and Roman literature mention the sounds created by the loom and by the weaver's tools, and distinguish a few different weaving sounds.[7] Latin poets throughout antiquity mention the sounds arising as a weaver works: they note both the sound of clattering clay loom weights and refer to the sounds of the pinbeater used to settle the weft.[8] The assumption is readily made that authors also use sound play to imitate such textile making sounds in passages about weaving, incorporating the making of textiles into the making of poetry.

Tibullus' comments about the song of clay loom weights (*tela latere sonat* in Tibullus 2.1.66-67) or the resounding heddle rods of Lucretius (*scapique sonantes* in *De rerum natura* 5.1353) unfortunately give no details on what these sounds were like.[9] Therefore, I rely extensively on experimental archaeology to assess how the sounds of weaving are mimicked in poetic texts. I draw on video- and audio recordings, as well as observation, of weaving experiments conducted at the Centre for Historical-Archaeological Research and Communication at Lejre and the Centre for Textile Research (CTR) in Denmark.[10] As this

[5] Corbin 1998, cf. also Picker 2003; Fritz 2015.

[6] On resonance in Vergil's *Eclogues*, e.g. Fitzgerald 2016.

[7] Literary reflection on sounds generated by weaving has received more attention in relation to Greek literature, cf. Restani 1995; Tuck 2006; Tuck 2009; Nosch 2014; Heath 2011. For discussion of passages in Latin literature, cf. Restani 1995; Kissel 1980.

[8] Noises generated by heddle rods, Lucr. 5.1353; loom weights, Tib. 2.1.65-66; pin beaters, Verg. *Georg*. 1.294; *Aen*. 7.14; Macrob. 5.12.7.5; *Cod. Iust*. 11.9.4, weaving pins: Symphos. 17.2; Claud. *Carm. Min. App*. 5.48.

[9] For *scapus* as 'heddle rod', Johncock 2016, 254 with further bibliography.

[10] Such experiments are conducted in line with the methodological principles of documentation, craft experience,

project is concerned with the generic sound of weaving rather than any one specific setting, I have compared features from several different experiments, all of which used period-appropriate materials and reconstructions and saw the work carried out by experienced weavers.

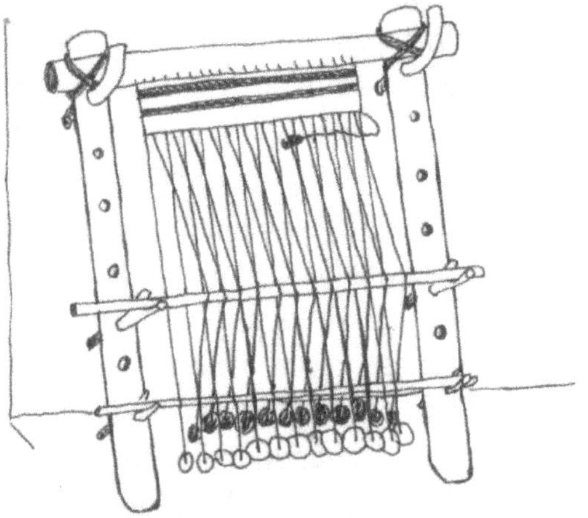

FIG. 1. Overview of a warp-weighted loom with a tabby weave. The heddle rod is resting on its supports to keep the desired shed open while weft is inserted.

Documentation has been collected from weaving on both warp-weighted and two-beam looms, capturing sounds generated by different types of weaves.

In set-ups on the warp-weighted loom, the shape and weight of the loom weights have also been varied. The analysis in this paper is based on the experiments on the warp-weighted loom, as this loom type is most relevant to the periods from which my two literary examples derive.

The following sound events characterize weaving in a warp-weighted loom:

Beating: The sound of the weaving sword beating in newly inserted weft is dull and regular. The width and density of the weave influence the number of repetitions and pauses. After beating, faint scratching sounds are sometimes audible as the weaver moves the pin beater through the warp to fits the next weft thread.

Shed change: When the weaver moves the heddle rods to change the shed, distinctive sounds arise from wooden parts of the loom moving against each other.

Warp noises: Noises arise from the warp as the weaver unsnags threads and clears the new shed opening with their hands. This may also generate chiming from the weights as the movement of the warp causes them to move against each other.

These sounds recur cyclically as the weaver continues their work. A basic acoustic spectrogram of illustrates the strong rhythmic qualities of the soundscape of weaving on a warp-weighted loom, as well as its key sound marks, shed change and beating.

This spectrogram is based on a recording from an experiment at Lejre reconstructing the functional parameters of heavy, pyramidal loom weights.[11] The sounds of shed change appear as regular, tall turquoise spikes (00.05 – circled in yellow, etc.). After each shed change, higher-frequency sounds generated by the loom weights are notable as weaver clears the shed by hand, joggling the warp (e.g. 00.05-00.15).

and close replication of ancient tools based on archaeological finds set out by CTR (Mårtensson, et al., 2009: 379-380).

[11] This experiment was designed for Dr Anna-Rosa Tricomi (Padua) in collaboration with Ida Demant (Lejre) and Eva Andersson Strand (CTR), who generously allowed me to record their work. On the loom weights and their context, Tricomi 2012.

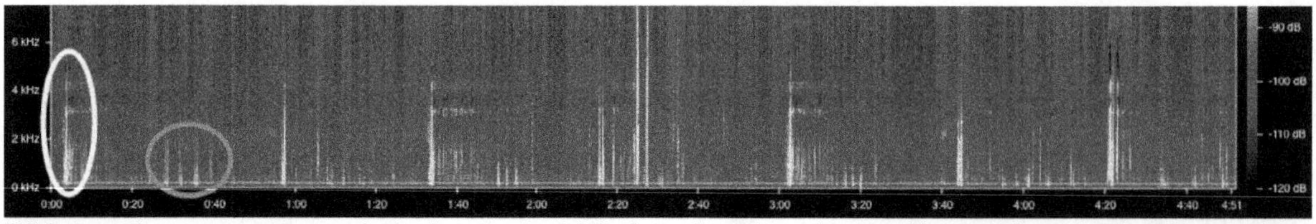

FIG 2. Acoustic spectrogram generated using Spec (version 0.8.2) of experimental weaving of a tabby weave in wool, using 26 loom weights. Sound frequency (kHz) is indicated on the vertical axis, whereas sound intensity (dB) is expressed through colour, with bright green and bright turquoise indicating sounds of 70 dB and 80 dB respectively. Time is indicated on the horizontal axis.

The sounds of beating appear as short series of finer spikes, clustered together after each shed change (00.29-00.38 – circled in red, etc.). After beating, there tends to be a short interval of comparative silence as the weaver fits the weft (e.g. 00.40-00.55): other set-ups exhibit thin sound spikes, corresponding to the low scratching of the pin beater used to fit the weft before changing the shed, but these sounds are noticeable only at a very close range.

Modern sound studies remind us that "[s]ound has a strongly tactile aspect, particularly with lower frequencies, which we feel as much as we hear".[1] A weaver will be acutely aware this: the sounds of parting fibres in the warp with one's hand are distinct but only audible immediately next to the loom, and the experience of them blend with the feel of the threads against one's hand. A text that mimics sound is uniquely placed to merge their sonic-linguistic representation, so that sound play renders both sound *and* tactile experience. This is prominent in many poetic descriptions of weaving.

Systematic tracing of one specific activity (in this case, weaving) allows us to identify commonalities in its sonic-literary representations in Latin poetry. Consistent with their prominence in recordings of experimental weaving, beating of the weft and sounds arising from loose, wooden parts of the loom moving against each other are the features of the soundscape of weaving to which Latin poets allude most frequently. Clanking noises from the loom frame and/or beating are frequently mimicked through clusters of voiceless plosive consonants (p, t, and k-sounds). In isolated, more elaborate passages, a distinction is also made between the clanking of wood and the noise from beating. In these cases, voiced plosives (d, b) are used in combination with open vowels to mark the duller sound of the weaving sword hitting the weft.[2] Warp noises, from parting the shed and unsnagging threads, are often represented in poetic texts by means of rhotic and sibilant sounds (r, s).

Of course, some words likely to occur in the context of weaving have inherent, onomatopoetic qualities (e.g. *texo* 'weave'), but how such words are deployed in context is far more important for the effectiveness of poetic sound mimesis: for example, do they appear as part of a cluster of certain sounds, or are they placed in an position which is emphasised through syntax or metre?[3] Rhythm is a vital component in poetic sound-mimicking of textile work. Roman poetry depends on rhythmical sequences: details can be adjusted and varied only within the framework of metrical rules. Audiences are trained to listen keenly for how poets engage with set sequences, and for

[1] Czink 2010.
[2] Ov. *Fast.* 3.820: *erudit et rarum pectine denset opus* ('she teaches [them to run the weft through the warp] and she presses together the remarkable weave with a pin beater') and Claud. *Carm. Min. App.* 5.47: *densentur pectine texta* ('the weave is packed with a pin beater').
[3] Cf. Lateiner 1990, 204-206.

how they satisfy or defy audience expectations of the chosen rhythm to underline the content of the poem.

The Pseudo-Vergilian *Ciris* exemplifies some of these features.[4] References to spinning and weaving illustrate how the character Scylla has abandoned all her regular pastimes due to her lovesickness. The allusions to the soundscape of weaving contribute to an image of female industriousness in the reader's mind, a counterpoint to Scylla's own inactivity:

an Libyco molles plauduntur pectine telae 179
...nor is the soft weave beaten with ivory pin.

In his influential commentary on the *Ciris*, Lyne uses this passage to illustrate the density of literary allusion, highlighting the prominent play with Vergil: the three lines on music and weaving in the *Ciris* bring together Vergil's description of a weaver noisily running a pin beater through the warp[5] with his description of Orpheus striking the strings on the lyre with an ivory plectrum in *Aeneid* 6.647.[6] These literary parallels underline that music and woolwork go together, a connection which encourages us to listen for the music of weaving itself.

The single most noticeable sound-mimicking feature of the line is the centrally placed verb *plauduntur* ('is beaten'). While it has inherent onomatopoetic qualities, the *Ciris* poet has integrated it into a larger, sound-mimicking context and thus amplified its effect. Sensitised to the presence of weaving sounds through the emphasis on *plauduntur* and the intertextual presence of texts connecting music and textile work, one may associate the line's opening plosives (b and c in *Libyco*) with the clanking sounds of wooden loom parts moving against each other during a shed change. Lyne attributes the poet's unusual choice of *Libyco* for 'ivory' to a desire for variety in line with the literary aesthetics espoused by the *Ciris* as a whole,[7] but it also supports the allusion to the soundscape of weaving better than the alternatives, especially the poetically traditional *eburno* (with the same meaning), which lacks appropriate 'clanking sounds'.[8] Next, the adjective *molles* ('soft', describing the weave) with its liquids and final s recalls the sound and feel of threads moving smoothly against the weaver's hand as they clear the shed before beating in the new weft. The brief silence arising when the weaver picks up the weaving sword is mirrored by the penthemimeral *caesura*, a regularly expected pause occuring in the middle of the third metrical foot. Through its dark vowels and a slow, spondaic rhythm, alluding to the dull resonance of the growing weave, *plauduntur* captures the beating of the newly inserted weft.

Then, a new weft thread is inserted and positioned with the pin beater. This lighter work element, and its brief, scratching sounds, is represented through the combination of plosives and close vowels in *pectine*. *Telae* as the subject of *plauduntur* grammatically signifies the growing weave, but it also evokes the loom frame itself: elsewhere, the word is used for the whole of the loom or its frame.[9] In this way, the plosive t

[4] For a survey of the discussion of the dating of the *Ciris*, Kayachev 2016, 1-7.
[5] Verg. *Georg.* 1.204: *arguto coniunx percurrit pectine telas* ('the wife runs through the weave with a chattering pin beater'; Aen. 7.14: *arguto tenuis percurrens pectine* ('she runs the fine weave through with a pin beater').
[6] Lyne 1978, 37-39.
[7] Lyne 1978, 38. Prevosti 2013 offers several typical examples of bone pin beaters from the Roman period
[8] Cf. Mart. 14.150 *Niliaco pectine* ('with a pin beater from the Nile') for another location adjective (equally rare), which also contributes to a clustering of plosive consonants mirroring the sounds of beating the weft.
[9] *telae* referring to the loom (frame), e.g. Lucr. 5.1351; Verg. *Georg.* 1.286; Tib. 1.6.79. Cf. Öhrman 2017.

in *telae* points forward to the movement and sounds of heddle bars moving against the loom during shed change. Thus, the *Ciris* traces the sounds arising from one working sequence, from shed change (*Libyco*) to shed change (*telae*), in a single line.

FIG. 3 A pin beater is used to position the weft. Detail of experiential weave at CTR, using white weft for illustration purposes.

One of Tibullus' elegies about the ideal life in the countryside provides our second example (Tib. 2.1.63-66). This passage includes both a direct reference to the sound of loom weights and an indirect representation of the soundscape of weaving.

hinc et femineus labor est, hinc pensa colusque, 63
fusus et adposito pollice uersat opus: 64
atque aliqua adsiduae textrix operata mineruae 65
cantat, et a pulso tela sonat latere. 66

...from here comes also the woman's work, the daily allotment of wool and the distaff, from here also the weaver singing as she busies herself with constant craft, and the loom resounds with struck clay loom weights.

Tibullus explicitly mentions that the loom itself creates noise (*tela sonat* 'the loom resounds'). He also highlights the part of the loom from which the sound derives (*pulso latere* 'with struck clay weights'). There are no other mentions in Latin poetry of sound arising from loom weights, and until recently, the term *later* in this sense was only securely attested in this passage.[10] The find of an inscribed loom weight in excavations of Caesar Augusta near Zaragoza, bearing the legend *ama lateres* [...] ('Love [your] loom weights!') now provides evidence of the term's usage in Tibullus' period.[11] Maltby argues that Tibullus' use of this and other technological terms expresses his Hellenistic preference for displaying specialist knowledge,[12] but this knowledge is also clearly rooted in the practice and vocabulary of those engaged in textile production.

It is fitting, therefore, that Tibullus' text also replicates the sounds of weaving with considerable detail in an indirect way. The two lines on weaving open with a half-line characterized by a-alliteration, stressing the unity of the three initial words, and plosives emphasized through their position at the beginning of each metrical foot (*at<u>que</u> ali<u>qu</u>a adsiduae*). The clanking impression of this half-line corresponds well to the sound of shed change, where wooden parts of the loom clank against each other. The persistent elision of ending vowels in *atque aliqua adsiduae* blurs the boundaries between words, mirroring how the clanking of heddle rods drowns out other sounds. The sibilant in *adsiduae* expresses the swishing of the warp pulled back or forth by the heddles.

Next, a shift in perception of where the weaving sounds come from is created by syntax and the audience expectation of a prominent pause in the middle of the line: when the word *textrix* ('weaver') stands emphasized after the mid-line pause and as the

[10] Murgatroyd 1980, 55-56; Maltby 1999, 246; Maltby 2002, 378. Maltby notes that Lucil. 681M also uses *later* in the context of looms, but the meaning is unclear.

[11] Beltrán Lloris and Beltrán Lloris 2012, also noting that *later* occurs in *Inscr. Aquil.* 3444, from roughly the same period.
[12] Maltby 1999.

subject of the sentence, the focus shifts from the noise of the loom itself (i.e. its heddle rods, as we just saw) to the *textrix*, the weaver herself, who generates sound by moving her hands through the warp. The repeated rhotic sounds in *textrix*, *operata*, and *minervae* bring the low-frequency crackling sound of unsnagging warp threads to mind. The combination of close vowels and plosives in *textrix* may also allude to the light chiming of loom weights that accompanies this work element, developed fully in the next line. Thus, the sound play of the line beginning *atque aliqua adsiduae* generates a sonic zoom-in from sounds audible across the room (the clanking of heddle rods) to sounds perceptible to the weaver only (warp and weight noises).

The final line is initially about the weaver's song, opening with *cantat* ('she sings'), but later shifts to how the loom resonates with the sound of loom weights striking against each other during beating. Plosives (c, t, p, and t) and open vowels (a, u, o) merge with the sense of the verb *pulsere* ('beat' or 'strike') to recall the sound of regular beating of the weft along with the explicitly mentioned chiming of loom weights. A pronounced rhythmical pause (*diairesis*) falls after the word *pulso* ('struck'). Although sound is explicitly mentioned here through *sonat* ('resounds'), it is the light, falling rhythm[13] that best captures the sound of loom weights settling into place after the louder noise of beating has ceased: the sequencing of *tela sonat latere* ('the loom resounds with … clay') draws the reader's attention gradually away from loom sounds in general to those arising from the clay loom weights. The word *latere* ('clay') opens with two light syllables, suggesting that the clatter of loom weights is barely audible, and then fades on a heavy, emphasized e, just as the light sound of loom weights peters out as the movement in the warp gradually stops.

Based on this work-mimicking sound play, Tibullus' couplet on weaving encapsulates one repetition of the weaver's work sequence (shed change, weft insertion, and beating with accompanying chiming of loom weights). Through its syntax and sequencing, the passage also highlights how sounds of different quality and origin are perceived, making a distinction between loud noises and noises perceptible only at close range and noises strongly supplemented by tactile experience.

These are but two examples of weaving sound play in Latin poetry: there are at least twenty passages that exhibit sound-mimicking features in their descriptions of weaving. [14] They are evenly divided from the late Republic into Late Antiquity and occur in authors from different parts of the Empire. They vary in level of detail and length, and their level of engagement with the sounds and rhythms of textile work differs, but features such as those I have discussed here are typical. Even though experiments show that many sounds arising from weaving would be easily obscured by other sounds and do not carry beyond the space immediately around the loom, clearly these sounds were distinctive and meaningful even to those not directly involved in weaving themselves. Thus, the sounds of textile production emerges as an intrinsic part of elite domestic soundscapes from the Republic into Late Antiquity.

[13] Morgan 2010, 352-359.

[14] Cf. esp. Verg. *Georg.* 1.285-286; Tib. 1.6.79 (on which Öhrman 2017); Ov. *Met.* 4.275; 6.576-577; Ov. *Fast.* 3.819-820; Luc. 10.142-143; Sil. It. 14.656-660; Juv. 9.30; Aus. *Epist.* 22.14-16; *Epigr.* 28.1-2; Prud. *Ham.* 291-292; Claud. *Rapt.* 1.276-275; Claud. 1. 224-225; 8.594-595; *Carm. Min. App.* 5.44-48; Symph. 17.2; Paul. Petric. 320-321; Sidon. Ap. 15.154-161.

BIBLIOGRAPHY

- Beltrán Lloris, F. & M. Beltrán Lloris (2012) "Ama lateres! Sobre una pesa de telar cesaraugustana relativa al lanificium." *Sylloge epigraphica Barcinonensis: SEBarc* 10, 127-148.
- Betts, E. (2011) Towards a multisensory experience of movement in the City of Rome. In R. Laurence & D. Newsome (eds.), *Rome, Ostia and Pompeii: Movement and Space*, 118-132. Oxford.
- Bjørnebye, J., S. Malmberg & I. Östenberg (eds.) (2015) *The Moving City: Processions, Passages, and Promenades in Ancient Rome*. London.
- Corbin, A. (1998) *Village Bells: Sounds and Meaning in the Nineteenth Century French Countryside*. New York.
- Courtil, J.-C. & R. Courtray (eds.) (2015) *Sons et audition dans l'antiquité. Pallas : revue d'etudes antiqué*. Toulouse.
- Czink, A. (2010) Sounding Interiors: Daydream, Imagination, and the Auscultation of Domestic Space, *Soundscape. The Journal of Acoustic Ecology* 10, 11-13.
- Edwards, C. (1996) *Writing Rome. Textual Approaches to the City, Roman Literature and its Contexts*. Cambridge.
- Emerit, S., S. Perrot & A. Vincent (eds.) (2015) *Le paysage sonore de l'Antiquité. Méthodologie, historiographie et perspectives. Actes de la journée d'études tenue à l'École francaise de Rome, le 7 janvier 2013*. Paris.
- Fitzgerald, W. (2016) Resonance: The Sonic Environment of Vergil's *Eclogues*, *Dictynna* 13, 1-11.
- Francois, P. (2015) Clamore sublato: le bruit de la guerre. In J.-C. Courtil & R. Courtray (eds.), *Sons et audition dans l'antiquité*, 89-114. Toulouse.
- Fritz, J.-M. (2015) Littérature médiévale et sound studies. In S. Emerit, S. Perrot & A. Vincent (eds.), *Le paysage sonore de l'Antiquité. Méthodologie, historiographie et perspectives. Actes de la journée d'études tenue à l'École francaise de Rome, le 7 janvier 2013*, 63-85. Paris.
- Grand-Clément, A. (2015) La paysage sonore des sanctuaires grecs. In J.-C. Courtil & R. Courtray (eds.), *Sons et audition dans l'antiquité*, 115-130. Toulouse.
- Gurd, S. A. (2016) *Dissonance, auditory aesthetics in ancient Greece*. New York.
- Hartnett, J. (2016) *The Roman Street: Urban Life and Society in Pompeii, Herculaneum, and Rome*. Oxford.
- Hartnett, J. (2016) Sound as a Roman Urban Social Phenomenon. In A. Haug & P. Kreuz (eds.), *Stadterfahrung als Sinneserfahrung in der römischen Kaiserzeit*, 159-178. Turnhout.
- Heath, J. (2011) Women's Work: Female Transmission of Mythical Narrative, *TAPhA* 141, 69-104.
- Hemelrijk, E. & G. Woolf (eds.) (2013) *Women and the Roman City in the Latin West*. Leiden.
- Johncock, M. (2016) Life Hanging by a Thread: The Weaving Metaphor in Lucretius. In G. Fanfani, M. Harlow & M.-L. Nosch (eds.), *Spinning Fates and the Song of the Loom. The Use of Textiles, Clothing and Cloth Productioin as Metaphor, Symbol and Narrative Device in Greek and Latin Literature*, 253-270. Oxford.
- Kayachev, B. (2016) *Allusion and Allegory: Studies in the Ciris*. Leiden.
- Kissel, W. (1980) Horaz c. 3,12 - Form und Gedanke, *WS* 14, 125-132.
- Lateiner, D. (1990) Mimetic Syntax: Metaphor from Word Order, Especially in Ovid, *AJPh* 111, 204-237.
- Laurence, R. (2017) The Sounds of the City: From Noise to Silence in Ancient Rome. In E. Betts (ed.), *Senses of the Empire: Multisensory Approaches to Roman Culture*, 13-22. London.
- Laurence, R. & D. Newsome (eds.) (2011) *Rome, Ostia, and Pompeii: Movement and Space*. Oxford.
- Lyne, R. O. A. M. (1978) *Ciris: a Poem Attributed to Vergil, Cambridge Classical Texts and Commentaries*. Cambridge.
- Maltby, R. (1999) Technical Language in Tibullus, *Emerita. Rivista de linguitica y filologia classica* 67, 231-249.
- Maltby, R. (ed.), (2002) *Tibullus: Elegies. Text, Introduction and Commentary*. Cambridge.
- Morgan, L. (2010) *Musa Pedestris. Metre and Meaning in . Roman Verse*. Oxford.
- Murgatroyd, P. (ed.), (1980) *Tibullus I. A Commentary on the First Book of the Elegies of Albius Tibullus*. Pietermaritzburg.
- Nosch, M.-L. (2014) Voicing the Loom: Women, Weaving, and Plotting. In D. Nakassis, J. Gulizio & S. A. James (eds.), *KE-RA-ME-JA. Studies presented to Cynthia W. Shelmerdine*, 91-102. Philadelphia.
- Öhrman, M. (2017) Listening for Licia. In S. Gaspa, C. Michel & M.-L. Nosch (eds.), *Textile Terminologies from the Orient to the Mediterranean and Europe, 1000 BC to 1000 AD*, 278-287. Lincoln.
- Picker, J. M. (2003) *Victorian soundscapes*. New York.
- Prevosti, M. (2013) A Textile Workshop from Roman Times: The Villa dels Antigons, *Datatextil* 28, 10-18.
- Restani, D. (1995) Il suoni del telaio. Appunti sull'universo sonoro degli antichi Greci. In B. Gentili & F. Perusino (eds.), *Mousike. Metrica ritmica e musica greca in memoria di Giovanni Comotti*, 93-109. Pisa.
- Tricomi, A. R. (2012) Archeologia della lana in età romana. Dati preliminari dalla provincia di Rovigo. In M. S. Busana, P. Basso & A. R. Tricomi (eds.), *La lana nella Cisalpina Romana. Economia e società. Studi in honore de Stefania Pesavento Mattioli. Atti del convegno (Padova-Verona, 18-20 Maggio 2011)*, 587-598. Padova.
- Tuck, A. (2006) Singing the Rug: Patterned Textiles and the Origins of Indo-European Metrical Poetry, *American Journal of Archaeology* 110, 539-550.
- Tuck, A. (2009) Stories at the Loom: Patterned Textiles and the Recitation of Myth in Euripides, *Arethusa* 42, 151-159.
- Veitch, J. (2017) Sounscape of the Street: Architectural Acoustics in Ostia. In E. Betts (ed.), *Senses of the Empire: Multisensory Approaches to Roman Culture*, 54-70. London.
- Vincent, A. (2017) Tuning into the Past: Methodological Perspectives in the Contextualised Study of the Sounds of Roman Antiquity. In E. Betts (ed.), *Senses of the Empire: Multisensory Approaches to Roman Culture*, 147-158. London.
- Weaver, P. (ed.), (1997) *The Roman Family in Italy: Status, Sentiment, Space*. Oxford.

Sound, Cognition, and Social Control

Vincent C. Paladino

VINCENT C. PALADINO is a Certified Audio Engineer, a Cognitive Anthropologist and an independent researcher.

ABSTRACT: The question of intentionality is central to interpreting our findings in Archaeoacoustics, the question of whether or not ancient peoples, literate and pre-literate, possessed detailed knowledge of acoustic phenomena tasks us to search for answers that help contextualize the subject. Additionally, the issue of social control has been addressed within archaeoacoustics, with questions arising from the notion that knowledge of acoustic phenomena might add to the prestige and power of an elite. The intention of this paper is to contribute to the establishment of a theoretical context within which our observations and experimental results can be framed, positing the notion that anthrophonic sound is a mediator of social control and its intentional use is the result of adaptive cognitive processes. The use of sound in the construction and maintenance of social systems allows for the effective manipulation of the environment. Social control is defined here as the necessary establishment of structure and form which assists in the establishment of more predictable outcomes. The use of sound in a conscious fashion is understood here as an inevitability, driven by the cognitive and physical needs of humans.

Humans use their voices and bodies to transmit and receive information, and a great deal of it is in the form of sound. This information has explicit and implicit components, some conscious and others unconscious. We inquire of our territory and inform our environment, interrogate our surroundings and fill space through our use of sound.

Sounds of all kinds emanate from humans, anthrophonic sounds like grunts, screams, groans, speech and music. There are work sounds, play sounds and the sound of people sleeping. Their rituals, rites and ceremonies need sounds to empower them, and the use of sound even empowers the application of silence. Quoting Miles Davis: "It's not the notes you play, it's the notes you don't play." Thus all sacraments and traditions, sacred and secular, acknowledge sound's importance, even those performed in silence. There is the persistent image of the one-eyed witchdoctor, but he's never deaf "A noteworthy fact is that a deaf ritual specialist is an anomaly whereas a blind one is common enough" (Jackson, pp296). Sound is crucially important, because it is a means of contact and control.

The purpose for which humans make sound is unique. We do more than signal some present condition, we indicate the past and future as well. We pattern it into languages spoken, written and musical that broadcast feelings and incite the formation of ideas. These patterns transmit knowledge, accumulated facts and insights that are passed on through generations, driving technological and social developments that would not be possible without such collected information.

Sound resonates with great power in our lives. It has been a vitally important component of our development, its power within the developmental dynamic well demonstrated. Throughout our evolutionary process we have had an ongoing exchange with and through sound, and it is reasonable to conclude that people have used sound to great effect for a very long time. Did pre-literate people render paintings in resonant cave spaces intentionally? Did their leaders or their shaman exploit acoustic characteristics as a component of their schemes to control and direct the actions of the group?

The position taken here is that sound has been used with knowledge and intention throughout human history, with the degree of knowledge and technical skill varying through time and territory. Furthermore, people use sound as a tool of social construction and maintenance, a set of actions that arise from the functional characteristics of human consciousness. Significantly, the idea of intentional action may have to take on a dimension of inevitability or a deterministic quality given the structure and function of the human mind/body/brain.

We use and have used sound for social control because it is one of the vibrational components of an energetic exchange within our environment that we can contribute to directly using our bodies. That complex relationship of energy exchange and reorganization of matter manifests with unknown and hidden causes that drive common events, and our survival depends on an ability to construct models of the world and make accurate predictions. Controlling the structure and function of social groups is a method of ensuring predictable outcomes based on defined and established relationships. Social control is an effort toward survival, an adaptive behavior that includes within its myriad entanglements the emergence of a powerful elite and a definition of people as a resource to be utilized in the pursuit of other scarce resources. The minimization of prediction error entails the construction of a controlled environment which requires restriction and control of individual and group behavior, a goal served well using language and music. This process can be examined in a structured way using theoretical constructs formulated to explain and predict the functions of human cognition and its action in the environment.

Free Energy

When utilizing the information based paradigm of human cognition as formulated by Karl Friston (1), communication through sound is understood as a tool used to minimize free-energy. Free-energy is a measure of surprise experienced when observations diverge from one's mental model of the environment. This model is expressed by the equation (1):

$$F = -\ln p(\tilde{s}|m) + D(q(\vartheta|\mu)\|p(\vartheta|\tilde{s}))$$ Free-Energy = Surprise + Cross Entropy

F	Free-Energy
$-\ln p(\tilde{s}\|m)$	The Upper bound of surprise associated with receiving a sensory input: $\tilde{s}$
m	Model of the world
ϑ	Unknown Causes, or unknown quantity that caused the sensory state
μ	The internal states of the brain
$q(\vartheta\|\mu)$	Recognition Density
$p(\vartheta\|\tilde{s})$	True Distribution of Unknown Causes: ϑ
$D(q(\vartheta\|\mu)\\|p(\vartheta\|\tilde{s}))$	The divergence between the Recognition Density and the True Density

Cross entropy is the divergence between the recognition density and the true distribution of the causes of a sensory state (3).

The minimization of prediction error is an adaptive characteristic of consciousness, a functional quality that plays a crucial role in the organism's survival within its environment. The construction of an accurate mental model of the world, along with active reduction of danger, is an ongoing process. The environment is continually sampled through the senses, the model adjusted, and/or the sensory input is changed through action within the world (2). Such actions include the creation of social systems that control people and their environment.

Using tools to transform the material world and transduce surrounding energies to meet their needs is a power unique to humankind. These processes begin with sound, which is used to create social cooperation and generate effective group activity. Myths, religions, science, laws, taboos and customs are all social constructs which make life a more understandable and safer experience. Human culture is a network of systems that serve the purpose of creating stability and predictability. Vocal sounds, music and language transmit the knowledge and emotional drive to create the needed social and physical structures that sustain us.

Internal to External Structures

Through an information exchange between the internal and external environments, there is mutual construction and modeling taking place. Social constructs and systems are reconstructions within the material world of hierarchical structures in the brain, which are modeled on the surrounding environment. An example of this can be found in the shape of neurons. Action at a distance, such as gravitational influence between objects in space, is modeled by the long slender nerve body of a nerve cell, which connects axon to dendrite. The characteristics of the universe in which the brain exists are cast within its structures. Social structures model the brain's communication network, moving information between people as the brain does between neurons. Here, brain functions are modeled in the surrounding environment when the system is driven with the intentionality of sound and symbols.

Sound Transmits Brain Activity

Sound is a medium that couples our brain activity to the external world. Music and spoken language carry vibrational information from the brain to the outer environment, intended to order that environment in a manner that allows us to continue the feedback process of experimentation, construction, learning and growth. When we speak, talk, sing and play instruments, we are reducing free energy (surprise) by acting on the environment to change its states, which then changes our sensory input. We change the world around us to get from it the feedback we expect, on a consistent basis.

Oscillations in the brain are periodic, and the periodic waves that emanate from us as sound are used to create social cohesion. Periodicity is experienced as regularity resulting in predictability. Where periodic phenomena occur, a conscious agent can predict occurrences more accurately. People communicate to share feelings and thoughts, to form alliances and groups that act as force multipliers. They make music, a rhythmic set of periodic impulses, to express themselves and the experiences they share with their group. This builds cohesion and fosters effective cooperation to build society and its structures. In this way, the regularity of periodicity is used within the system to reduce uncertainty. What beautiful symmetry!

It is well known that periodic waves can interfere with each other destructively, cancelling each other out. We know that social interactions can be harmonious or disharmonious. Evolution itself may be characterized as a progression driven by constructive and destructive interference. Those changes-mutations-that are harmonious result in constructive interference propagating in a wave-like fashion through the biosphere, and those that are dissonant result in destructive interference and are cancelled.

Beliefs, Myths, Social Control

In caves, reverberant sounds might be interpreted as emanating from within a rock, and myth supporting that would reduce uncertainty and fear of the unknown. A model is thus provided that reduces surprise. The same would be true of resonant spaces within buildings of religious worship, such as cathedrals. The glory, magnificence and wonder of the building, and by extention the universe, can be attributed to the workings of the worshiper's deity. Those who worship in such a place would be receiving confirmation of their beliefs with every encounter. Their model of the world is reinforced each time they experience the sounds and sights in the cathedral. The energy input the system (cathedral + people) needs to continue reassuring worshipers that their world model is correct is largely in the form of sound. Sermons, songs and the anthrophonic sounds of people in groups supply auditory and physical sensations that facilitate bonding, which reinforces shared beliefs.

Conclusion

Did people of ages past use sound knowingly, and as a means of social control? Utilizing these models of cognition and brain activity, it appears that the answer is yes. Initially, they may have done so without conscious awareness, driven by the need to reduce free energy through manipulation of their mental models and their sensory input data. However, as their awareness grew, their conscious power of manipulation would have grown. During the initial condition, intention may be defined as actions taken as the result of internal cognitive processes, separate from the consciousness occupied with everyday tasks. During later conditions, as awareness developed progressively, this intention became the actions taken as the result of observation and calculation within the conscious mind of everyday life. Social control and the intentional use of acoustic effects emerge from a system that seeks to verify its beliefs, operating internally at frequencies that can be externalized as sound.

REFERENCES

1. Friston, Karl J., and Klaas E. Stephan. "Free-Energy and the Brain." *Synthese*159.3 (2007): 417–458. *PMC*. Web. 9 Jan. 2018.

2. Friston K. Nat Rev Neurosci. 2010 Feb;11(2):127-38. doi: 10.1038/nrn2787. Epub 2010 Jan 13. Review.

3. Collell G and Fauquet J Brain activity and cognition: a connection from thermodynamics and information theory. *Front. Psychol.* (2015) **6**:818. doi: 10.3389/fpsyg.2015.0081

4. Atasoy, Selen, Isaac Donnelly, and Joel Pearson. "Human Brain Networks Function in Connectome-Specific Harmonic Waves." *Nature Communications* 7 (2016): 10340. *PMC*. Web. 9 Jan. 2018.

5. Jackson, Anthony. "Sound and Ritual." *Man* New Series, Vol. 3, No. 2 (Jun., 1968): 293-299.

On Foundations of Archaeoacoustics

Iegor Reznikoff

IEGOR REZNIKOFF: Emeritus Professor at the University of Paris (Université de Paris Ouest) is a specialist of the foundations of Art and Music of Antiquity and particularly of Early Christian Chant for which his performance, based on deep comparative studies, is known worldwide. This speciality has led him to study the resonance of Romanesque and Gothic churches and on practice of ancient scales, as well as the resonance of prehistoric painted caves for which he has shown the relationship between paintings and acoustics. He is a specialist in Sound Therapy and more generally in Sound Anthropology.

It is now a few years that Archaeoacoustics has emerged as such, with this very name, although the field has existed since long ago, e.g. since Vitruvius. Archaeoacoustics is now very prolific, and it is time to make an appraisal in order to improve studies in the field. There are many different subjects, such as studies of ancient instruments (flutes, lithophones, etc.), closed spaces (caves, temples), open spaces and sites. Methodology has to be elaborated for each of these subjects; particularly the adequacy of different means of measurements to the field of study; but also one has to reflect on the kind of results one is looking for. Often conclusions are wrong: for instance, it is not because a space, now, sounds well that it was intended to be such or really used for sounds. The anthropological point of view has always to be kept in mind. This is implied by the meaning of *archaeo*, since in ancient times it was not just for the sake of acoustics that instruments or temples were built and used. The acoustic approach was – be it in means or aims – anthropological. Finally there is the problem of backgrounds and training needed for good studies in archaeoacoustics: apart from some basic studies in archaeology and obviously in acoustics, certainly good knowledge in ethnology, particularly in ethnomusicology is necessary, also in music and – most important for the study of resonant spaces – some elementary vocal and singing practice is needed, as well as some knowledge in compared religions and rituals. However all this may remain superficial without a deep consciousness of sound and a trained ear in fine perception of sound waves, as we can learn from Masters of the past, from Vitruvius to Helmholz.

A Little History

Before reflecting on Foundations for the present and future o Archaeoacoustics, it is wise to look into the past and on how, what we call nowadays acoustics appeared and were practiced. Let us be reminded first that the word *acoustics* proceeds from the Greek *akoustikos* which derives from *akouein* = to hear or to give attention for. Clearly, it refers to perception, since there was no other possibility to realize or express the matter, but it doesn't mean that sound and acoustics were not perceived and used in a very subtle way. We will see many convincing examples of the use of fine perception on which acoustics have necessarily to be grounded. The modern introduction of the word *acoustique* (in French) is due to the French natural philosopher Joseph Sauveur (1653-1716) in the year 1700; Sauveur physically founded *musical* acoustics. Now speaking of *archaeoacoustics*, the first important text on *ancient* acoustics, ancient meaning relatively to the writer's time, is in the first century BC: the well known *De Architectura* by Vitruvius (see below). It is interesting that already in the time of Plato and Aristotle, it was understood that sound is vibration (e.g. of a string) which is transmitted by air. Later, medieval texts on acoustics are mostly devoted to musical intervals and their numbers expressed in length of strings, repeating what was already known from the time of Pythagoras, and devoted to forms of scales more or less inherited from the Greeks, that they understood no more, as for the *De Institutione Musica* of Boethius (c.510).

The first approach of *acoustics* as we understand it now but also of *archaeoacoustics,* is the *Harmonie Universelle* (1627 and 1636) by Marin Mersenne (1588-1648), a French Franciscan, who can be considered as the founder of these studies because he considers, as it was always done, the knowledge of the past, but makes precise the ancient notions, particularly in what concerns harmony. Moreover Mersenne makes measured experiments, namely on the swiftness of sound, measuring it by using echo effects so that the same clock measures time between emission and return of the sound. His experiments gave the speed of about 310 m/sec., which is a remarkable result for that time (the speed under normal conditions is 340 m/s.). In the same period, we may also mention Galileo's measures (1638) of vibrations of strings, and mention, from a philosophical point of view, the great astronomer Johann Kepler (1571-1630). Apart from his major scientific discoveries in astronomy, Kepler was also concerned by the metaphysical notion of Celestial Harmony, making the distinction (which Galileo, who refused metaphysics, couldn't make) between the Visible World – the astronomical – and the Invisible in which, for him, everything was tuned in Harmony following the laws of acoustics and music; Kepler, in this respect, calling himself a Pythagorean. Still in the 17th c., one has to mention Anastasius Kircher and his *Musurgia Universalis* (1650) where he discusses many traditions of the past e.g. the Egyptian and the Greek. He has the honour of being the very first to give a transcription of a Greek hymn of which he says that he found it in Italy (in a papyrus which eventually disappeared after his discovery). Not going further in the details of history, we have however, in our perspective of foundations, to mention the name of Hermann von Helmholtz (1821-1894), the great scientist and acoustician who promoted and developed a theory of *sound perception* and tried to find relationships between physics and the physiology of this perception.

The Importance of Fine Perception

It should be clear that for studies in fields related to perception, be it of colours, sounds, etc., the corresponding perception (and its refinements) is the absolute base on which these studies must be founded. We have to insist here on this because on another hand it is not true. For some aspects of modern acoustics that are based on pure physics, as for instance underwater or ultrasonic acoustics. And therefore, there is nowadays an obvious temptation to rely only on modern electronic devices and to forget about the reality of perception. Certainly computers and software for sound are fascinating and so easy to use, but for serious studies in archaeoacoustics, a good perception of sound is necessary. Indeed, all ancient often advanced data were discovered by hearing, listening, echolocation (particularly for progression in obscurity), or by direct perception of vibrations, be it in music, singing, imitation of animal sounds for hunting, or in building instruments, temples and theatres. All ancient practical but very efficient data and achievements in music, construction and tuning instruments or in architectural acoustics, were founded on hearing in a very subtle way. It is completely misleading to believe that now with all our sophisticated machinery, we do better. If we limit ourselves to electronic devices, without a fine consciousness of sound, we may even not notice some essential aspects of ancient data and miss the real point, so that conclusions may be superficial and of little interest.

Of course, for verifications, improvements and accurate analysis, modern machinery and computers can open new horizons and discoveries, but the first approach must be through trained perception. Exactly as, say, in a prehistoric cave, it is first by seeing and scrutinizing the walls that one discovers paintings and which animals are painted or engraved. Then, and only then, modern chemistry can show what minerals were used for colours, and atomic physics give a reliable dating. If you go into a cave you know nothing about, with a computer, but blind and deaf to the surrounding space, what's the point? You have no chance for a discovery. It is only because pictures and resonance effects were discovered previously that chemical, physical and acoustic studies were initiated. But while pictures are seen with eyes, it is amazing that for sounds people who measure are not interested in hearing and ignore how to listen seriously and even more how to make consistent singing sounds.

All prehistoric civilizations were founded on very subtle perceptions, particularly of sound. At that time, as in other hunter-gatherer societies, one had to learn to walk absolutely silently to approach animals for hunting, sometimes by night; one had to listen to their noises or cries and possibly imitate these cries with the voice or with implements like bird calls which required an extraordinary gift for hearing and making sounds. Let us remember here the traditional example of an American Indian on his knees, one ear close to the ground in order to hear or rather perceive the noise of the hooves of a herd of horses or bison and estimate their number? It is the same kind of gifted people who, long ago, explored in almost complete darkness and certainly using echolocation, the dangerous caves that they eventually decorated and painted.

But even more: how were all the human spoken languages elaborated, if not with subtle hearing and the exceptionally refined possibility of making so many different sounds, vowels and consonants? In the Caucasian area, one of the richest areas in phonetics, a language (now extinct) had more than a hundred phonemes. (Phonemes are different sounds of the language; about 36 in French and 44 in English.) Certainly human language and speech is an extraordinary achievement if not a miracle, that computers are not likely to create.

Now, concerning musical instruments, despite our modern acoustics and possibility to analyze vibrations of various materials, our contemporary realizations have not yet prevailed over incomparably fine constructions of violins, cellos or viola da gamba elaborated by the subtle perception of musicians and instrument makers of 17th and 18th c.. Nor have they prevailed over the elaboration of mouth-pieces for oboes or clarinets made traditionally from bamboo reeds (best from Provence). For violins, the name of Antonio Stradivarius (1644-1737), the Italian *luthier* (a French word), is still very famous. Here, it is worth telling a personal anecdote; a friend of mine, an architect, decided at the end of his career to build new instruments. He decided that shapes of classical string instruments were no more needed and that time has come for creation of new forms: why peculiar shapes of the past and not simple rectangular ones? So that he built some cellos with "rational" forms, squared, parallelepipedic, cylindrical, conic, etc. and mixed them. He was very proud of the idea and asked me (why me!) about the sound of these strange objects. Well – I told him – you have to listen and of course train a little your ear; it is not a simple matter of shape. Sound runs by waves and not in a linear way, so that if violins or cellos have a very specific, fragile, incredible shape, it is just because the Masters of the past were listening very carefully and discovered progressively that such and such shapes, woods and special varnishes were probably the best. It is the perceived sound that dictates how to improve the shapes. Finally, concerning instruments, less known is the extraordinary craft of melting and casting bells, which climax belongs also to the 17th – 18th c. We still do not know exactly their techniques for alloys and our bells do not ring as well. I, myself, have worked on vibrations in solid bodies with an extremely precise electronic device, an accelerometer. However in what is relevant for human beings, perception is incomparable since the machine doesn't know, feel or hear what is good and meaningful for humans, body, senses and consciousness.

Concerning theatres, we give here the remarkable lesson of oral practice given by Vitruvius in *De Architectura*, Book V, chap.8 (my translation from Latin and French *L'Architecture de Vitruve*, translated by Ch. L. Maufras, Paris, 1847), which shows how refined was the perception in architectural acoustics.

Attention must be bestowed on the choice of a place where the voice may fall smoothly, and be repulsed so as to keep a distinct meaning to the ear. There are some places which naturally hinder the movement of the voice: the dissonant, the circumsonant, the resonant and the consonant places. The dissonant places are those in which the first part of the voice, when increasing, strikes against solid bodies that repulse it so that the voice impedes its following utterance. The circumsonant are those where the voice moves around, is collected in the middle, and dissipated, dies away in sounds of indistinct meaning. The resonant are those in which the voice, striking against some hard body, is reverberated and makes last syllables appear doubled. Finally, the consonant are those in which the voice, reinforced from below [in the amphitheatre], rises to the ear with great clarity of words. Hence, if due care be taken in the choice of sites, the effect of the voice will be improved, and the good quality of the theatre increased. The differences of the figures consist in this, that those formed by means of squares are used by the Greeks, and those formed by means of triangles by Latin. He who attends to these precepts will be enabled to erect a theatre in a perfect manner.

If we take as an example the Greek amphitheatre in Epidaurus (4th c. BC) which was certainly built following the instructions inherited by Vitruvius, we have remarkable proof of how powerful and subtle are human perceptions. This architecture was renowned from antiquity and is still in our modern times as an exceptional architectural and acoustic achievement. The words of the half singing / half speaking actors' voices were heard from everywhere in the theatre space. It must be noted that Vitruvius speaks rather of a singing or recitative singing voice since elsewhere he speaks of musicians. Moreover, in the same chapter, concerning acoustic vases, Vitruvius speaks of *musical intervals* (fifth, fourth, etc.) and *modulations* of the voice which are relevant only for chant[1].

In the same paper (see footnote 1), the precisely intended acoustical architecture of Romanesque churches (11th – 12th c.), is discussed as well. This architecture was progressively elaborated by centuries of singing practice and hearing what kind of vaults, spaces, apses and resulting resonance, answer the best to the voice and makes the words understandable in one place or, from a different place, the chant melismas more beautiful. It is important to emphasize that in ancient times, and until the 18th c. included, singing was performed in natural scales (not *tempered* modern ones), scales of which intervals are the same as intervals appearing from the harmonic laws of resonance. Otherwise, if singing in tempered scale, the beauty of the resonance is blurred and erased.

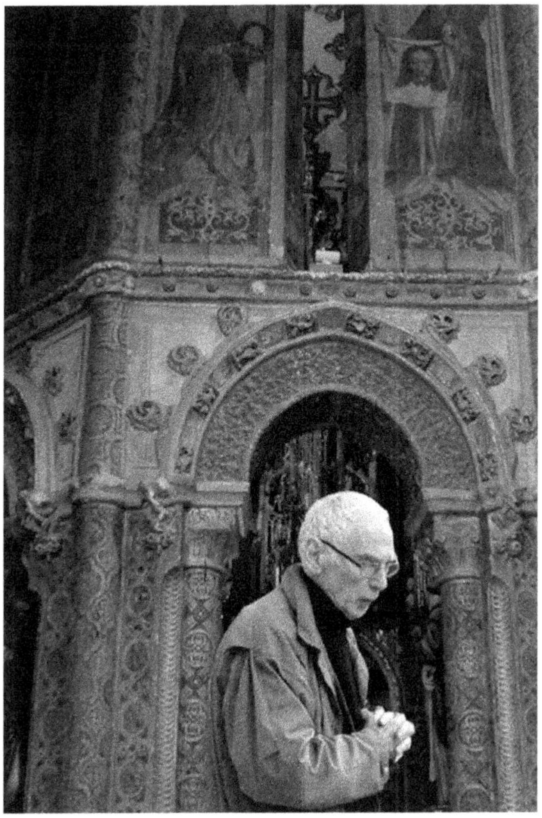

The author and the retunda of the Knights Templar, Convent of Christ, Tomar, Portugal. Image: OTSF

[1] For a detailed commentary about acoustic vases, see I. Reznikoff The evidence of the use of sound resonance from Palaeolithic to Medieval times, *Acoustics, Space and Intentionality*, Proceedings of a conference held in Cambridge, 27-29 June 2003, Lawson, G. and Scarre, C. eds., Cambridge (McDonald Institute for Archaeological Research, Monographs), 2006, p.77-84.

It is out of the question to sing in a Romanesque resonance in *bel canto* or lyrical style, as however almost all ensembles do! The reader may listen to the CD *Le Chant du Mont St-Michel* (ADF-Bayard, France) in order to hear and feel what indeed a great Romanesque resonance is, or, for those who were there, they may remember the remarkable concert at the Convento de Cristo in Tomar.

Now, when you have started to discover and rediscover the various harmonic sounds produced by a trained voice in resonance, and all its surprising effects – sometimes bad effects e.g. when the resonance is unequal, possibly with holes depending on the pitch of the voice, as for instance in the Romanesque Cistercian church of Sylvacane (Provence), contrary to the Cistercian church of Le Thoronet (also in Provence) which has the most wonderful sound quality – when you have progressed in mastering the resonance you may start first a careful *musical* analysis (see below) and then use software for a complementary approach, but as far as we know, there exists yet no adequate software for revealing the essence of resonance, since available software is conceived for acoustics of rooms or concert halls which have nothing to do with rich resonances such as in some caves or churches. We explain this below.

Practice and Measurement

The fine perception and consciousness of sound we spoke about above, yields progressively an *interaction* with the studied space. This is particularly true because perception of sound is the main perception – body and hearing – we have: the whole body is taken by sound and vibrations as are the deepest levels of consciousness[2]. Gradually, in a prehistoric cave, a Romanesque church or even in an open space with echoes, this interaction helps to discover still more subtle acoustical qualities of the studied site. It is important to be involved in the intensity of this interaction, as certainly were prehistoric men when discovering caves that they eventually adorned after proceeding mostly by echolocation – actually making sounds for echoes – in a frightful dark space which appeared as a living being or deity that answers to calls, echoes and chant[3]. This call-and-answer play depends on the route, and one has often to repeat the experience along the path to discover the best answers of the resonance. This intimate relationship is clearly proved by the decorated recesses and small niches painted with red dots we find in many painted caves. It is impossible to do this when moving with a cumbersome electronic material. By using voice, one can feel directly a meaningful, possibly attractive effect, or an impressive humming effect inside a niche and around [see photo 1 and below about *Bison effect*].

Moreover, in some noteworthy parts of the cave the process can be continuous in order to discover possible connections between different parts of the cave while progressing. Discovering the sound dimension of a cave is a real and intense performance.

[2] For the relationship sound / body and sound / consciousness see I. Reznikoff, *On Primitive elements of Musical Meaning*, http://www.musicandmeaning.net/issues/showArticle.php?artID=3.2.

[3] I had myself a rather successful experience in blind echolocation, but see *World Access For the Blind*, https://visioneers.org/, or the incredible https://www.youtube.com/watch?v=YBv79LKfMt4.

PHOTO 1 – Producing a *Bison effect* by humming in a niche in Kapova cave (Ural).

PHOTO 2 – Trying the resonance of small recesses (Kapova cave, Ural

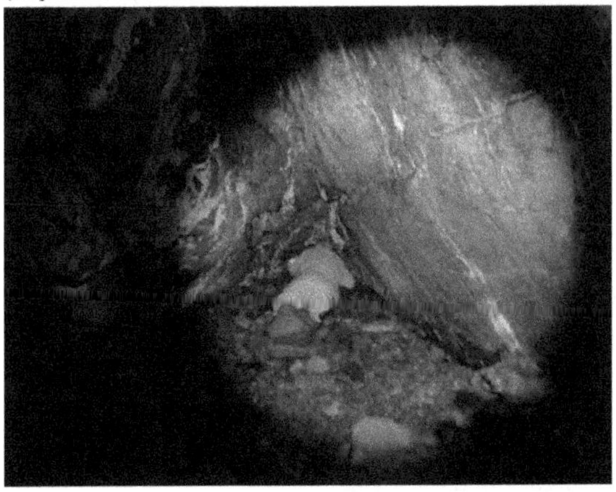

PHOTO 3 – In Solsemhula cave (West coast of Norway)

With electronic equipment, only partial and point-wise measures can be done, measures that you discover only *post factum* in a studio, so that you cannot realize the richness of resonance and, for instance, the correlation with other echoing parts in the cave.

While with electronic devices and software you can certainly check some aspects of cave acoustics, this material is not manageable for the *discovery* of the resonance complexity in the cave which can beapprehended only as described above. As it is indeed the case for the pictorial discovery which can be done first only by seeing carefully the walls, niches and low or higher parts in the cave, and not with electronic tools and screens. The same is true for sounding temples and churches, although the route inside is regular and therefore simpler. And it is true also for the discovery of resonance in an open space, studying correlation with painted rocks. The incomparable advantage of using the voice for echoes is that in case of a rich echo, you can sing (not shriek) short melodies in the right way i.e. following resonant harmonic intervals. Melodies are returned by echo and are repeated many times, possibly overlapping and giving a harmonic magic impression. This is a purely anthropological approach which, moreover, can give an evaluation of the quality of echoes. The simplest and measurable evaluation is given by the number of echoes. A resonance giving 5 or more echoes is a rich one, while one echo is certainly meaningless because it appears almost everywhere in caves or open spaces with rocks or cliffs.

The best way for measuring resonance is the well-tested musical way: making vocal sounds – homogenous sounds of around 70-80 dB (and 90 dB for echoes), during 1-2 sec. – and measuring:

1) time with a chronometer (e.g. a good watch),
2) pitch with a tuning fork (usually tuned on A = 440 Hz),
3) intensity with a sonometer,
4) and richness of the resonance by the number of echoes.

This simple manageable approach used by all musicians, tuners and orchestras to define sounds precisely, gives all needed information. Moreover, a trained hearing may reveal also the main harmonics given by the resonance. Point 4) is relevant only for caves, etc.

Some beginners in archaeoacoustics say presumptuously that their studies with electronic systems – computer, software, etc. – are 'more robust' than the musical study. But this is ridiculous pretention. An electronic device such as that ordinarily used by acousticians, in situations and spaces as described above, gives neither in pitch nor in duration or intensity more useful information than the musical measurement needed for human perception. For instance, let us consider the device given by software EASERA (AFMG Technologies, with sonometer Brüel and Kjaer 2250 (class 1), sound card, amplifier HPA D604, omnidirectional source and microphones, producing continuous sounds from 45 to 20.000 dB during 5 sec.), This electronic system is not really adequate for measurements in resonant spaces like the caves or churches we are actually interested in because while sounds are produced, resonance and echoes of previous sounds are hidden in or erased by the mass of oscillographic information. Moreover, while using the device, you cannot realize (in 5 seconds and so many sounds!) what happens on the spot, so that you cannot improve right away the measurement e.g. the direction or intensity, needed to deeply discover the acoustic quality of the studied space, be it a large hall, a narrow tunnel or a small recess where the machinery has no access without saturation phenomena. The discovery of what we call *Bison effect*[4] would have been impossible. ; it is a very powerful resonant phenomenon obtained in a niche, by a simple humming performed at the right pitch, the main pitch of the resonance which amplifies the humming so that it imitates extremely well roars of bison or other big animals. The imitation of animal cries is essential in the shamanic process.

After the study and a post factum analysis, it may be complicated or too late to return to the studied spot, since time, particularly inside painted caves, is strictly limited. In this case, the study remains superficial. Let us recall that such acoustic studies with complex devices were motivated by discoveries made already 30 years before by the musical alive approach.

[4] See *Prehistoric Paintings, Sound and Rocks*, in Studien zur Musikarchäologie III (Colloque international, Michaelstein, sept.2000), E.Hickmann *et al.* éd., p.39-56, Rahden (Westf.), Allemagne, 2002. In this paper, this effect I called *Camarin effect* because it was first discovered in a niche called Camarin in Le Portel cave. See also I. Reznikoff, *On the Sound Related to Painted Caves and Rocks*, Sounds Like Theory. XII Nordic Theoretical Archaeology Group Meeting in Oulu 25.–28.4.2012, J. Ikäheimo *et al.* (eds), Monographs of the Archaeological Society of Finland 2, Helsinki 2014, p.101–109; *Bison effect* are mentioned on p.102.

One has to know and hear first. If one starts with a device as described above, which doesn't show clearly echo effects, and therefore a correct account of resonance, one can miss discoveries and announce false results. Actually what is a *scientific* approach or study? It has to give results measured in a standard way; namely here in frequencies, duration and intensity. The experience must be repeated and of course must have the property of being possibly repeated several times.

The musical approach as presented above has all these properties and moreover is much simpler and more subtle as well. Of course, afterwards, a complex technical study is welcome to possibly specify the results.

It is one of the major weaknesses of our time: the belief that eyes are more reliable than ears. But why would seeing a needle on a screen be more objective than hearing a sound? Such an amazing statement was heard at a lecture given at our Congress in Tomar; but all human knowledge is founded on speech and words we say and hear, and only after on written sources! We progress in learning with words *yes* and *no*. The sound perception is the first and main perception, and this is quite important for our subject. The very first consciousness of space is given by perception of sounds and not by sight, since the baby in the womb doesn't see while he (or she) has an already structured consciousness of sound (see footnote 2).

Needed Background Knowledge

I have worked with acousticians belonging to the renowned School of Acoustics founded by Guy Busnel (1914-2017), specialist in animal acoustics and whistled languages. One of these acousticians told me that it is not necessary for him to go inside the cave I was interested in: "You simply give me the map with all dimensions of the cave obtained by a laser technique and the nature of the rock; immediately the computer will give all acoustic data you need regarding the cave." Clearly he was thinking of ordinary rooms and concert halls. I told him however that it was much more complicated because ground, ceiling and walls are usually of different and changing nature. It can be rock, clay, earth, water or mud for the ground; and there are often many recesses and niches that cannot be noticed by dimensional laser technique, therefore the acoustical data would be inaccurate. After a while he admitted this and understood that the point was also anthropological: that we want to understand how prehistoric tribes progressed in a given cave, possibly using echoes, and what could be the relationship between paintings (or engravings) with different points of resonance. The aim of such studies is, indeed, through careful visual and aural survey to get some information on rituals or processions that certainly were performed in some resonant parts of the cave. One has to be really involved in the study; as we stressed above: involved in body and consciousness.

A purely acoustic study limited to measurements of duration and frequencies is not in itself sufficient to understand the acoustical meaning of a monument. If the results are negative, maybe it is because you missed some hidden points. But if the results are positive it doesn't yet mean that the sound quality of the monument was calculated intentionally or even used at all. Many other aspects, particularly anthropological for which some background knowledge is needed, have to be integrated for reliable conclusions. Among needed backgrounds, we can mention the following.

Trained Hearing

Before working seriously in archaeoacoustics, obviously some basic knowledge is necessary. Why are learning and training needed in archaeology and not in archaeoacoustics? In archaeology, traditionally, for excavations one works slowly and very delicately with fine brushes and training is compulsory. It is out of the question to start on the spot with computers first. A refined training using voice and hearing is clearly needed as a first step in archaeoacoustics. There are workshops for such training. However, some more knowledge is needed.

Ethnology and Ethnomusicology

First, in ethnology; how can one pretend to study ancient monuments, their meaning and use, without knowing about ancient societies? For this, particularly for economically primitive societies, a comparative knowledge must be acquired; above all in ethnomusicology. One has to know that in such societies, *singing is practised day and night, all lifelong*: songs before birth for the baby, sounds during delivery, lullabies, working songs at home or outside in the fields or on rivers, war and peace songs, and chants for various rituals addressed to ancestors, spirits and to the Invisible World. How can one pretend to understand these societies without even be able to sing a simple O or AA for two seconds! And why are people in archaeoacoustics so afraid by their own singing voice? An elementary training is needed. It is not a question of *bel canto* and opera singing. The appropriate training is simple and very rewarding.

Contrary to what people often believe, namely that the more primitive a society, the more primitive the music, ethnomusicology teaches that these societies e.g. Pygmies, South African Bushmen, Australian aborigines, have a very subtle approach of sound and music, proportional to the fine perception of sounds they need to have, as we have seen above. Therefore, it is wrong to believe that sounds and chants practised by Prehistoric tribes were 'primitive', although they may appear simple if compared to our western classical music since say, the 17th c. Usually monodic, indigenous music is very subtle in various intervals and intensity. Some people with short imagination believe that ancient tribes could only make clicks with two stones and shriek like monkeys. But let us look at their paintings: they had indeed great masters. Why not masters in music, as we can hear in Pygmies chants or in other ancient oral traditions? Moreover, ethnology teaches that their way of living was, and – sometimes for extant societies with primitive economy – still is, superior to what we can do now with our learned technologies. Who could, without heavy logistics, survive for thousand years in the hard conditions of life of Eskimos in Greenland, of Bushmen in deserts of Africa or of tribes in deep forests of Amazonia or Congo?

The same kind of error one hears concerning Prehistoric flutes: if such a flute has, say, three holes, some naive prehistorians say that people who used this flute had only three notes making primitive melodies. But with three holes, a good player can play a lot of different but also subtle intervals and melodies. Once, a Japanese musician (I think it was in Paris) came to me with a peculiar flute, rather of ocarina shape, but with just one hole: a unique hole both for mouth and perforce only one finger. By closing more or less the hole with the finger or the lips, he made a lot of various sounds in different ways, light or strong, continuously or staccato. I could only marvel. At the Congress, in Michaelstein organized in September 2000 by Ellen Hickmann (see footnote 4), some evenings, 'concerts' were given by amateur musicians playing on copies of prehistoric instruments, mostly flutes, doubtful rattles and bullroarers (rhombus). They played 'prehistoric' i.e. in a very childish way; it was pitiful. And then, not at one of these concerts, but during a session at the Congress, a German scholar took a flute, a copy he made of a prehistoric one, and played. He played really, musically, gently, melodiously. The audience was fascinated and kept a long silence after his performance. This reminds of reed *nasal flutes* played e.g. in Philippines and some Pacific Islands, by blowing into the flute through the nose. The sound, of course, is very soft and subtle, a kind of wind poetry.

For ethnomusicology, the best is to listen to recordings of genuine oral traditions. Nowadays, there are ensembles or persons who imitate traditional music, e.g. a shamanic ritual, pretending moreover that after a workshop one can become a shaman; but this, of course, is just fake tradition. To avoid such a falsity, one may listen to renowned Collections, as the UNESCO traditional music series or the Ocora Radio France one.

Addressing the Invisible

Now, ancient monuments, temples, churches, theatres, painted caves, remarkable sites, were devoted to worship, praise, prayer, rituals or liturgy, performed essentially by singing or vocal uttering. How is it possible to try to understand the meaning of these monuments when one avoids completely the notion (not speaking about practice) of prayer and ignores a possible relationship with the Invisible World. In any case, such studies, even if one has to be agnostic for the results, have to take into account that ancient people, concerned by temples etc., believed in invisible forces and that it was the very aim of the extraordinary monuments they erected for millennia, overmastering enormous difficulties. These monuments demand respect, and beyond chant, a contemplative silence.

In all ancient or so called primitive societies -- be it tribes of Indians in Amazonia, Jewish or Islamic communities, or in Buddhist temples and Christian Monasteries or churches -- rituals and liturgies, first prepared and then celebrated, were sung almost continuously for centuries and all lifelong. Ethnology teaches that there are no rituals and celebrations without singing. It is not because all these traditions have disappeared almost everywhere and particularly in our modern West, that one has to ignore such a fundamental reality

To acquire a deep knowledge of what Addressing the Invisible may be, the best thing is to attend, for instance, a Christian orthodox or Eastern Church liturgy, a Hindu *Puja*, Buddhist ceremony or genuine ancient but still alive oral worship, and to reflect upon the notions of offering and sacrifice.

Studies in Prehistory and Archaeology

This should be obvious for those who are interested in acoustics of prehistoric caves or marked open spaces, or in acoustics of ancient monuments, temples, churches, amphitheatres, etc. Fortunately, such knowledge is usually given in corresponding Institutions.

Acoustics

Finally, a minimal knowledge in this field is of course necessary, at least on an elementary level in the physics of sound. Unfortunately, in Acoustic Departments there is usually no teaching in resonance phenomena. It is why acousticians are usually lost in resonant spaces and why there are no appropriate devices (see above). Some Departments should remedy this, but the field of archaeoacoustics is not so well known and not as rewarding as other parts of acoustics related to modern industrial or research technologies.

Of course, the more knowledge and experience one has, the best it is, but the fields of knowledge mentioned here give already a large foundational base for Archaeoacoustics.

Conclusion

Although the number of studies and papers in archaeoacoustics has enormously increased these last years, many studies remain at an amateur level, be it by a lack of precision or comprehension concerning resonance and its complexity, or lack of background knowledge and anthropologic understanding. Too often, people think that with a standard device and software, they are immediately able to study and discover some mysteries of ancient monuments or sites. As it would be in archaeology or in prehistoric studies, such an approach is not serious. It is time to put archaeoacoustics at a university level with serious background knowledge that the practitioner has to acquire. Provided such backgrounds, we wish all trained researchers beautiful discoveries: perception of sounds is the deepest you have and certainly there are, in archaeoacoustics, many discoveries to be found.

Researching the Epistemological Paradigm of Archaeological Bone Flutes' Sound-Studies

Etienne Safa

ETIENNE SAFA, Researcher and craftsman, PhD student in history and sound-archaeology at the University of Burgundy. Fields of research: history, archaeology, musicology, sociology, cognitive sciences, University of Burgundy, France

ABSTRACT: Recent results in this research have revealed our tendency to consider flutes' sounds as an absolute reference while they should be considered as one set of possibilities amongst others: flutes may give multiple results depending on the experimenter's culture, know-how, technicality, state of mind, motivation, etc. However, studying archaeological flutes is not only about creating sounds: it rather starts with the observer's cognition, as who and what we are definitely influences the way we perceive and process information. Typology itself isn't neutral and may be analyzed from sociological, biological and cognitive points of view to reveal its cultural and personal subjectivity. This paper is an attempt to synthesize both cognitive and experimental approaches, aiming toward the production of an essay for a global epistemological paradigm of archaeological bone flutes' sound-studies.

KEYWORDS: Epistemology, Acoustics, Cognition

Introduction
Context and Other Works

Archaeological flutes remain silent. Researches have to face difficulties (or even biases) amongst which some are more subtle than others and have succeeded to elude investigation for decades. Since the early 1950's, archaeomusicology has thus been illustrated by different methods which lead sometimes to very different results for the very same objects, thus evoking potential lacks, shortcomings or veiled biases either in methods, study processes, technical means or knowledge of the object's specificities (Brade, 1975; Safa, 2018b).

A good illustration of this state of research lies in the study-case commonly referred as the "Divje-Babe cave 1 flute" which has been largely studied (D'Errico et al., 1998; Kunej & Turk, 2000; Horusitzky, 2003; D'Errico, 2002; D'Errico & Lawson, 2006; Turk et al., 2006; Morley, 2006; Tuniz et al., 2011).

The most interesting fact here is not about archaeological methods, neither about their results, but rather about the researchers' approaches to sounds and shapes: whatever the published papers, some people always seem tempted to "see a flute in it" because it both looks and sounds like a flute.

Considering the fact that up until now few works have warned against the researcher-related cultural and technical biases (Buckley, 1994; Homo-Lechner, 2001, 2002), learning craftsmanship proved to be a fruitful way for enhancing the understanding of those special objects through embodied and empirical experience (Safa, 2015, 2016, 2018a) as well as for raising new epistemological issues related to the object's ontology and phenomenology as well as to the observer's understanding of it.

A Cognitive Approach to Object Surveys

The ways we look at objects never are and never will be neutral. They are built upon, maintained by and developed through the diversity of our intellectual and practical experiences as well as of our memory. Both of those are deeply altered by the social and cultural context we grow and live in. To be more specific, some cognitive instances such as typology typically are first and foremost social-learnings which will be subsequently extended as part of technical/specialized trainings – such is the case for archaeologists and craftsmen.

From this point, we raise the following questions: up to which point our modern, partial and socially-influenced mind may contribute in the understanding of the cultural complexity of (sound) artifacts from the past? To which extent an experimenter's ears and eyes may deliver biased, altered sounds and drawings, thus mirroring his thoughts rather than the object as it actually was?

This paper will attempt to rise epistemological issues regarding those influences in order to question the methodological reach of archaeologists in their understanding of ancient, exotic, strange archaeological bone flutes. In the end, this paper aims to offer the beginning of an essay for an epistemological paradigm of archaeological bone flutes' sound-studies which would synthesize the phenomenological situation of this kind of research, thus raising questions about the range and consistency of our actions as either classical or experimental researchers.

A Biased and Partial Gaze; A Typological Story of Visual Learning

The very first lessons of typology take place during early childhood: firstly by living in a home which has its own organization system and, secondly, by being taught how objects of specific shapes/types/functions are being stored and used in this context. Let us consider a child growing up and learning how to store and how to use objects of his everyday-life. Basically, this child will learn to understand how things work together: the meanings, functions and use-cases of each and every object he encounters. He'll learn to put them into small conceptual boxes that will allow grouping, differentiating and linking assessments. He will also learn to create semantic or functional connections between different objects from the same box or between different boxes. All of those knowledge – and know-how – acquisitions are related to the definition of typology itself:

From CNRTL french dictionnary, typology is defined as a "science of analysis and description of typical shapes from a complex reality that enable classification systems" (Science de l'analyse et de la description des formes typiques d'une réalité complexe, permettant la classification) (CNRTL http://www.cnrtl.fr/, 2012).

In the end, this child will grow up and have to choose a specialized activity. In a western context, this event is mostly related to professionalization, either by short or long studies, or by professional trainings. Any specialized activity generates specialized cognitive tasks including shapes recognition and functional understanding:

Archaeologists will be able to differentiate technical or social variations and preferences in a group of a priori identical objects by spotting small differences in their shapes, technical choices and/or tool marks,

Flute-makers will be able to determine the quality of a raw material from its look or its touch, as well as to identify technical and musical choices from geometrical and morphological characteristics in order to estimate the sound a flute should produce.

How and what we learn during a specialized training depends on our own background. Thus, if a specialized training gives us a common skill-set amongst peers, each and every one of us has its own understanding and its own manners when he uses it. To be short and explicit: there is no specialized context in which everyone agrees in either how to do things, or what this or that means.

Primary and Secondary Epistemic Knowledges

Assembled memorized information that contributes to the understanding of an object is called Epistemic knowledge (Jacob & Jeannerod, 2003, 148–162 cited in Levitte, 2010, 31–33). It is split in two levels of understanding:

Primary (or direct) epistemic knowledge assembles every memorized information (shapes, thickness, volumetric configuration, raw material's physical properties, etc.) which enable the understanding of the object itself (its shapes, its function, its use).

Secondary (or indirect) epistemic knowledge assembles correlated memorized information which allow qualitative deductions regarding indirect manifestations, influenced environment and extended deductions.

In other words, epistemic knowledge is about knowing what is seen and deducting what is not. In this theoretical scope, the volumetric knowledge and understanding of an object (trypology, metrology and geometry) are stored in the primary epistemic knowledge, while the meanings of the latter are stored in the indirect epistemic knowledge (fig. 1).

A Theory of Human Relationship to Objects

Any considered object comes with an associated culture. Before being brought to the world by hand and mind, it needs to be designed, then crafted. There is a need before any object – even a shallow one – that makes the whole scene of its creation happen. And, in the end, both archaeological and handcrafting researches aim to understand the culture (and thus men) underlying the objects of their interest. According to J. J. Gibson's theory of perception, knowledge is what reduces semantic and empiric distances between a person and the complex reality (Braund, 2008; Levitte, 2010, 52), meaning its role is to enhance the connection between what we think/know and what actually is/was.

Differential Relationship to Typology and Objects

An archaeologist makes a living of specialized, analytical, scientific typology. He thus have to train and extend his basic typological know-how in order to turn it into a scientific, systematized expertise tool for his research. His typological activity may address either shapes (geometries), aesthetics or even tool marks if they remain. He knows how to qualify typology, metrology and geometry both for classification and documentation as well as to create connections with contextual information coming from historical data. The result of his work is often illustrated by technical drawings, pictures and images.

In this scope, the archaeologist's role is to understand the evolution and the social complexity of objects by creating connections between shapes and contextual information (history, geology, etc.).

A Flute-maker makes a living of manufacturing his products and thus needs a particularly deep, profound understanding of them. Hence he needs a more empirical, personal, embodied approach in order to increase his senses and enable more sensory inputs to manufacture, verify and finish his making: visual inputs of course, but also haptic, auditory, olfactory and eventually gustatory ones. All of those will contribute in his understanding of his making process as well as of the interactions between gestures, shapes and sounds.

In this scope, the flute-maker's role is to understand and influence those interactions through meticulous hand-work, which means creating meaningful connections between shapes and their meaning in sound-producing.

Phenomenology and Ontology of Humans and Objects

The diagram (fig. 1) is an attempt to bind different aspects of objects' ontology as well as to consider both archaeological and handcrafting's range of action in order to consider the human–object relationship.

In this depiction, the object is defined by four main aspects of its existence:

Metrics is about observational "raw" data coming from both the object itself (as a combination of geometrical spaces) and from the observer (how he understands and processes data). By definition, parts of an object's metrics remain out of perception's reach and thus out of cognition's scope (any measurement is only an estimation of the object's geometrical reality).

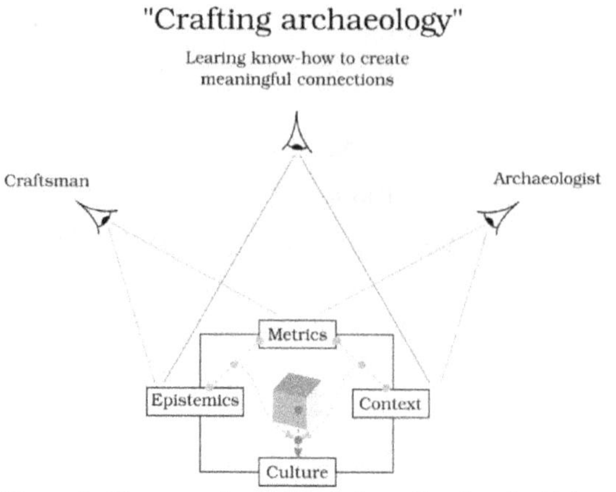

Figure 1: Phenomenological depiction of archaeologists' and flute-makers' scopes of action

Context is about any kind of informational data coming from either historical, archaeological, ethnographical or geological researches (amongst others) which help to complete and enlarge the scope of our understanding.

Epistemics is about meanings of each part of an object, both independently and in their relationship to the object as a whole.

Culture is about what was actually there, existing and underlying the design of an object while the creative process was ending, at a specific time and place. By definition, it is already gone as soon as the time goes on, meaning that one may only tend toward a better understanding of it without ever being able to reach it. In this phenomenological and philosophical scope, both archaeologists and flute-makers have a different but complementary role to play. Both of them strive for the same goal: understanding the bond connecting an object with its associated culture.

Anatomy of a Partial Gaze

Partiality and subjectivity originate from a complex set of causes – as mentioned above: society, culture, psychology, life experiences, etc. However, quite a substantial part of it originates from anatomical conditions and may be approached through a more practical perspective.

The Precise Vision Area

Anyone passing any object to different people and asking them to observe it precisely in order to describe it afterward would be able to observe three interesting events:

The very first spot targeted by the eyes varies from one individual to another,

Quick saccadic eye movements happen, of which the followed path varies from one individual to another,

The level of noticed details varies from on individual to another.

Those movements are called "saccades" and are inherent to any visual analysis task. Even if we tried, we could not prevent them for long, which is due to a simple fact: the range of precise vision in our visual field is particularly narrow (about 15°). This situation originates from an anatomical condition of the eye: the retina, which contains numerous photo-receptor cells, possesses a specific spot where their density is highly increased and which is called "fovea centralis".

Sophisticated Saccades and Visual Paths

Visual paths are an instinctive but quite complex phenomenon which involves several factors such as anatomical condition, attention, intention and memory in the effort of incrementing visual data in order to lead toward understanding, then action (Jacob & Jeannerod, 2003). As we're blind while the eye is moving, the next target is selected by anticipation – a process which is also influenced by artificially and naturally salient objects in the view field, as well as routines.

Memory's influence: stored knowledge (expertise) will help the observer to create convenient visual paths, processes and patterns that will be selected because of their efficiency (Yantis, 2005).

Routines: those patterns may turn into routines which are here for later energy savings. They represent both a risk to eliminate potential discoveries and an efficient tool for adaptive cognition (Quéré, 1997, 172–176).

Objectives of the perceptual activity: whether the intention aims to a cognitive or behavioral objective influences the way we perceive and process visual informations (Serences & Yantis, 2006).

Salience of objects: the salience of an object characterizes its relative remarkableness compared to other objects from the same scene. Objects may either have a "natural" salience (flashy colour in the middle of gray objects) or an artificial, conceptual one (personal interest and knowledge may create an overlay of salience to an observer's point of view).

Attention: the observer's ability to focus on the perceptive task is deeply influenced by both the situation itself and the previously mentioned factors. One may "ignore irrelevant details regarding his needs" (Pacherie, 2003, 269)

Eyes-movements' control is thus an intelligent process as it needs to simultaneously call on visual perception, on every kinds of memories (sensory, short-term, semantic, episodic, long-term) and on several cognitive systems such as curiosity, imagination and deduction (Henderson (2003, 500), cited in Levitte (2010, 45)). Hence it is deeply influenced by life experiences, trainings and personal expertise.

Neurology of Perceptions
Different Kinds of Perceptions

Perceptions are complex mechanisms meant to give us information about the world we live in. Our brain is an incredible processing machine which allows us not only to perceive pieces of information, but also to assess them, to judge them and to get inspired by them.

There are several kinds of perceptions, all of which being involved in visual analysis of scenes and objects. While semantic and visuo-motor perceptions respectively allow identification and functional understanding of an object, egocentric and allocentric perceptions are involved in the cognitive process of visual data (Levitte, 2010, 67–69). Egocentric perception is the most intuitive one as its reference point is the observer itself: it allows direct actions and direct distance/measurement estimations. On the other hand, allocentric perception is about understanding a space (or a geometry) from an external point of view, using another object as the reference point. Such is the case when the observer tries to understand his position in a space through map-reading or to understand how the geometries of an object fit together by mentally projecting himself into the latter.

To a certain extent, analyzing an object is pretty much the same as analyzing a space, as both of them are defined by geometrical shapes which are mentally conceptualized.

Visual Percepts, Representations and Consciousness

Perception should not be narrowed down to visual inputs. As stated previously, every senses contribute to the understanding of a scene, especially kinesthetic inputs (motor, haptic and proprioceptive ones) which stimulate the attention of the observer as he is more involved in the perceptual task (Wexler & van Boxtel, 2005), thus taking advantage of increased attention and consciousness degree.

This situation may be illustrated in the everyday-life, when people instinctively try to grasp an object or to get closer to it (for example in a museum, in which fences and glasses may sometimes create audible frustration). It also directly refers to the cerebral mechanisms related to visual percepts and representations.

On one hand, visual percepts could be defined as a mental database of standard objects which are only characterized by raw data (color, lines, shapes, global aspect) in order to allow extremely quick (almost instantaneous) visual identification (Pacherie & Proust, 2004).

On the other hand, visual representations could be defined as a visual catalog picturing different realities of objects grouped into the same visual percept, which increases through time and experiences.

The cognitive process toward an eventual identification of the object depends both on the existence of an associated visual percept in the observer's experience of life, and on the degree of consciousness of the perception process regarding this stimulus (fig. 2).

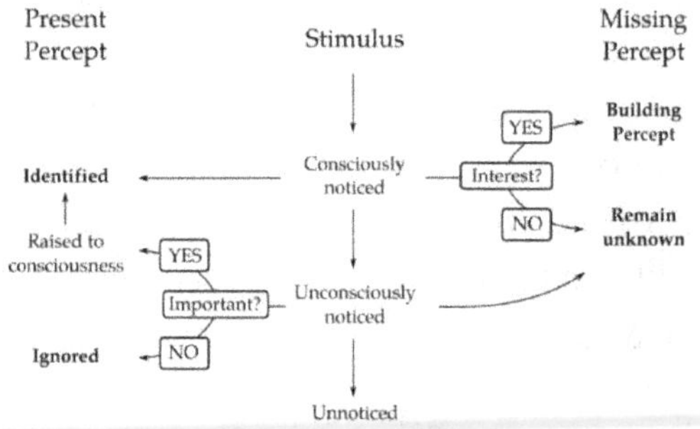

Figure 2: Stimulus processing diagram

Percepts and representations are therefore both precious tools and abrupt biases in the sense that they may either serve or mislead the observer. For example, possessing a totally different percept and/or representation than those corresponding to the object in the present scene often leads to a kind of "virtual blindness", where the observer is unable to spot the object (Levitte, 2010).

As a consequence, the way a specialist expertize an object is thoroughly connected to the way he experienced it beforehand.

Differences in the Specialists' Cognitive Scopes

An archaeologist's experience tends to lead to an analytical mind in which measurement tools occupy a prominent place (calipers, compasses). Archaeologists' objectives are mostly of cognitive nature: every effort aims to contribute to the understanding of the studied object. However, the visual process will most probably change according to the specific task he is conducting (for example: systematic measurement for 3D reconstruction versus targeted measurements for typological classification). Those two cases are a perfect illustration of visual paths variations simultaneously induced by the archaeologist's intention, interest and expertise. They furthermore evoke the salience variation relatively to the same factors.

A flute-maker's experience, in the other hand, tends to lead to an empirical mind for which memory and routines (and thus expertise) will this time induce multi-sensory information recalls which are connected to both conscious and unconscious manufacturing experiences. Measurement mostly aims here to verify a manufactured geometry and rework it if necessary. This specific situation generates both a gestural knowledge of measures and an embodied experience of the relationship between shapes and sounds. It also entails a differential salience of a flute's shapes, which are overlaid by new informational dimensions coming from sound-working experience.

A craftsman's understanding of a flute's geometry thus fundamentally differs from the archaeologist's point of view as every geometrical component of a flute echoes both with sound specificities and embodied measurements. If studied in dedicated experiments involving eye-tracking devices, this difference of approaches and experiences would probably result in two main kinds of visual patterns.

Epistemology of Experimental Sound-Archaeology

In the previous pages we covered the study of cognitive biases to visual analysis of archaeological bone flutes, hoping to have successfully illustrated how influenced one may be in this task. However, these archaeological remains also possess a sound dimension which entails its own specific issues and potential biases. Related to cultural, social and aesthetic considerations (amongst others), those have become key issues for questioning experimental research when it leads to archaeological reconstruction, acoustical estimations or musical performance (Safa, 2016; Safa et al., 2016; Safa, 2018b).

As stated previously, handcrafting proved to be an excellent analytical tool for reconsidering academic assessments on handcrafted objects from the past, especially as in our case those are sound-instruments, which raises the question of the ear's subjectivity. The following pages are dedicated to this issue.

Bone Flute's Acoustics and Handcrafting

Handcrafting bone flutes for one year alongside a traditional flute-maker and thereafter for experimental researches contributed to reveal different acoustical specificities to these instruments (idem). To keep it synthesized:

The inner cavity of a bone (its volume, shapes, surface and volumetric configuration) defines the sound basis of the flute, thus giving to each bone family its own sound potential.

The morphological variability of bones has a significant influence on the inner cavity of each individual, thus making each bone different from others.

An adaptive sound-geometry (carved shapes) is the only way for the flute-maker to craft a functional bone flute as each bone comes with its own acoustical characteristics, making it impossible to copy the geometry to another bone without avoiding a set of technical issues due to non-respect of the bone's individuality.

Meaningful details are much tiny, most of flute-makers considering that cutting even 0.1 mm may generate, change or kill the sound of a flute.

Experimental Approach to Bone Flutes' Sounds

Experiments conducted since 2015 delivered several interesting results while raising much more questions. They were designed to progressively enhance the systematization of their own process while revealing in the meantime as much biases and sources of errors as possible. We decided to use and combine technological processes, handcrafting know-how, sound studies and statistical analysis in the process. Their main objective was to explore bone flutes' acoustics and sound behavior in order to highlight eluding biases in current and past methods used for sound-studies of archaeological flutes.

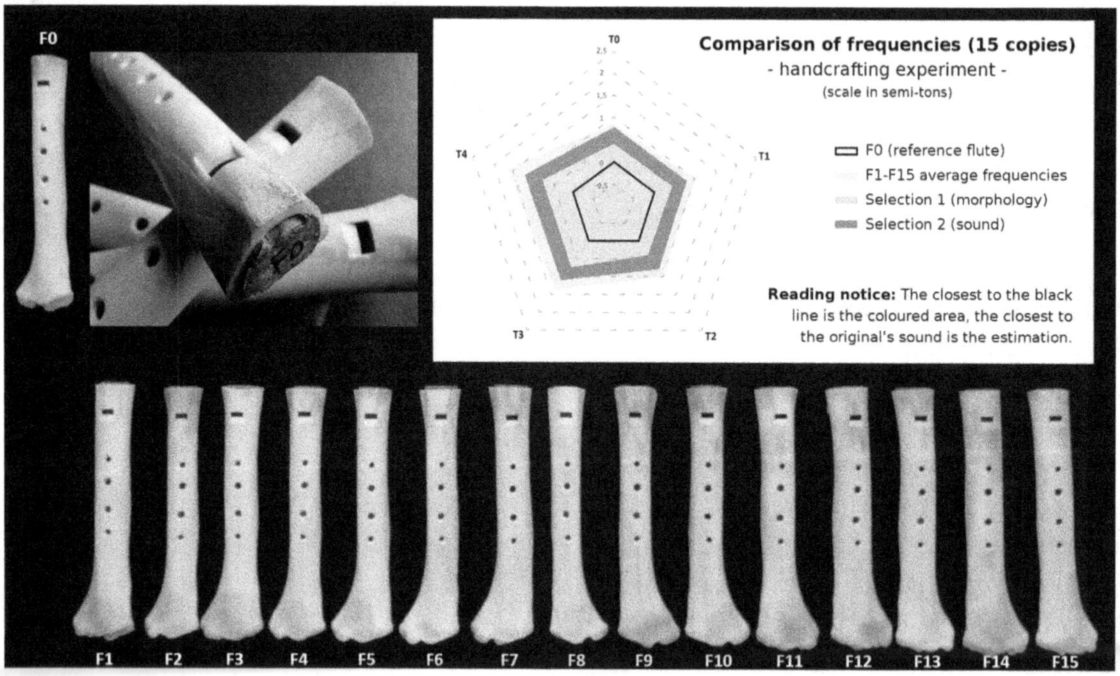

Figure 3: Frequency comparison and illustration of the sound influence of the morphological variability – from (Safa, 2018b).

During this process, we thus had the opportunity to confront 3D printing to handcrafting. Both results' analysis and interpretation's scopes evolved with more time and experiences, from first clues arguing for equivalence (Safa, 2016) to broader and more critical analysis arguing against. In the end, thanks to statistical analysis, handcrafted replicas revealed themselves as untrustworthy due to the weight of morphological variability (fig. 3) – an ineluctable, inherent factor of bone flutes' sound definition – while 3D copies seem to offer better results if printed in high definition resin printers (Safa, 2018b). However, while this conclusion seems convincing, a lot of unknown factors remain to be investigated.Considering that when played in the past those flutes were played by humans, natural blowing technique was preferred to the wind-tunnel artificial blowing device commonly used by acousticians, thus inputing its variability into the statistical analysis. Though this choice reduced the acoustical reach of our studies (because of its non-repeatability), it was consistent enough as long as it was consciously considered and treated as a variable. Moreover, it enabled the identification of two epistemological milestones directly connected to natural blowing techniques:

Sound-flexibility: correlated by craftsmen's and musicians' experiences (and experimented during the apprenticeship) is the fact that bone flutes possess a particularly important range of emitted frequencies for each finger-hole, depending on the blowing strength and pressure, and thus on the blowing technique (Safa, 2018a).

Sound-subjectivity: when averaging the notes played by the experimenter on a 2 minutes-long track, those differ from the original tuning of the instrument (fig. 4), thus revealing the influence of the musician on what we call "frequencies" and "notes", and consequently questioning the intrumental tuning as an absolute reference (Safa, 2018b).

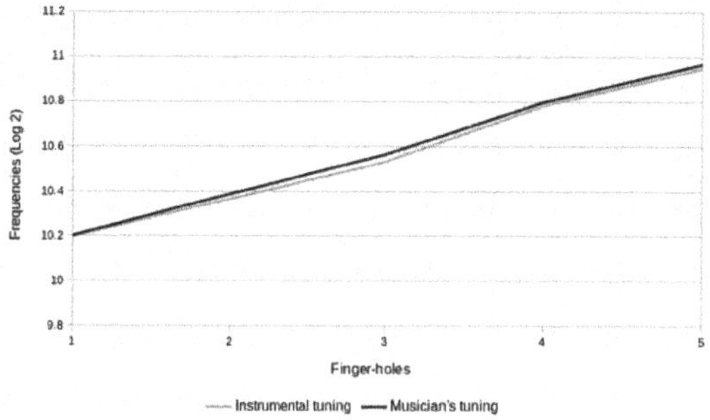

Figure 4: Depiction of the frequency differences (in logarithmic scale) between a bone flute's instrumental tuning (light gray) and its technical tuning (average frequencies played by the musician, dark grey) – from (Safa, 2018b).

A lot more remains to be done in the investigations of both 3D and handcrafted copies, as well as regarding the frequencies emitted by the musician. However, based on those considerations, the conclusion of these experiments is the following: trying to discover the exact frequencies and pitches of ancient bone flutes will most probably lead to either misunderstandings or lacks of interpretations, while orienting our efforts toward the understanding of their acoustical and musical potential as sound-tools seems much more consistent.

Essay for a Sound-Archaeology Paradigm

In the phenomenological scene of an observer and an observed object, we have to consider both entities as well as the four cases of perceptual results: consciously/unconsciously processed, ignored and unnoticed (fig. 2). The following diagram (fig. 5) and paragraphs are an attempt to collect and to bind together different phenomenological aspects of such a scene in which the observer would be either an archaeologist or a flute-maker, and the object an archaeological bone flute. In this scope, every constitutive element is an object inside an object (the flute) and may itself be composed by multiple sub-elements. All of them have to be considered as different objects composing the scene (or space) which will here be named "the flute".

As time passes, the flute's geometry is progressively eroded by its journey through the soils, while both its technical and cultural meanings are progressively forgotten. The "definition" of the object lowers even more due to human actions, as excavating and study processes may both damage the archaeological remain (broken during excavations, marked by excavating or measurement tools, over-manipulated, etc.). Even restoration processes will alter its integrity, as any restoration is first and foremost an interpretation, e. g. an estimation of the object's original state. In the case of archaeological bone flutes for which geometry and morphology are both meticulously intertwined into the sound-producing process (Safa, 2018a), the slightest lag or error in the reconstruction choices and operations will definitely influence the sound of eventual copies if based on these altered shapes.

Even the minimal lag induced by the glue's additive thickness may change the sound characteristics of the flute if operated in sensitive sound location such as the mouthpiece.

We won't give here too much attention to the observer's sight condition which obviously alters the quality of his perceptive experience. However, the previously highlighted cognitive factors and phenomena undoubtedly have a role to play in this phenomenological scene.

First, visual paths are one (if not the most) subjective element of this perceptual analysis: they result both from our conscious and subconscious abilities, interests and curiosity, both in general and regarding the considered object. They furthermore rely on memory: the reiterative practice will ultimately lead to the selection of preferred processes and patterns. They are influenced by the observer's intention, but also by the task itself as well as by its nature and its multiplicity (the more tasks simultaneously handled, the more chances there is to induce attentional blindness to smaller details). Routines will induce more straightforward processes, patterns and paths. Fatigue finally will alter the general state of the observer by lowering his attentional abilities.

But those visual paths will also depend on the observer's experience of the object: whether he's used to keep his distances in a strictly observational relationship (usual archaeologist) or to ground his feelings into multi-sensory perceptual and deduction processes (usual craftsman) will most certainly influence the construction as well as the use of specific visual patterns.

A strictly analytical experience of the object allows the archaeologist to keep a constant, consistent and critical distance between him and the flute. It ensures the will of diminishing his subjectivity in the collection and connection of informations. It is meant to bypass his own beliefs in the search of facts that would illustrate the diversity of the object's reality in its historical, archaeological, sociological and ethnographical dimensions (amongst others).

An embodied experience on the other hand allows the flute-maker to connect gesture to measurements, thus creating both episodic and semantic memories that will ensure the link between the flute's sound-characteristics and their influence on the final sound. From this point, one may understand that in

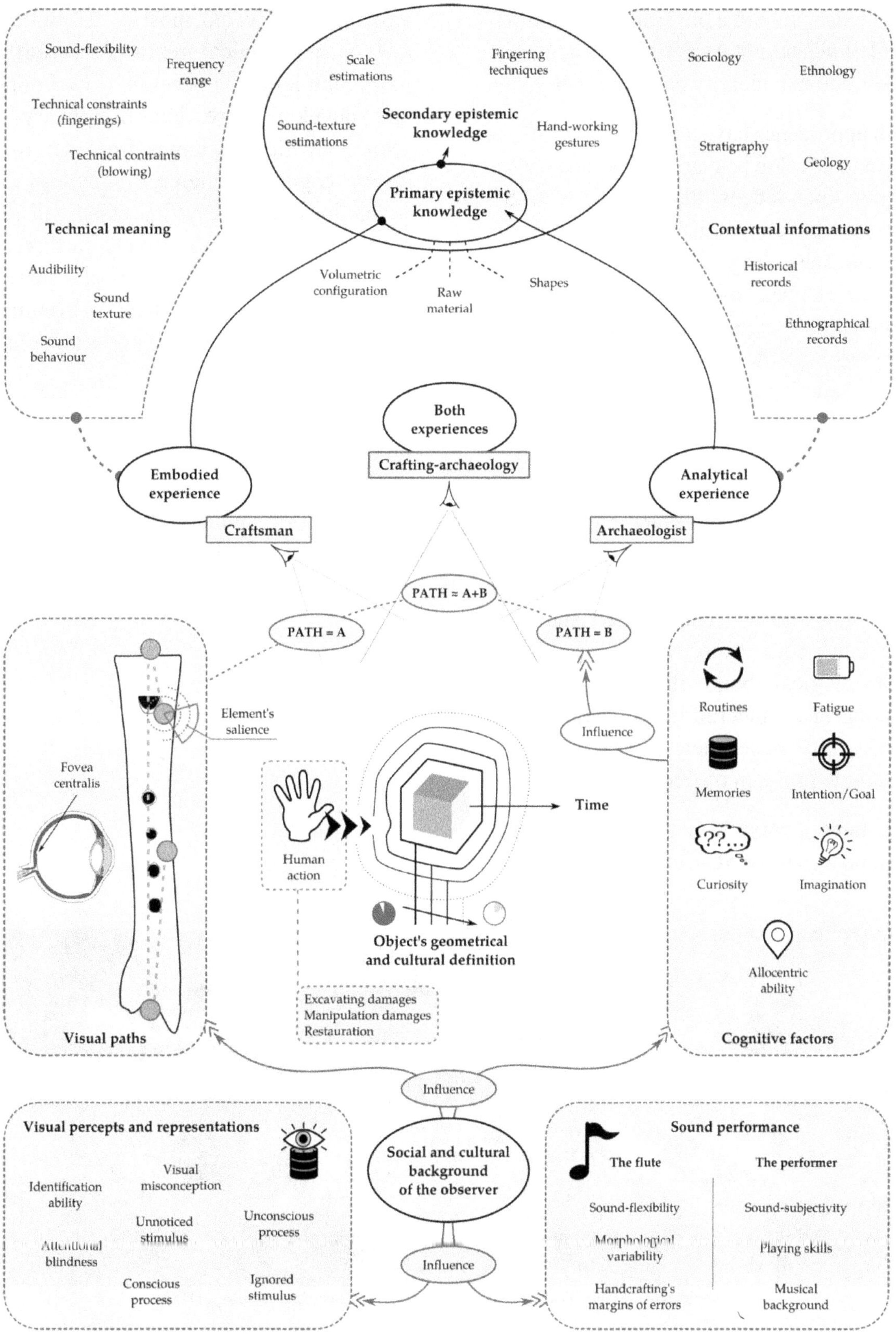

Figure 5: First glimpses of an epistemological paradigm for archaeological bone flutes' sound-studies.

the mind of a flute-maker, a volumetric element exists in both a physical and an immaterial dimension: it casts whispers about the flute's sounds, identity and personality.

Both approaches have a different way to access and develop primary and secondary epistemic knowledges, thus leading to a differential development of the observer's consciousness regarding the complexity and the richness of the considered instrument. In this scope, "crafting-archaeology" offers to bind those two experiments in order to benefit from both approaches (Safa, 2018a).

Conclusions

This work, as it is, is nothing but a starting point for more intense and immersive researches: a lot remain to be done. If we're starting now to get a better understanding of the main issues related to sound-studies of archaeological bone flutes through 3D printing and handcrafting processes, those results have raised new questions beyond the single problem of frequencies.

The new inputs from cognitive sciences have proven to be of an extreme importance regarding the epistemological issues of this work, as they would most certainly be for any epistemological research. It furthermore highlights the necessity to encourage and develop more interdisciplinary researches with new inputs from neuro-sciences, cognitive sciences, sociology and psychology. This very first essay will most certainly evolve while diving further into the cognitive approach to object's surveys. Chances are this first sketch will eventually reveal to be incomplete, if not erroneous.

Hopefully, it will one day fully include a more important variety of experiences, factors and biases that would help in the creation of a comprehensive epistemological landscape for sound-archaeology and archaeological bone flute's studies.

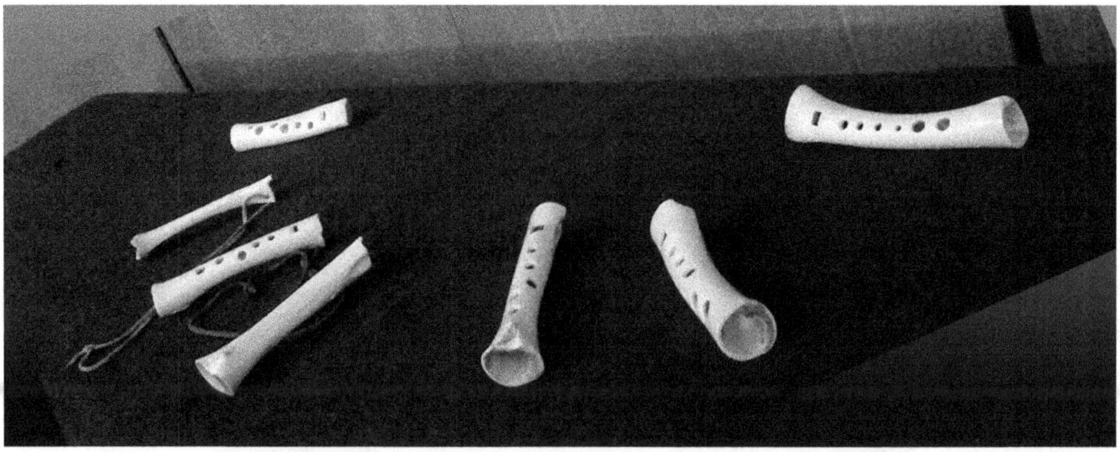

Samples provided by the author at the Archaeoacoustics III conference. Image OTSF

REFERENCES

Brade, Christine. 1975. Die mittelalterlichen Kernspaltflöten Mittel-und Nordeuropas, Ein Beitrage zur überlieferung prähistorischer und zur Typologie mittelalterlicher Kernspaltflöten. Vol. Band. 14. Neumünster: Wachholtz.

Braund, Michael James. 2008. The Structures of Perception: An Ecological Perspective. Kritike, 2, 123–144.

Buckley, Ann. 1994. L'histoire de la musique médiévale : la musicologie, ethnomusicologie, ou archéologie de la musique ? Quelques questions de notre époque. Pages 503–512 of: Homo-Lechner, Catherine, & Bélis, Annie (eds), La pluridisciplinarité en archéologie musicale. IVe rencontres internationales d'archéologie musicale de l'ICTM, Saint-Germain-en-Laye, 8-12 octobre 1990. Saint-Germain-en-Laye: Maison des Sciences de l'Homme.

CNRTL http://www.cnrtl.fr/. 2012. Centre National de Ressources Textuelles et Lexicales.

D'Errico, Francesco. 2002. Just a bone or a flute ? The contribution of taphonomy and microscopy to the identification of prehistoric pseudo-musical instruments. Pages 89–90 of: Hickmann, Ellen, Kilmer, Ann D & Eichmann, Ricardo (eds), Studien zur Musikarchäologie. Archäologie früher Klangerzeugung und Tonordnung, vol. 3. Rahden: Verlag M. Leidorf.

D'Errico, Francesco, & Lawson, G. 2006. The Sound Paradox. How to assess the acoustic significance of archaeological evidence? In: Scarre, Chris, & Lawson, Graeme (eds), Archaeoacoustics. Cambridge: Mc Donald Institute Monographs.

D'Errico, Francesco, Villa, Paola, Pinto Llona, Ana C., & Ruiz Idarraga, Rosa. 1998. A Middle Palaeolithic origin of music? Using cave-bear bone accumulations to assess the Divje Babe I bone "flute". Antiquity, 72, 65–79.

Henderson, J. M. 2003. Human gaze control during real-world scene perception. Trends in cognitive sciences, 7, 498–504.

Homo-Lechner, Catherine. 2001. L'interprétation des découvertes archéologiques. Pages 18–21 of: Laloue, Christine (ed), Archéologie et musique : actes du colloque des 9 et 10 février 2001. Paris: Cité de la Musique.

Homo-Lechner, Catherine. 2002. L'Archéologie musicale. Pages 17–22 of: Clodoré-Tissot, Tinaig, & Lecrerc, Anne-Sophie (eds), Préhistoire de la Musique : sons et instruments de musique des âges du Bronze et du fer en France. Nemours: Musée de préhistoire d'Ile-de-France.

Horusitzky, François Zoltán. 2003. Les flûtes paléolithiques: Divje babe I, Istállóskö, Lokve etc. Point de vue des experts et des contestataires - Critique de l'appréciation archéologique du spécimen. Arheološki vestnik, 54, 45–66.

Jacob, Pierre, & Jeannerod, Marc. 2003. Ways of Seeing: The Scope and Limits of Visual Cognition. Oxford University Press.

Kunej, Drago, & Turk, Ivan. 2000. New perspectives on the beginnings of music: Archaeological and musicological analysis of a middle Paleolithic bone "flute". Pages 235–268 of: Wallin, N L., Merker, B., & Brown, S. (eds), The Origins of Music.

Levitte, Agnès. 2010 (oct). La perception des objets quotidiens dans l'espace urbain. Ph.D. thesis, Ecole des Hautes Etudes en Sciences Sociales (EHESS).

Morley, Iain. 2006. Mousterian musicianship? The case of the Divje babe I bone. Oxford Journal of Archaeology, 25(4), 317–333.

Pacherie, Elisabeth. 2003. Modes de structuration des contenus perceptifs visuels. Philosophies de la perception. Phénoménologie, grammaire et sciences cognitives, 263–289.

Pacherie, Elisabeth, & Proust, Joëlle (eds). 2004. La philosophie cognitive. Paris: Ophrys, Maison des Sciences de l'Homme.

Quéré, Louis. 1997. La situation toujours négligée ? Réseaux. Communication - Technologie - Société, 15(85), 163–192.

Safa, Etienne. 2015. L'Archéologie du son : transmettre les connaissances et les savoir-faire de la France paléolithique à la France médiévale. Pages 23–46 of: Moretton, Carine (ed), Polymatheia - Transmission et musique, vol. 1. Vanves: Tourmaline.

Safa, Etienne. 2016. Handcrafting in archaeomusicological research: record of a one-year apprenticeship alongside a traditional-flute-maker and its application to sound archaeology. In: Emerit, Sibylle, Perrot, Sylvain, & Vincent, Alexandre (eds), Sound Making : Handcraft of Musical Instruments in Antiquity. Paris: IRCAM.

Safa, Etienne. 2018a. Artisans-chercheurs – Récit d'une expérience d'apprentissage en lutherie des flûtes populaires et de ses implications dans la recherche archéomusicologique. In: Clément-Dumas, Gisèle, & Meegens, Rachel (eds), La flûte, de la Préhistoire au Moyen Âge – In Press. Montpellier: PULM.

Safa, Etienne. 2018b. Commentaires sur l'utilisation des copies de flûtes archéologiques en os – Enquête méthodologique et épistémologique. In: Clément-Dumas, Gisèle, & Meegens, Rachel (eds), La flûte, de la Préhistoire au Moyen Âge – In Press. Montpellier: PULM.

Safa, Etienne, Barreau, Jean-baptiste, Gaugne, Ronan, Duchemin, Wandrille, Talma, Jean-daniel, & Arnaldi, Bruno. 2016. Digital and Hand-crafting Processes Applied to Sound-Studies of Archaeological Bone Flutes. Pages 184–195 of: Ioannides, Marinos, Fink, Eleanor, Moropoulou, Antonia, Hagedorn-Saupe, Monika, Liestøl, Gunnar, Fresa, Antonella, Grussenmeyer, Pierre, & Rajcic, Vlatka (eds), Digital Heritage - Progress in Cultural Heritage: Documentation, Preservation, and Protection, vol. 2. Nicosia: Springer.

Serences, John T., & Yantis, Steven. 2006. Selective visual attention and perceptual coherence. Trends in Cognitive Sciences, 10, 38–45.

Tuniz, C., Bernardini, F., Turk, I., Dimkaroski, L., Mancini, L., & Dreossi, D. 2011. Did Neanderthals Play Music? X-Ray Computed Micro-Tomography of the Divje Babe "Flute". Archaeometry, 54, 581–590.

Turk, Ivan, Blackwell, B. A B, Turk, Janez, & Pflaum, Miran. 2006. Résultats de l'analyse tomographique informatisée de la plus ancienne flûte découverte à Divje babé I (Slovénie) et sa position chronologique dans le contexte des changements paléoclimatiques et paléoenvironnementaux au cours du dernier glaciaire. Anthropologie, 110, 293–317.

Wexler, Mark, & van Boxtel, Jeroen J.A. 2005. Depth perception by the active observer. Trends in cognitive sciences, 9, 431–438.

Yantis, Steven. 2005. How visual salience wins the battle for awareness. Nature Neuroscience, 8, 975–977.

Of Stars and Bells: the Sound of Light

Christiaan Sterken

CHRISTIAAN STERKEN is Emeritus Research Director and Guest Professor at the University of Brussels. His principal field of research is observational astronomy, with focus on pulsating stars and on comets. His research and teaching also include specialized subjects in the history of nineteenth-century astronomy.

ABSTRACT: This paper is about the oldest orchestra that we know of: our universe. Many stars appear to behave like bells or like wind instruments or drums, and reveal multiply-periodic fluctuations in brightness and in local velocities. These oscillations can be observed and understood by measuring and by analyzing the light that we receive on Earth. Some similarities between astronomy and archeoacoustics are outlined, and the relevance of asteroseismology for contemporary astro-music is documented.

KEYWORDS: asteroseismology, constellations, Music of the Spheres, stars, comets

The Stellar Universe

Stars are huge spheres of glowing gas that produce energy in their cores through nuclear fusion via the conversion of hydrogen into helium.

Stellar distances are gigantic, and starlight takes years to hundreds and thousands of centuries to reach us.[1] As such, the nightscape that we contemplate is a projection of age-old light (actually, photons), and the "archaeo" perspective that we see is thus entirely illusive.

Likewise, our exalting constellations are equally delusory: some stars move at velocities of several hundred kilometres per second, and this effect causes visible changes in the constellation patterns when considering archaeological time scales. Not to speak of the rotation of the Earth's axis that revolves in 26,000 years as if it were a spinning top.

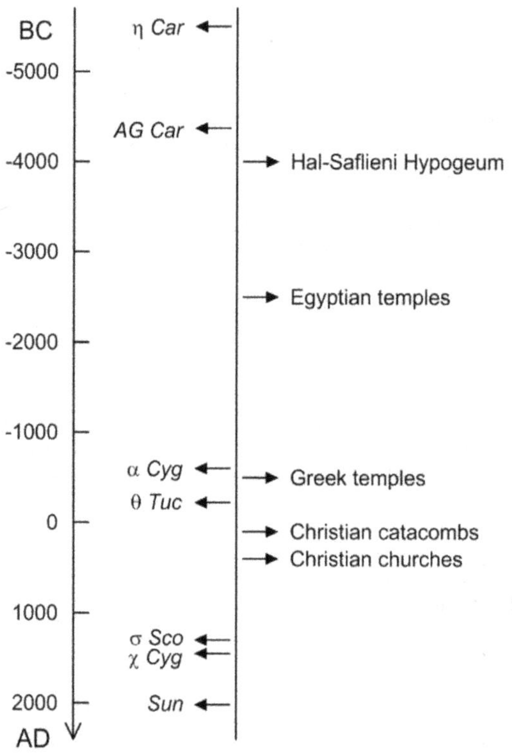

Figure 1: Comparative timeline: archeological and architectural key dates versus the time that light from a specific star needs to travel to reach us.

[1] Our nearest star is the Sun, at a distance of 150,000,000 km, or only 8 light-minutes.

The effect of this precession is that the twelve zodiac signs of western astrology have shifted by nearly a complete sign in 2000 years, so that at the moment when the Sun is now in Sagittarius, it was in Capricorn at the beginning of the current era. Figure 1 shows a comparative timeline that serves archaeology and astronomy.

Stars are born, live and die, and their life cycles are of the order of millions to billions of years. During their lifespan, stars seem in quasi-equilibrium, but some events and processes may cause sudden changes that can even lead to explosive cataclysms. Some of these variations – as astronomers call them – have been observed over several centuries as cyclic changes in the light output, as is illustrated in Figure 2.

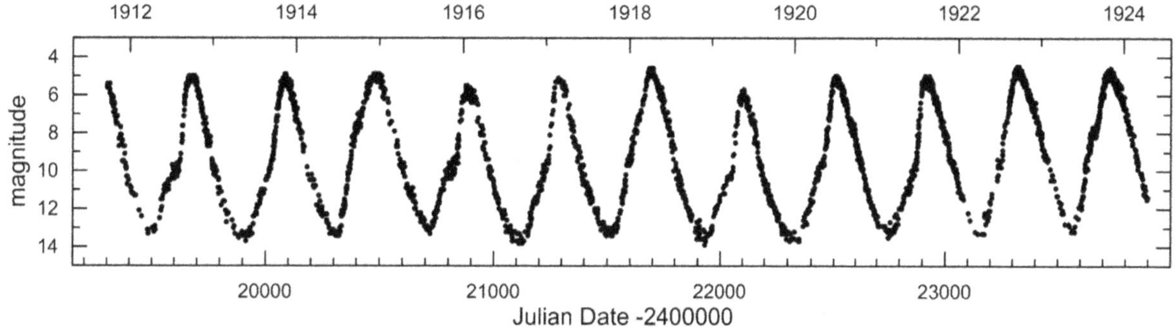

Figure 2: Cycles of brightness (i.e., magnitude) changes of χ Cygni 1912–1924. The lower time axis is in Julian Day Numbers. Note that the magnitude axis is reversed. Based on data from Broens, E., Sterken, C., McAdam, D. & Watanabe, M. 1998, The Journal of Astronomical Data, 4, 8.

Astronomers measure stellar brightness in "magnitudes". Originally, magnitudes were not more than estimates of a visual sensation, without quantitative definition. The brightest stars in the night sky were considered to be of first magnitude, whereas the faintest were of magnitude six – the limit of visual perception.[1] The response of the eye to illuminance is logarithmic, and in 1860 Georg Fechner[2] asserted that a logarithmic law also holds for sensations of pressure, noise and taste. The step of the logarithmic magnitude scale was experimentally determined to be about 2.5, but Norman Pogson[3] selected the value 2.512, the fifth root of 100. Thus a difference of one "mag" corresponds to a change in brightness by a factor of 2.512. The magnitude scale resembles the tonal scale that was created with equal distances in an octave, the so-called Equal Tempered Scale that uses the base 12 system and octave ratio 2, and in which the selected semitone ratio equals 1.0595, the twelfth root of 2.

The Vibrancy of Gaseous Spheres

Some properties of gaseous spheres are not unlike those of musical instruments: stars also possess natural frequencies, as well as overtones. Physical processes bring forth sound waves in their interiors, and that

[1] That is, for a dark-adapted eye.
[2] Fechner, G.T. 1860, *Elemente der Psychophysik.* Leipzig: Breitkopf & Härtel.
[3] Pogson, N. 1856, Monthly Notices of the Royal Astronomical Society 17, 12.

sound makes the star resonate[4] as if it were a violin, a kettledrum or an organ pipe.[5] Since the speed of sound depends on temperature and on the density of the ambient gas, golf fronts become curved, and the propagating sound waves reflect and interfere.[6] This leads to surface gravity waves (as we can see on the surface of a pond) and to interior acoustic waves (like we can hear when underwater). Just like in an organ pipe – i.e., a column open at one end and closed at the other – waves in a gaseous sphere reverberate and manifest themselves as Chladni-type patterns formed by the node lines along which there are no pressure changes.[7] An example of such spherical node lines is shown in Figure 3.

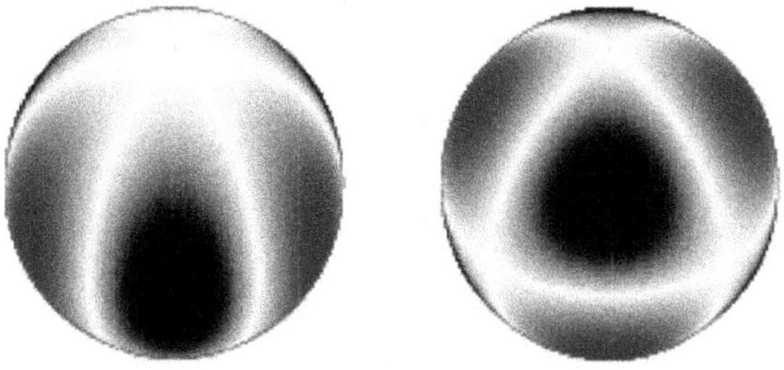

Figure 3: Node lines on a gaseous sphere. Alternating sections of the sphere are moving in opposite directions.[8]

Some stars possess a mechanism that can generate and maintain steady standing waves inside their interior, and that mechanism makes the star expand and contract, and renders the star cyclically brighter and fainter.[9]

The frequency of stellar pulsations depends on the size and density (thus mass) of the star, and the brightest stars have the longest vibration periods. The fundamental oscillation is quite often the strongest mode, just like it is for musical instruments. In astrophysisist's scientific jargon, overtones are often called "harmonics", a somewhat misleading terminology that comes from the fact that these terms have the same meaning for modes on a uniform string. The topic of harmonics and overtones in stars (and wind instruments) is more complicated than in the case of string instruments. In Cepheids[10], for example, the ratio of the first overtone to the fundamental period is 7/10, and is quite different from the ratios in organ pipes and violin strings.

Just like we have classes of musical instruments, we also have classes of variable stars. Expanding and contracting stars are called pulsating stars. The scientific discipline that studies the internal structure of such stars by the interpretation of the frequency spectra of these cyclic variations is called "asteroseismology", and uses astronomical measurements of brightness and velocity, together with stellar models. This endeavour includes the detection of frequencies of sound waves in stars.

The study of pulsating stars is not just a pastime for academics. Henrietta Leavitt (1868–1921) discovered the so-called period–luminosity relation for Cepheids, and in 1908 she came to the conclusion that "It

[4]A body that can oscillate has a natural frequency of oscillation. Resonance occurs when an oscillating system experiences a periodic force with a period near or equal to the natural period of that body.

[5]Archeoacousticians know of ringing stalactites and stalagmites issuing pure bell-, drum- or gong-like notes when struck.

[6]At the open end of a woodwind instrument, a sound wave encounters a pressure drop and reverberates back to the embouchure. In the same vein, stellar golf fronts reflect when reaching the surface layers.

[7] Kettledrums make for 2-D resonance forms: for a visual impression of drumhead oscillations, I refer to Dan Russell's dedicated webpage http://www.acs.psu.edu/drussell/Demos/MembraneCircle/Circle.html

[8] Graphics adapted from Kurtz, D.W., 2006, *Stellar Pulsation: an Overview*, ASP, Vol. 349, 101, Astrophysics of Variabie Stars, Eds C. Sterken & C. Aerts, with permission from the Astronomical Society of the Pacific.

[9] A driving process feeds mechanical energy just like a mechanical driving mechanism keeps a pendulum clock going. In a violin, the driving force is the scraping of the taut strings by the bow, and the vibrations of the string are conducted to the body that, in turn, transmits at natural frequencies while it damps the others.

[10] Cepheids are a class of pulsating stars.

is worthy of notice that … the brighter variables have the longer periods".[11]

This assertion makes a most direct link to the acoustics of bells. Bells and chimes, like stellar bodies, are resonators. The tone of a church bell is inversely proportional to the cubic root of its mass (thus halving the frequency requires increasing the mass by a factor of eight). Western bells have single tones, and the partials are not simple arithmetic ratios. Since the tone of a church bell does not depend on the place where the body is struck, a clapper is conveniently used to ring such a bell.

Figure 4: Zhong bell. The lenticular section measures 14 x 8.5 cm. Photo C. Sterken.

The Chinese zhong bells,[12] on the other hand, have an oblate form and can produce two separate pitches. The zhong bell has an oval cross section, and each side is cast with bands of nipples (Figure 4). The two sets of modes are triggered by striking either the sui position (front) or the gui (side). Nipples are placed so as to produce the desired tunes, and separate the sui and gui pitches.[13]

Figure 5 shows the relationship between size and frequency for the sui and gui tones: it is a fairly simple function, just like the relation between a trombone's tube length and its pitch, or between the period of a pendulum and its length.

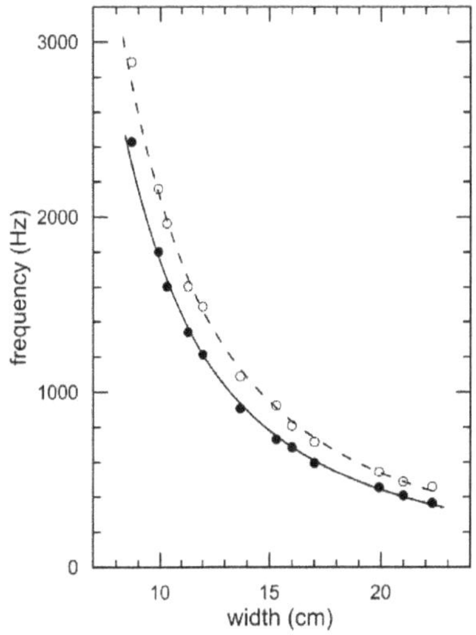

Figure 5: Relation between size (bottom tip-to-tip length) and frequency for the sui tones (front strike note, filled symbols) and the gui tones (side strike note, open symbols) for a set of 12 zhong bells. The curves are power law fits. Based on data from Takahashi, J. 1989, J. Acoustical Soc. Japan (E) 10, 5.

The Music of the Spheres

Star-music is commonly understood in terms of "Music of the Spheres", a metaphor that links cosmology and music, and that is credited to Pythagoras of Samos (c. 570–475 BC). The Pythagorean concept of celestial harmony puts forward that the putative solid crystalline spheres that carry the planets produce sounds during their orbital revolutions. For the stationary and geocentric Earth, these spheres were, in the order of distance, the orbs of the "planets" Moon, Venus, Mercury, Sun, Mars, Jupiter and Saturn. The outer orb, the sphere of the

[11] *Leavitt, H. S. 1908,. 1777 variables in the Magellanic Clouds.* Annals of Harvard College Observatory 60, 87.

[12] Zhong bells date from around 1500 BC.

[13] Shen S. 1987, Scientific American 256, 94.

fixed stars, was immobile. Planets were assigned notes in seven intervals that were linked to the relative – but constant – distances between the celestial bodies.

Lorenzo, in The Merchant of Venice, says: “There's not the smallest orb which thou behold’st but in his motion like an angel sings”[14]

Ordinary mortals were not supposed to be able to hear these sounds, except the Master himself, as well as those belonging to the unique populace described by Jonathan Swift: “…the people of their island had their ears adapted to hear “the music of the spheres…”[15]

Koestler[16] points out that this Pythagorean harmony is not the pleasing effect of simultaneously-sounded concordant strings, but the concept of armonia, or the attunement of the strings to the intervals in the scale, and the pattern of the scale itself.

Johannes Kepler (1571–1630) describes his own view of the harmony of the heavens in Book V of Harmonices Mundi under a title that begins with “De harmonia perfectissima motuum coelestium …”.[17] The radical departure from the Pythagorean celestial “music” follows from his explicit acceptance of the heliocentric universe,[18] wherein the planets revolve around the Sun in elliptical orbits with variable speeds.

Kepler considered that the diurnal arc of Mars, when the planet is closest to the Sun, is about 38 arcminutes, whereas the diurnal arc of the Earth when it is farthest away from the Sun (when it travels at its slowest) is about 57 arcminutes, hence the ratio of these numbers is 3/2. In other words: a perfect fifth, the musical interval between one note and another five tones higher. Since the velocities of the planets in their orbs vary periodically, they will produce glissandos in their “songs”. Figure 6 shows Kepler’s musicographic presentation of the planetary motions. Each body starts from the note that corresponds to its motion when it is farthest away from the Sun in its orbit. Kepler remarks that Venus remains almost on unison.[19]

The music-of-the-spheres worldview has now totally vanished from the scene of modern science.

Star Music

World literature incorporates very few other references to an “audio” cosmos. Blaise Pascal (1623–1662), one of France’s most famous philosophers even refers to the absence of sound: "The eternal silence of those infinite spaces terrifies me" [20]

[14] William Shakespeare (1564–1616) *The Merchant of Venice*, Act 5, Scene 1, lines 60-61.

[15] *Gulliver's Travels*, Part III. *A Voyage To Laputa, Balnibarbi, Luggnagg, Glubbdubdrib, and Japan.*

[16] Koestler, A. 1959, *The Sleepwalkers*, Penguin Books, London, Part I, p. 29.

[17] An excellent translation of Kepler’s opus was published by Aiton; E. J., Field, J. V. & Duncan A. M. 1997, The Harmony of the World, Memoirs of the American Philosophical Society vol. 209. Their translation of the opening sentence of Book V reads: “ON THE MOST PERFECT HARMONY OF THE HEAVENLY MOTIONS, and on the origin from the same of the Eccentricities, Semidiameters, and Periodic Times. According to the precepts of the most thoroughly corrected astronomical teaching of the present day, and the hypotheses of Copernicus, but also those of Tycho Brahe, one or the other of which are today publicly accepted as true, superseding those of Ptolemy.”

[18] Kepler’s “universe” consisted of the solar system (Sun, Moon, and six planets) and the celestial vault of the “fixed stars”.

[19] Translation from Aiton et al. (1997).

[20] *Le silence éternel de ces espaces infinis m'effraie ;* Pensées, 1669.

Figure 6: Excerpt from Kepler's Harmonices Mundi (1619) Liber V. The lower line says Hic locum habet etiam ☽: "Here the ☽ also has a place". Mercury's song has the highest pitch, because it is the closest to the Sun and thus the fastest moving, and makes the steepest glissandos. Terra (Earth) and Venus with their near-circular orbits have a smaller range. The frequency of Jupiter is sometimes below the level of human hearing.

Two positive sources may serve as an example, though:

"Where wast thou when I laid the foundations of the earth? … When the morning stars sang together,…" [1]

"At night she took to walking out into the pampa and lying on her back to look at the galaxy above,…and this star-music was as close as she came to joy." [2]

Obviously, we cannot hear any of the celestial sounds: hearing (as well as smell) depends on the existence of a medium between the "sender" and the receiver. In addition, as illustrated again in Figure 7, the frequencies of "cosmic sounds" fall entirely outside the audible region of humans.

Absence of that medium has another consequence: astronomers are entirely dependent on the light, i.e. the photons, that they capture with their telescopes.[3] Therefore, astronomers cannot manipulate nor control the objects of their study, nor can they exactly duplicate an experiment since the object may have changed as time goes by.

Many stars have been observed to exhibit this "stellar music", and some of the "songs of the stars" that we capture today were produced 5 to 10 millenia ago. A vast range of frequency spectra have been collected so far. The periods of stellar oscillations cover a time interval from minutes to years, thus the range of frequencies is about thirty octaves compared to the ten octaves covered by human-audible sound. The highest stellar frequencies are more than ten octaves lower than A4 (440 Hz), see Figure 7.

[1] Job 38:7, King James Version. Any translation of the Bible can be considered as a landmark of literature.

[2] Rushdie, S., 1988, *The Satanic Verses*, Vintage.

[3] The human eye functions as a telescope with an 8-mm aperture.

Some Examples

In early July 1686, Gottfried Kirch (1639 – 1710) observed a star in the constellation Vulpecula, and at this occasion he compared the surroundings with the charts in some celestial atlases. He could not find the star that was known as χ Cygni. Kirch consequently kept this region on close surveillance, and found on 19 October 1686 χ Cygni again as a star of fifth magnitude. Figure 2 shows a section of its light curve: this star vibrates monoperiodically with a period of 450 days, and emits a dull monotone chord since centuries – sounding, in fact, like a Pythagorean planetary crystalline sphere, or it can be thought of as a ringing, undamped, church bell. χ Cygni is a so-called Mira star: its diameter varies over several hundred solar diameters through its cycle, and the range in light variation covers 10 magnitudes, i.e. a 10,000-fold fluctuation in light output.

Another interesting pulsator is α Cygni,[4] one of the visually most luminous stars in the Galaxy: a 20-solar mass star whose light, seen today, was emitted in 1500 BC. A very massive star similar to α Cygni is η Carinae. The star underwent a massive explosion in the 1840s when the gas lobes, visible in the Hubble Space Telescope image (Figure 9), were formed. This is one of the stars that combines several physical processes: pulsation, binarity, percussive activity and irregular noisy[5] behavior.

The constellation Scorpius includes the star σ Scorpii, a member of the family of β Cephei type stars. Unlike χ Cygni, it pulsates in two nearly equal modes simultaneously: one period is 5h 45m, the second is about 10 minutes longer. Such nearby frequencies give rise to beats (see Figure 10), just like when combining two violin strings that vibrate at slightly different frequencies give rise to a slow beat wave (at an inaudible frequency, though with an evident effect in intensity).

The constellation Toucan, in the southern hemisphere, includes θ Tucanae, a binary star that pulsates in 10 frequencies corresponding to periods between 39 and 71 minutes in equally-spaced groups (Figure 7). These frequencies are located in a 411-cent wide interval (a major third), while three additional frequencies are 4–6 octaves lower. This star is a member of the family of δ Scuti type stars.

Our nearest star is our Sun. The solar sound waves are of the order of millihertz. Sound waves are trapped in the solar atmosphere, and the pattern they make tends to repeat itself at intervals of about 5 minutes, with distortions of more than 2000 km, and leads to a 5-minute periodicity. The Sun generates millions of simultaneous oscillation modes, and the entire Sun acts like a gigantic organ.[6]

Listening to Planetary Music

Quite a number of compositions have been created for rendering the "Music of the Spheres". A most interesting project was carried out by Willie Ruff and John Rodgers: "The Harmony of the World: A Realization for the Ear of Johannes Kepler's Astronomical Data from Harmonices Mundi 1619". Ruff & Rodgers programmed tone generation for a music synthesizer.[7] Their

[4] Also known as *Deneb*.

[5] The correct term is stochastic behavior.

[6] The Sun also has an 11-year periodicity in the appearance of sunspots. That frequency (of the order of 4 microhertz) is not considered here and is not included in Figure 7.

[7] The code was written on punched cards for an IBM 360/30 mainframe computer.

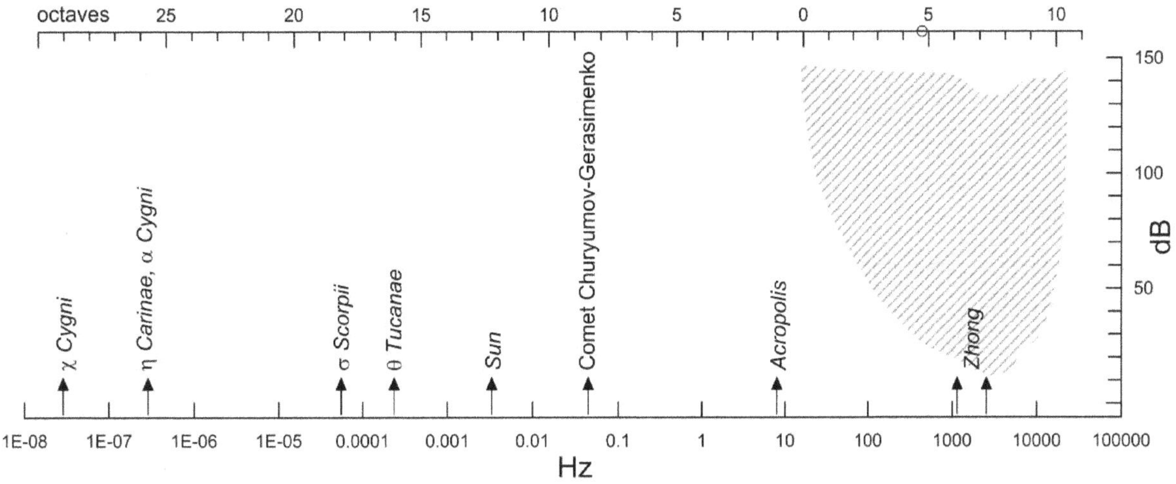

Figure 7: The human hearing region (hatched area) in the hertz/decibel domain. The quietest sound has a sound pressure of 20 micropascal, while the loudest sound that a human can hear without pain or permanent damage has a pressure of several hundred pascal. The upper horizontal axis is in units of octaves, the zero point coincides with the lower limit of human hearing (C0). The o symbol denotes the position of A4 (440 Hz). Vertical arrows indicate fundamental frequencies of vibration of some selected celestial objects, as well as a subsound frequency measured in the Alatri Acropolis (see Paolo Debertolis' presentation in this meeting), and an example of the dominant sui and gui frequencies of a Zhong bell.

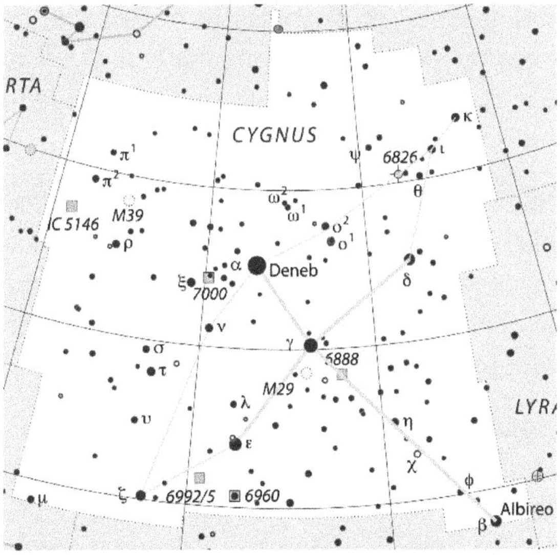

Figure 8: The constellation Cygnus. The thick lines show the principal lines of the asterism that leads to the constellation's nickname "Northern Cross". Faint stars are represented as small points, brighter stars are shown as larger dots. α Cygni (Deneb) is near the center of the image, χ Cygni is in the lower right quadrant. North is up and East is left. Credit International Astronomical Union and Sky & Telescope magazine.

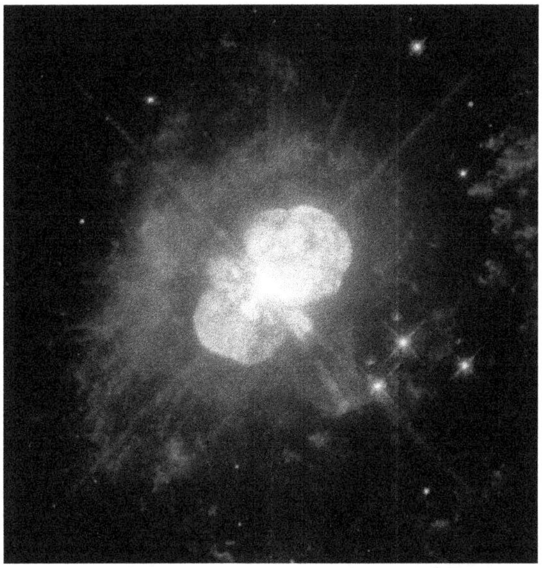

Figure 9: η Carinae as seen by the Hubble Space Telescope. Credit: J. Hester/Arizona state University, NASA/ESA [1]

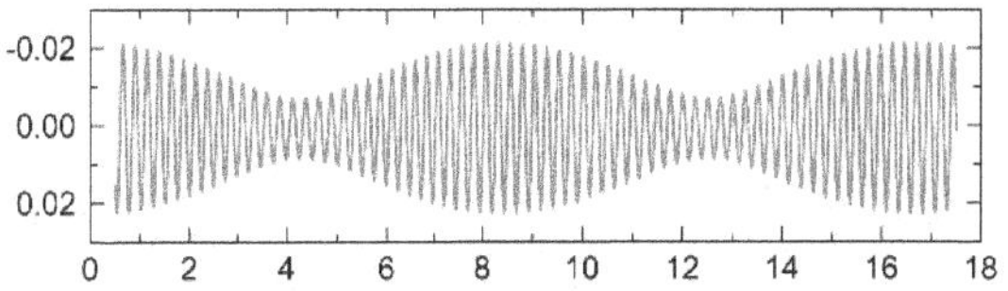

Figure 10: Cycles of brightness changes of σ Scorpii over 17 days (computed light curve). The range in brightness variation is a mere 4%.

[1] https://www.spacetelescope.org/

work[2] resulted in an electronic interpretation of the movement of the planets around the Sun for a period of several centuries, and can be described as a "planetarium for the ear", an aural picture of how the solar system works.[3]

Another composition, though much less known, is "Die Harmonie der Welt des Jupiter"[4] that was commissioned by the Münster Astronomical Society to commemorate the 350th anniversary of the death of Kepler. This organ work "visualizes" the orbits of Jupiter's four largest moons. This project was patronized by two astronomers: Waltraut Seitter (1930–2007) and Hilmar Duerbeck (1948–2012).[5] The composer created his work in less than eight months, quite a challenge not because of the frequency ratios of the second and third to first moons being, respectively, 2:1 and 4:1, but because the ratio between moons IV and III is 7:3.

The four satellites of Jupiter (Io, Europa, Ganymede, and Callisto) have an interesting property: adding two times the daily motion of Ganymede to the daily motion of Io, we get three times the daily motion of Europa.

Even comets emit "songs": Comet 67P/Churyumov-Gerasimenko produces oscillations in its magnetic field environment. A sonification of these magnetic data was compiled by composer Manuel Senfft (www.tagirijus.de), and the resulting sound tracks can be enjoyed on selected internet sites.[6]

Listening to Stellar Music

Pulsation periods of days to years correspond to frequencies below a millionth of a hertz, and that requires us to transpose these frequencies over a wide range to make them audible. Note that transposition of frequencies to the audio range is also done in archeoacoustics, as explained in this meeting by Paolo Debertolis who renders audible the infrasounds measured in the Acropolis at Alatri by transposing the ultrasonic signal into the audible band: frequencies in the infrasonic range between 8 and 20 Hz[7] are transposed over 20 Hz to bring them into the audible range. Ruff & Rodgers also "transformed" frequencies: since the outer planets[8] – unknown to Kepler – move too fast for us to perceive the subtleties in their relationships, the pitch had to be lowered by slowing down their computed motions in such a way that one Earth year is reduced to ten seconds of real-time listening.

Such frequencies transposed by a key change can also be used for composing for an ensemble of "stellar instruments". Some astronomer-composers have used such virtual instruments for music composition.[9]

Zoltán Kolláth and Jeno Keuler demonstrated that sounds conceived according to the principles of stellar physics, and designed to the nature of the processes inside stars, can be used as a new basis for music composition. They developed an alphanumeric score similar to the C-sound score. They also created a Stellar Music Project

---

[2] 1979, Label: Kepler – LP 1571, now available on audio CD.

[3] See www.youtube.com/watch?v=ArXrDAlGlYU

[4] Günter Bergmann 1981, *The Harmony of the World of Jupiter*, Verlag Schwann GmbH Düsseldorf.

[5] For additional background information, see Sterken, C. 2012, *The Eagle and the Dove*, JAD, www.vub.ac.be/STER/JAD/JAD18/jad18_1/jad18_1.pdf

[6] https://soundcloud.com/esa/a-singing-comet https://sploid.gizmodo.com/scientists-reveal-the-sound-of-the-comet-67p-churyumov-1657727449

[7] Frequencies above 20 Hz were cut.

[8] Uranus, Neptunus and "ex-planet" Pluto.

[9] Kolláth, Z. & Keuler, J. 2006, *Stellar acoustics as input for music composition, Musicae Scientiae* 10, 161.

that includes the first piece of music composed for "stellar instruments": Stellar Music No. 1. This work can be downloaded in mp3 format.[10] The website also provides a video file that illustrates the sound generated by a pulsating star in a binary system.

Transcending Mensuration

Several inspired speakers at this meeting offered stimulating blends of quantitative science and qualitative contemplations related to their archeoacoustical research activities. Ancient structures as listed along the time line in Figure 1 were described as ritual places, and there was mention of visitor's body resonance and of relaxation, of "seeing" sound, of reaching a state of ecstasy and of visual experience that transcends what one had anticipated to encounter. There was mention of "elevation", the "celestial", extra-sensory perception, and the impact of silence. In other words, cognitive content – though not necessarily sacred – of the visiting person twinned with the experiences of the visiting scientist.

So, what about my own personal experience as an observational astronomer who has spent more than a thousand nights observing the skies? I must admit that the quantitative aspect of this kind of observational work is[11] tedious and stressful night work: stars rise and set and determine the ever changing observing window, clouds come and go, and technical problems come as well, but do not go in a natural way.

But there were always moments for contemplation, though (in my case) not intense enough to see the night sky as a ritual place or to experience a state of ecstasy with extra-sensory perception. But during every single observing mission I have always felt the impact of the silence, and the impact of the darkness that reigns at those remote observatory sites – two deep feelings that I dearly miss now in my urban life. And there is a third sensation that I dare describe as exaltation: seeing in real time the dynamics of the naked-eye cosmos, exactly the way our ancestors contemplated the celestial vault: the daily motion of the Moon and the planets projected on the rock-steady and magnificent panorama of our Milky Way. Remember that the Moon moves around the Earth in approximately 27.3 days, and from one night to the next it moves almost one hour[12] eastward. For an observer on a mission of one month or longer, the daily view of the motion of our near companion is a delight.

Epilogue

The morning after the meeting I wanted to leave Tomar with an early train, so I was the earliest bird in the breakfast room, and I took a solitary table along the Eastern window wall. Very soon I got the company of Slobodan Dan Paich, and then Paul Oomen also joined us. "No Venus today", Slobodan said. Indeed, for the duration of the conference we had enjoyed magnificent views of the morning star,[13] just before sunrise, from our adjacent hotel rooms that looked East. After a while the fog cleared, and the three of us were given a splendid and sharp ruby-red Sun instead, but only for a short while, because nebulous cirrus clouds were forming.

Two hours later I was in the train to Lisbon, and to my delight, Nancy and Alvin Holm joined me on the opposite seats. "How

[10] www.konkoly.hu/staff/kollath/stellarmusic/
[11] Actually "was", as many observing routines have now been automated.
[12] Actually, 13 degrees.
[13] *Morning star* is commonly used as a name for Venus when the planet appears in the East before sunrise. When Venus is visible in the West after sunset, the term *evening star* is applied.

beautiful the Sun is", Nancy said. I mentioned that, earlier in the morning, we did not contemplate a muted orange Sun, but a vivid red disk with a sharply-edged rim.

So, why was it so different, I was asked. Well, the molecules in the air, as well as the very small dust particles, scatter the shortest wavelengths, and because at sunrise the beam of light takes a much longer path through the atmosphere, a sunray will render a much redder visual impression than when the Sun stands higher in the sky. The diffuseness that we saw was caused by the layer of cloud that had suddenly appeared.

While explaining this, I became all of a sudden aware of the fact that I was "unweaving a rainbow" in the sense as conveyed in John Keats' poem *Lamia*[14]:

Do not all charms fly
At the mere touch of cold philosophy?
There was an awful rainbow once in heaven:
We know her woof, her texture; she is given
In the dull catalogue of common things.
Philosophy will clip an Angel's wings,
Conquer all mysteries by rule and line,
Empty the haunted air, and gnomed mine–
Unweave a rainbow, as it erewhile made
The tender-person'd Lamia melt into a shade.

The question is whether a scientific explanation of a phenomenon entails a reduction in the mind's sensation of beauty?

But have we not been unweaving the arches of sound all these days in Tomar? From my personal experience I am rather inclined to follow Michael Shermer, who argues that scientific understanding brings on an addition rather than a reduction to aesthetic appraise.[15] After all, Shakespeare wrote:

Beauty is bought by judgement of the eye.

And, as I have seen and heard all throughout this meeting – in the *conference room*, in the synagogue and in the Convent of Christ:

Beauty is in the ear of the beholder.

[14] John Keats (1795–1821). *Lamia.*

[15] *The Believing Brain*, 2011 Macmillan.

Ħal Saflieni Hypogeum: An Introduction to the Site and its Acoustics

Katya Stroud

KATYA STROUD is the Senior Curator of Prehistoric Sites within Heritage Malta. She is an archaeologist and has a particular research interest in the history of conservation of archaeological sites and public perception of the Neolithic.

ABSTRACT: The Ħal Saflieni Hypogeum in Malta is a Neolithic site dating to between the 4th and 3rd millennia BC. It has a particular acoustic phenomenon whereby when someone with a deep voice utters some words or chants into a niche in a particular chamber, the sound is heard to resonate around the entire site. A number of studies have been carried out about it, but these have really just touched the tip of the acoustic properties of the site. This article summarizes the results of these studies and starts paving the way for future research on the acoustics of this unique site.

Welcome to the Ħal Saflieni Hypogeum

The Ħal Saflieni Hypogeum is found on the island of Malta, home to a number of unique monuments built between the 3rd and 4th millennia BC which have been recognised by UNESCO as sites of World Heritage value. The Ħal Saflieni Hypogeum was inscribed on the UNESCO World Heritage List in 1992 as a site which bears unique testimony to a cultural tradition which has disappeared.

Ħal Saflieni is found close to other contemporary Neolithic sites such as the Tarxien Temples, and Kordin Temples, and although the area was largely built up in the 20th century, in the Neolithic these sites would have occupied a prime location, being found at the edge of agricultural plains while having easy access to the sea; the factors that very likely made living in Neolithic Malta sustainable.

The Ħal Saflieni Hypogeum is an underground, rock-cut, funerary site, with structural extension and development probably reflecting developments in the rites and rituals that accompanied the burial of the deceased. It is set on three levels, the Upper, Middle and Lower Levels, with the deepest part of the lower level reaching around 10m below the modern street level (Figure 1). The complex consists of a number of interconnected oval chambers which were excavated out of the living rock with stone and bone tools. No metal tools were available at the time and the population's familiarity with the local geology appears to have been used to their advantage in creating the site, with a number of chambers being excavated following natural faults and fissures in the rock. The layout of the different levels also suggests that very little natural light would have originally filtered into the site beyond the Upper Level.

Site Discovery

The site was discovered accidentally in the 1890s when the area was being developed to provide housing for the workers who were attracted to the region by the employment prospects of the new dry docks constructed in the nearby harbour. However, the site was only reported to the authorities in 1902 when housing construction in the street overlying the site was completed and parts of the Upper Level had in fact been integrated into the house foundations and basements (Figure 2).

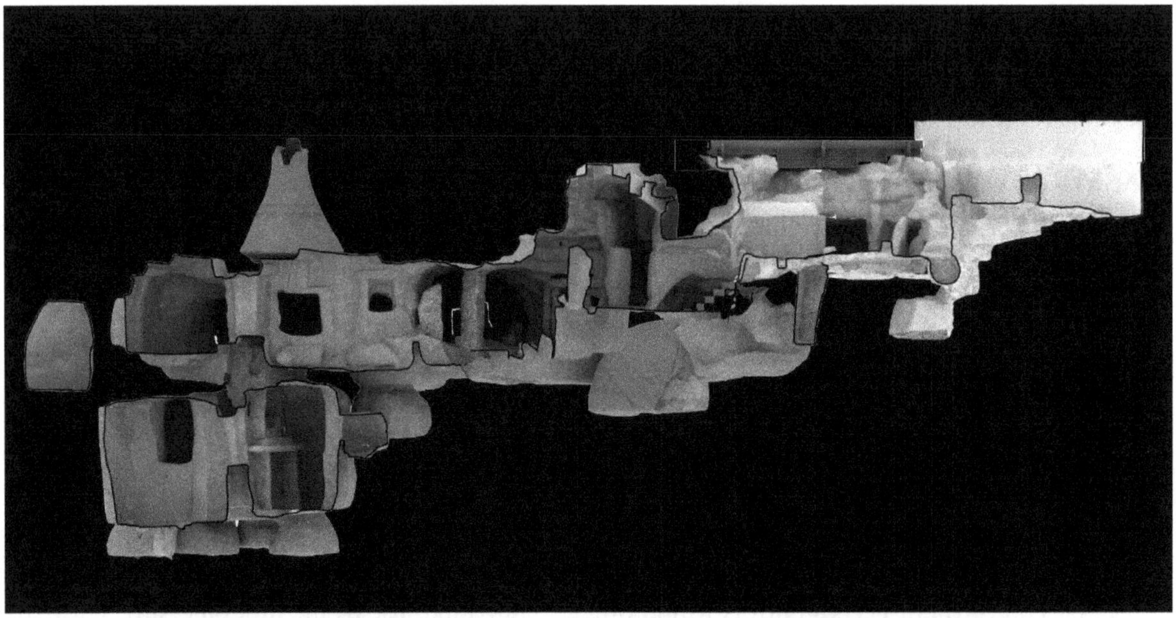

Figure 1 Cross-section of the digital model of the Ħal Saflieni Hypogeum showing its layout on three levels

Figure 2 Part of the Upper Level was integrated in the foundations of the houses built above the site at the turn of the 20th century (Heritage Malta archives)

Following the report of the site's discovery to the authorities, it started being excavated in 1904, and although it was opened to the public in 1908, investigations continued until 1911. The investigations yielded numerous bones; calculations indicate that there may have been up to 7000 individuals buried here. They were buried with personal ornaments such as beads and pendants. A number of statuettes were also discovered and these were either buried with the individuals or used in burial rites and rituals.

Features and Characteristics

The site itself, together with the remains found within it, build a rather bleak picture of the hypogeum in the Neolithic; burial rituals would have been carried out in near darkness, or to the flickering light of burning torches, and would have taken place in the inevitable smell of decomposing bodies. This is contrasted by the above-ground contemporary buildings which were also used for rituals but never for burials, and which in some cases were built to mark the position of the rising sun at the start of the seasons.

The Ħal Saflieni Hypogeum seems to offer a direct contrast to these above-ground structures, not only in its use, but also in its construction, as parts of the hypogeum were hewn out to resemble the above-ground structures by reproducing their architecture in the negative. This aspect of the site is not only of value as it demonstrates the Neolithic society's artistic prowess but it also sheds light on the way the temples were roofed over since the ceilings are carved to imitate a roof built of corbelled masonry, with one course overhanging the one below (Figure 3).

Figure 3 One of the chambers of the Ħal Saflieni Hypogeum with its ceiling carved in imitation of the roof architecture of the above-ground temples (Photo – Clive Vella)

Figure 4 One of the red ochre paintings which have survived at the Ħal Saflieni Hypogeum (Photo – Clive Vella)

Figure 5 The niche which appears to be at the root of the acoustic phenomenon of the Ħal Saflieni Hypogeum (Photo – Clive Vella)

A distinctive characteristic of the hypogeum is that some of the walls are curved, which not only adds to the structural strength of the site, but also makes the spaces look visually larger than they actually are. Another valuable element of this site is the survival of wall paintings which have disappeared from other local Neolithic sites. These paintings are made in red ochre and other natural pigments and can be found in wide bands, in a chequer-board motif, as well as more intricate designs such as a honey-comb design with spirals around its edge, and a more complex design of spirals and discs (Figure 4).

The latter is found in what is known as the Oracle Chamber, where a small niche appears to have very particular acoustic properties (Figure 5). This was not immediately noticed by the site excavators since when it was first reported, Temi Zammit, then curator of the museum, gave little attention was given to it saying:

"The niche was possibly intended to receive a statuette, a cone, a pillar, or a similar object."

It is only a decade later, in 1920, when writing for the National Geographic Magazine, that William Arthur Griffiths points out;

"Here it was noticed only a few months ago that any word spoken into this place was magnified a hundred-fold and audible throughout the entire underground structure. A curved projection is specially carved out of the back of the cave near this hole and acts as a sounding board, showing that the designers had a good practical knowledge of sound-wave motion."

This niche and the Oracle Chamber have ever since become the centre of much fascination, (mis-) interpretation and the focus of research.

Acoustic Research

There are limitations to any study carried out in the Ħal Saflieni Hypogeum, both due to intrinsic characteristics of the site, such as its location and complexity, as well as to measures that have been put in place to ensure that impacts on the site are kept to a minimum and where possible mitigated. These factors affect any study made on the acoustic properties of the site. These limitations include:

- The number of people and duration of access is limited since these affect the humidity and temperature of the site which in turn have a direct influence on its preservation, in particular the wall paintings. Therefore, in conducting research on, and recording, the acoustics of the site, the number of researchers and technicians has to be limited to the bare minimum while the economic use of time on site is also important, making preliminary preparation and panning essential;

- The site is extensive, being divided into a number of interconnecting chambers on various levels. This complexity has to be taken into account in planning, recording and interpreting results;

- A number of changes have been made to the site over time including the deterioration of any 'furniture', the excavation of soil and bones, introduction of a metal walkway and construction over the Upper Level which makes the replication of the acoustic effects as heard in the Neolithic near to impossible;

- Modern noise still reaches the lower levels of the site including voices and traffic, creating interference while recording acoustics on site.

A number of studies have nonetheless been carried out on the phenomenon in the Oracle Chamber of the Hypogeum:

Researcher	Year	Primary Research Aim
Paul Devereux Thomas Andersen	2007	Primary Resonant Frequency
Daniel Talma	2007	On-site recording
Reuben Zahra	2008	On-site recording
Paul Devereux	2009	Acoustically-related Imagery
Paolo Debertolis	2014	Archaeoacoustics Conference
Rupert Till	2014 (2017)	Archaeoacoustic Study as part of the first Archaeoacoustics Conference in 2014

Daniel Talma (2007) and Reuben Zahra (2008), carried out recordings on site that were done to develop a preset that can be used with specialised audio software so that one can reproduce the acoustics of the Hypogeum in a digital recording studio.

In 2007 Devereux and Andersen set out to address the following research questions:

- How does the site compare to other sites studied by the researchers in Ireland and England?

- How does the female vocal range fare in the site?

- Are the nodes in the Maltese temples in a meaningful pattern?

- Is there a link between the acoustic phenomena and the wall paintings, be it in their location as well as the representation?

Unfortunately technical difficulties did not make it possible to address the majority of these questions but the research questions raised were in themselves valuable results as they remained very relevant for even the most recent studies. Devereux did identify a possible correlation between the paintings and the niche at the root of the site's acoustic properties; particularly the fact that the wall paintings stop abruptly at the niche at the root of the phenomenon and also the way the discs grow larger the closer they are to it, with three large discs appearing in the niche itself (Devereux 2009).

During the first Archaeoacoustics conference held in Malta in 2014, a study led by Paolo Debertolis, focused on the Oracle Chamber and using voice as well as a number of instruments, concluded that a low male voice can stimulate the resonance of the structure at two frequencies 114 and 68-70Hz, while a drum stimulates the resonance at 114Hz. It also concluded that the sound generated, particularly a low voice at 70Hz can be heard across the entire site (Debertolis 2014).

At the same time during the conference of 2014, another team, led by Rupert Till also carried out their own separate studies and came up with complimentary results. This study however, was the first to not only focus on the Oracle Chamber, but to also consider the Main Hall (Figure 6) of the site (Till 2017).

Figure 6 The Main Hall of the Ħal Saflieni Hypogeum (Photo – Clive Vella)

The principal results obtained from that study were the following:

- The primary resonances of the Oracle Chamber were found to be 41, 72 and 76Hz (the 41Hz resonance was stimulated by a large drum or grinding stones) and the Main Hall had a resonance that is lower in frequency and more powerful,

- Reverberation time was up to 16 seconds at low frequencies (63Hz)

- Sound from the space (the Oracle Chamber) was heard, if not understood, throughout the Hypogeum and became up to 50% louder outside the chamber,

- It was difficult for listeners outside the chamber to identify the location of the singer or performer,

- Music that is rhythmic or with several components became confused, but a female voice which would have better articulated consonants and enhanced vowels would be musically more clear and defined.

Although the research conducted to date has just started discovering and defining the nature of the acoustics of the Ħal Saflieni Hypogeum, it has primarily aided in refining the research questions and the methodologies and technology required to address those questions. In a way, it has paved the way for future studies, not only through the results produced but also through the shortcomings encountered and the numerous accompanying discussions which are not all necessarily represented in the published material, but which are nonetheless invaluable in designing future studies.

Lessons Learnt

As described above, the nature of the site presents limitations to any study made on it. The principal factor that effected each study conducted was the site's complexity. This made it impossible to apply a pre-existing methodology and any future study would have to be tailor-made and then tested and tweaked for the site.

One question that kept cropping up while conducting studies on the site's acoustics and even in the discussion on results achieved was the issue of intentionality; were the acoustic effects of the site purposely created, or were they accidental? The studies conducted and discussions that accompanied them tended to veer towards the idea that although the origins of the acoustics here may have been natural or accidental (any hollow space with rounded walls will produce some form of echo), they were definitely recognised in the Neolithic as indicated by the location and nature of the wall paintings in the Oracle Chamber, and attempts may have been made to amplify the effect in prehistory. But again, further research is required to better understand this aspect.

There has been a long-standing pre-conception that the acoustic effect was mainly intended for a male voice, but as Rupert Till indicated "Previous male-focused interpretations may be the result of male authors testing male voices". During previous studies, female voices were in fact tested with promising preliminary results (see section on Acoustic Research above), but this aspect clearly needs further investigation.

One aspect that has become evident through the research campaigns held is the importance of allowing the observations of individuals familiar with the site and its spaces, including curators, tour guides, maintenance personnel and security officers, to feed into the research design. The value of these observations is often brushed aside with the result that certain intrinsic aspects of the site are not even considered when planning research about the site.

A factor that is often highlighted by archaeologists and other academics, but which has not been addressed by the studies that have been conducted, is the fact that the site has gone through many changes, in the Neolithic itself as it was developed over thousands of years, and later on when large deposits were removed from it and modern intrusions introduced. The acoustics of the sites have similarly undergone changes at each stage of the site's development. This element has not been investigated and the studies carried out have only considered the current status quo of the site. This is understandable since it is not possible to reintroduce the deposits that have been excavated without irreversible damage to the site, and the physical recreation of the various stages of structural development of the site is clearly unachievable. However, digital modelling allows for the virtual manipulation of the structure so as to create various possible scenarios and test their acoustics, thus providing a useful tool in further understanding the acoustic development of the site.

Most previous researchers focused their attention on the Oracle Chamber. As witnessed by visitors and particularly, maintenance workers on site, there are other parts of the site that have acoustic properties. One of these chambers is the so-called Holy of Holies where uttering a sentence appears to be instantly amplified and yet made unintelligible due to the echoing in the space. One of the previous studies also indicated the potential of studying the properties of the Main Chamber. An aspect of the site's acoustics that has yet to be addressed is the phenomenon that it is often hard to tell the location of the source of a voice or sound in many parts of the site, not only when it originates from the Oracle Chamber, but also when it comes from other parts of the site.

Therefore future research would need to take into consideration the whole of the site together with observations by stakeholders so as to understand its acoustics holistically, rather than be designed purely on the pre-existing formulae and perceptions set by previous researchers.

BIBLIOGRAPHY

Till R., 2017, *An Archaeoacoustic Study of the Ħal Saflieni Hypogeum on Malta*, Antiquity 91:355, 7489

Devereux P., 2009, *A Ceiling Painting in the Hal-Saflieni Hypogeum as Acoustically-Related Imagery;*

A Preliminary Note, Time and Mind Vol.2 Issue 2, 225-232

Debertolis P., Eneix L., 2014, *The Ħal Saflieni Hypogeum Research Project*, in Archaeoacoustics Conference Proceedings 2014

Wonders Decoded: Extrasensorial Looking at the Ancient and Its Meaning Toward Future Architecture

Hyun Soo Suh

HYUN SOO SUH: Bsc + MSc at Delf University of Technology, The Netherlands, Architect (registered in the Netherlands), Director at The Korean Society of Contemplative Science, Head of Research at Institute of Self-Realization, CEO, Proper Mediation Inc. (Architecture consulting firm)

ABSTRACT: The search in this paper is set on the authentic analysis of the ancient architecture. There are many sites and architectures that are considered to be sacred, spiritual or mysterious in a popular sense. Because they are outside of the current knowledge of science, people irresponsibly give them titles of 'wonders' or 'mysteries' and put them in a cast of untouchable 'sacredness' which in turn makes it harder to see the truth about the real causes of their positioning and design. When geared by 'Extrasensory perception(ESP)' which is the natural cognitive ability of every human being, the scenes that have gone unnoticed come onto the surface and unfolded as the accurate reflection of the real. With ESP, one is able to measure certain forces existing in nature and find out that those forces have been the major factors affecting the ancient architecture. Having seen the correlation between what ESP can reveal and how the ancients have managed their built environment, from site selection to spatial design, one comes to the conclusion that the architecture in an original sense is much deeper than the current apprehension and practice. Observing the ancient manifestation with intuitive eyes (ESP) thus discloses the proper pattern of spatialization and it is what the architecture of tomorrow should be based upon.

KEYWORDS: Extrasensory perception (ESP), dowsing, ancient placements

What Do We See? (The Problem of Perception)

"Why is it," Jonathan puzzled, "that the hardest thing in the world is to convince a bird that he is free, and that he can prove it for himself if he'd just spend a little time practicing? Why should that be so hard?"[1]

Everything starts with one single moment of engagement between the observer and the observed. A variety of views and opinions are piled up to form a so-called 'knowledge' which acts as a stepping stone for the next cycle of observation. It is this on-going process of accumulation which most of the people consciously or unconsciously base their system of cognization of the world.

It does not mean that this subjective way of getting to know things is wrong. If the universe is designed in a way that it operates at its own will (which is considered to be 'perfect'), this subjectively-engaged process of observation is also perfectly fine. Nothing

[1] Richard Bach, Jonathan Livingston Seagull, London, Element, 2003, p. 881

must be judged by and everything is in its own perfect phase of the time and place. But it is also to be said that even if there is no right or wrong in this realm of 'perfectness', there is indeed a difference of depth in relation to acknowledging the true picture of the observed. The one who sees more wins in 'knowing' which directly deepens the understanding of it. In the realm of perception, thus, what is essential is quality, not the quantity. One genuine insight over infinite numbers of subjective opinions.

This subtle mechanism of 'quality-oriented' observation can be relatively effortlessly operated but it also can be the hardest thing ever to be achieved. As it might have been hinted in the previous paragraph, the key factor in getting access to authentic awareness is a conscious act of engagement with a self to get rid of infinite numbers of subjective opinions. In short, when an individuality dies out, the veiled truth is disclosed. The individuality in this context is largely the sense-based perception based on the process of interpretation which involves person's learning, memory, expectation, and prejudices. As long as a view is rooted on such perception, the case is closed, leaving one behind with social doctrines and personal inclinations.

The solution to subjugate sense based perception is a mindful act of dismissing what I know now. It is an act of going beyond of a present knowledge, being on the outside of the current norms and looking without discrimination. It is an 'extra'-sensory perception[2] which is quite essential in acquiring true picture of reality. It is not only important for a domain of science but also a deep concern for many religious systems since they are trying to find the true self, true god and after all, a true reality as it is.

Extrasensory Perception

Let us go into more detailed pictures of the realm of 'perception'. A 'perception' can be operated in three classifications. First is connected to the analysis of 'sensed' information and is called 'Sensory perception (SP)'. SP involves, in a traditional sense, reception of information gained through the physical senses such as sight, hearing, taste, smell, and touch. The problem arising from this particular and yet the most generally performed perception is that it is always the case that there are predetermined values affecting the outcome through the processes of organization, identification, and interpretation of the acquired information. In other words, the sensory perception is, by its descent, however excellently keen and acute it may be, only an interpreted mental grasp which leaves it with a considerably huge limit.

The second is incorporated with a higher sensing ability in which higher and deeper coherency with the factual entity of a subject occurs. It is called 'Higher sensory perception (HSP)[3]. HSP concerns a dealing of

[2] Extrasensory perception, ESP or Esper, also called sixth sense or second sight, includes claimed reception of information not gained through the recognized 3 physical senses but sensed with the mind. The term was adopted by Duke University psychologist J. B. Rhine to denote psychic abilities such as intuition, telepathy, psychometry, clairaudience, and clairvoyance, and their trans-temporal operation as precognition or retrocognition. Source: Noel Sheehy; Antony J. Chapman; Wendy A. Conroy (2002). Biographical Dictionary of Psychology. Taylor & Francis. pp. 409–. ISBN 978-0-415-28561-2. However, in this paper, I argue for a different definition of ESP (See 2.Extrasensory perception).

[3] The Korean word for 'contemplation' is 명상 (Myung-Sang). It is a thorough and radical process of clearing up (=명) the solidified concepts (=상) of a self with a 2 single pointed focus. In other words, it is a process of concealing the preconceived notions and judgemental views so that truth can be

a larger amount of information from phenomena acquired by senses. Well known examples are the manifestations of psychic abilities such as telepathy, clairaudience, trans- temporal operations involving pre or retro-cognition and etc. In the course of spiritual or ascetic exercise, this perception occurs at frequent base depending on individual intention. People who exert such perception know things through the ways that are not normally perceived by the general people. Simply put, people with HSP experience the world more multi-dimensionally because they can cognize more information from the nature of totality. But the problem is, however that HSP mode still relies on the sensory organs, and the processing of organization, identification and interpretation of the brain. Due to such restraints, the results of cognition differ from individual to individual. In other word, people with HSP can experience a world more diversely and multifaceted but still within their own frame of reference, meaning it is still not in line with the highest form of knowledge, the knowledge that coincides with the subject as it truly is, the truth.[4]

The last is one that is surpassing the previous two. Seen from the arguments before, it is clear that sense based perceptive images are in most cases incorrect or insufficient to know things as they really are. 'Extrasensory perception (ESP)' is a way of perceiving things without any engagement with the faculty of bodily senses and workings of a brain. It is a form of direct perception of a truth or a fact, an immediate apprehension independent of any reasoning process. Simply put, it is a realm of 'I just know it.' In this sense, it might be understood as a form of knowing based on an intuition or one that is equated with a spiritual encounter. The explicit distinction is in the mechanism that five senses have no place to play a part at all. It does not involve any of bodily senses but something much broader and radical realm of human capability.

Evidence of ESP

However, is there any proof that this particular type of perception really exists? Does ESP reside in all human beings or is it something that only a number of gifted persons have? Can it operate anytime in any place? Many questions about its credibility and capability may arise. The answer might however be found in our body.

In the late 1970s, Dr. G. Goodheart published the results of his research on the application of kinesiology. He argued that food with harmful ingredients decreases the strength of muscle whereas with healthy ingredients causes the muscle to generate more power.[5] Similar research on bodily muscle and its association with a set condition was conducted by Dr. J. Diamond. From the research, he concluded the result that positive mind-set makes muscle strong

successfully brought forth. A philosophical inquiries on the ancient Greek understanding by M. Heidegger reveals the interesting account of the Greek word for 'truth' which is 'aletheia'. 'For the Greeks, truth, a-letheia, then means "a-concealment" or "dis-concealment" in the specific sense of "being dis-concealed," "that which has been dis-concealed"(not "that which we conceal")*. *Source: R. Rojcewicz "The gods and Technology_A reading of Heidegger"_Part 3. The danger in Modern Technology P.171~172_State University of New York Press, Albany_ISBN 0-7914-6642-6

[4] The definition of HSP in this regard coincides with the definition of ESP from J. B. Rhine. If the term 'Extra' denotes going beyond of the 4 general mode of perception, it must not comprise any involvements or association with the five sense organs.

[5] The implication was that at a level far below conceptual consciousness, the body 'knew' and through muscle testing was able to signal what was 5 good and bad for it*. *source: D. Hawkins, Power Vs Force, Hay House Inc., 2005, p.3

while with negative mind-set does the opposite.[6] The common denominator of these experiments was obtained by the careful observation on the almost imperceptible and yet distinctive and subtle change in muscle. Extending from these unusual experiences and critical observations of a human body, the significance must be drawn that no matter what, where, when, who and how, the human body is always in a cognitive state. What is compelling and remarkable is the fact that the brain is not in a state to cognize or to know such signals exerted on a body. What does it signify? What is this phenomenal inevitability that a body seems to know in advance? What is clear from this experience is that it shares the common feature with a perceptive mode of non- sensory processing which is 'Extrasensory perception'. Among other body-awareness mechanisms, a technique called 'dowsing'[7] is a relatively easy way to detect muscle changes. An L-shaped rod held in hand is a tool that amplifies subtle twitch in hand muscle into a noticeable movement so that it becomes visually distinctive. Numerous tests suggest that when engaged, a rod gives clear responses in the form of 'twist' at certain spots across the field. This 'response' is actually caused by a distinctive muscle twitch in the palms. Properly stated, it is a body which is responsive to something. It is like a body is somehow aware of something that the brain misses. And this 'body-awareness' is always working regardless of time, space, sex and age. It is an unavoidable feature that every human body exerts. What a dowsing projects is the invisible realm of 'Extrasensory perception' into the realm of visible clarity. If people do not want to use a rod but still want to engage in a similar attainment, they can use their body itself for the same purpose as Marko Pogacnik[8] claims, "the human skeleton is composed of many joints that are capable of acting as a rod or a pendulum. The body can be trained to react to invisible phenomena like a rod. Observe the movements of knees, hips, or hands."[9] In order to be able to use a body alone, however, requires a lot of exercise and self-observation.

This subtle phenomenon might be linked to great teachings of the religions, especially in the scriptures and the sacred books that recorded discourses and disciplinary principles. They have distinctive rituals and conventions of their own, a core value in their contents shares the one same basic essence. It is the idea of 'part in the whole, whole in part'. In Buddhism, many scriptures and orally passed down teachings contain in their core constitution many phrases denoting the concept of 'part and whole'. The following dictions from the Buddhist scriptures are a few of those that express the idea in very explicit and clear form.

[6] D. Hawkins, Power Vs Force, Hay House Inc., 2005, p.36

[7] When looked in the dictionary definition, dowsing is defined as a 'type of divination employed in attempts to locate ground water, buried metals 7 or ores, gemstones, oil, grave sites, and many other objects and materials, as well as so-called currents of earth radiation (Ley lines), without the use of scientific apparatus'. Reference: http://en.wikipedia.org/wiki/Dowsing). Similar applications such as BPE-Method, radiesthesie, rhabdomancy, bio-location and others may have different names but shares the same principles of operation. Tools for dowsing are several. Mostly L-shaped and Y-shaped rod is used frequently and in some cases a pendulum is also employed.

[8] Born 1944 in Kranj, Slovenia, graduated as sculptor on the Academy of Fine Arts in Ljubljana 1967. He is developing art active in the fields of 8 ecology, geomancy, social movements and individual spiritual development.

[9] Sacred geography by Marko Pogacnik. Part 4 Geomantic perception and exploration_4.2 Methods of geomantic perception with exercises_Aura 9 sensitivity, body reactions, Lindisfarne Books, 2007, P.190.

一微塵中含十方 Cosmos is in a tiny dust
一中一切多中一 One in All, All in One
一卽一切多卽一 One is All, All is One[10]

* English translation by the author

In the Bible, the similar expressions can be found throughout the texts. The verses like 'One God and Father of all, who is over all and through all and in all (Ephesians 4:6)' and 'On that day you will realize that I am in my Father, and you are in me, and I am in you (John 14:20)' all point to the fundamental principle that a part, when perceived right in its proper stance, is composed of infinite whole. The relation between the notions of 'Atman' and 'Brahman', the idea of 'Avatar' and the indicative meaning of 'Namaste' in Hinduism all show the authentic coherency with Buddhism and Christianity. The search through religious manifestations of the same notion can go on and on. In one of the Korean ethnic religious beliefs, there is an idea of 'Human being is the cosmos itself' and 'macrocosm and microcosm'[11] which are exactly the same notions compared to an ancient Greek Neo-Platonic approach to cosmos and man. All point to the one and the same belief on 'whole in part, part in whole' essence. What can be concluded is thus a picture that the religions have been standing on an accorded ground of the same idea toward the cosmic structure and within such belief, man has tried to live by the orders of nature in order to resemble or to equate self to it.

The realm of science can also contribute to unfolding the picture. A core value from the teachings of the religious systems mentioned above which is 'part in the whole, whole in part' can be explicitly found in the well-known scientific theory called 'Fractal'[12] and 'Holographic universe'. The key concept is an assumption that the cosmos is composed of infinitely many parts in which the cosmos itself is embodied. In other words, every entity is micro-cosmos itself, enclosing all attributes of the cosmic order in itself. Everything is cosmos and cosmos is everything. Extending this scientific hypothesis to the religious teaching of 'whole in part, part in the whole', it is very clear that two shares the same idea on the understanding of the cosmos.

What the exploration into the evidence leads to is a constructive understanding of the way of thinking, experiencing and being as an entity that was not ever imagined. If this structural concept of the cosmos is true after all, what is it like to be in the mode of being as such, understanding that I am the cosmos itself? What am I then? What can I come to know? What seems obvious is the fact that if such structure is the reality, what I am composed of is exactly the same as

[10] Exert from the Chinese version of 'The Avatamska Sutra'. The Avatamska Sutra is one of the most influential Mahayana sutras of East Asian 10 Buddhism. The title is rendered in English as Flower Garland Sutra, Flower Adornment Sutra, or Flower Ornament Scripture. The Avataṃsaka Sūtra describes a cosmos of infinite realms upon realms, mutually containing one another. Reference: http://en.wikipedia.org/wikl/ Avatamsaka_Sutra

[11] Macrocosm and microcosm is an ancient Greek Neo-Platonic schema of seeing the same patterns reproduced in all levels of the cosmos, from 11 the largest scale (macrocosm or universe-level) all the way down to the smallest scale (microcosm or sub-sub-atomic or even metaphysical-level). Reference: http://en.wikipedia.org/wiki/Macrocosm_and_microcosm

[12] A fractal is a mathematical set that has a fractal dimension that usually exceeds its topological dimension and may fall between the integers. 12 The term "fractal" was first used by mathematician Benoît Mandelbrot in 1975. In physical cosmol ogy, fractal cosmology is a set of minority cosmological theories which state that the distribution of matter in the Universe, or the structure of the universe itself, is a fractal. Reference: http://en.wikipedia.org/wiki/Fractal_theory

what others are composed of, the cosmos itself. What I am is what others are and everything is identified as everything in infinity. I am others, I am the cosmos and as an ultimate 'General' being, I reside in everything in every frame of space and time. In this constitution, as a universal being in totality, the act of 'real-knowing' becomes obvious and simultaneous embodiment. In this sense, ESP can be considered as an obvious way of perceiving the true picture of the world. Going beyond bodily senses might first be supposed as unreachable meta- physical realm but embracing the true totality of the cosmos and its working systems, it can show the concrete picture that ESP has always been at work but has gone largely unnoticed out of its 'omnipotent' actuality. Because it is in everywhere, it goes unnoticed and claimed to be a non- existing element. Like the meaning of a Chinese character '無', it does not signify the state of nothingness or zero, but it implies 'a state of being everywhere at every moment'. Because of its omnipresent quality, it is considered to be something that is not experienced or more accurately said, it is always being experienced but failed to be acknowledged consciously.

Forces in Nature

The angel who talked with me had a measuring rod of gold to measure the city, its gates and its walls... (Revelation 21:15, New International Version)

"Measure-taking gauges the between, which brings the two, heaven and earth, to one another. This measure-taking has its own metron, and thus its own metric."[13] (M. Heidegger)

How can 'Extrasensory perception' be applied in order to analyze the architecture from the ancient in its proper sense? As briefly mentioned, it involves the process of concealing and un-concealing self in the form of question and answer. But what kind of questions must be formulated to really know the true picture of reality? Is it concrete enough to pose questions such as "why is this site chosen?" or "Is this architecture proper for this site?" The answer to such questions will definitely be attained since truth already resides in us. Anything asked will be answered accordingly. But there is a twist. Because the answer is always in correspondence with the question, if the question is not explicit or accurate enough, the answer will also be implicit and ambiguous.

It is well known that throughout the history people have taken great interests in inquiring ways to harmonize the human dwelling with the surrounding environment. Notable cases are found in the studies of Feng shui[14], Vastu Shastra[15] and Tajul muluk[16] which

---

[13] Martin Heidegger, Poetry, Language, thought, trans. Albert Hofstadter, New York: Harper and Row, 1971, "...poetically man dwells...", P.221 13

[14] Feng shui is a Chinese system of geomancy believed to use the laws of both Heaven (Chinese astronomy) and Earth to help one improve life by 14 receiving positive qi. Depending on the particular style of feng shui being used, an auspicious site could be determined by reference to local features such as bodies of water, stars, or a compass. Source: Wikipedia (http://en.wikipedia.org/wiki/Feng_shui)

[15] Vastu shastra is an ancient doctrine, which consists of precepts born out of a traditional view on how the laws of nature affect human dwellings. The designs are based on directional alignments. It used to be applied in Hindu architecture, especially for Hindu temples, and covers other domains, including vehicles, vessels, furniture, sculpture, paintings etc. Source: Wikipedia (https://en.wikipedia.org/wiki/Vastu_shastra)

[16] Tajul Muluk is the most commonly used name for the Malay system of geomancy, comprising metaphysical and geomantic principles considered when siting or designing buildings to improve and maintain well-being. It was traditionally practiced by shamans (dukun or bomoh) and architects

are all various forms of doctrines that concentrate on the prosperity of human beings by managing and improving spatial treatment. A notion known as 'Geomancy'[17] also shares an associated essence and implication. It is not so surprising to witness an extensive influence of Feng shui in construction markets in China and other Asian countries and still there are social believes and pious pursuits to follow the rules of it in domestic interiors. It is not only the case for Feng shui, a Vastu tradition is still playing an important role in Indian cultures especially for the construction of sacred architectures. Geomancy is known to be an Islamic or Arabic origin. By acknowledging the great architectural achievements in America and Africa, it is quite reasonable to claim that though there are not eminent tracks not discovered, they must have some kind of science that concerns the building regulation. The common aspect of all such systems of spatial engagement lies in understanding the logics behind the workings of earthly realm and trying to reconnect to the dimension of a celestial or spiritual encounter so that a building can resonate with the order of the universe in conformity.

One particular concept springs from the vast record which is rendered as some kind of flow or force that carries positive or negative energies. Well known terms are 'Qi' in Feng shui, 'Pancha maya bhoota' in Vastu Shastra, 'Tiang seri' in Tajul muluk and the modern perception on the similar phenomenal evidence includes terms 'Ley line'[18], 'Song line'[19] and so on. These are the terms that are used to describe the cause behind the ancient heritages such as Stonehenge, Pyramids, dolmens and other sacred sites and buildings but only in theoretical assumptions. But if the records of the scriptures and teachings do claim the existence of such flow of forces in our environment, it is not to be ignored without trouble. In fact, when 'Extrasensory perception' is brought in to consciously acknowledge the reality as it is, what is vague and uncertain becomes an incontrovertible picture that we have been missing habitually.

The forces are found to be pervading on the surface of the earth. Precise causes of their formation are not determined objectively but the possible explanation might be that the earth, within and without, is in constant interaction between entities of various scales from natural topographical features to planets and stars in outer space. If there are forces of interaction between celestial bodies at work, it is natural to infer that the earth which is only a tiny spot in the vast cosmos cannot possibly escape from this network and therefore always finds itself in this subtle flow of matrix. What 'Extrasensory perception' reveals through the exploration on the ancient heritages makes a clear-cut case that there are indeed certain forces present in space.

from Malaysia and Indonesia. Source: Wikipedia (http://en.wikipedia.org/wiki/Tajul_muluk)

17 Geomancy, from Ancient Greek geōmanteía translates literally to "foresight by earth". Source: Wikipedia (http://en.wikipedia.org/wiki/Geomancy)] Geomancy defined by Marko Pogacnik is an ancient word denoting the knowledge of the invisible and the visible dimensions of the Earth and its landscapes. He refers to geomancy as sacred geography which is not only about one side reality but the multi-dimensional eternity. Source: Sacred geography by Marko Pogacnik, preface.

18 Ley lines are supposed alignments of a number of places of geographical and historical interest, such as ancient monuments and megaliths, natural ridge-tops and water-fords. Source: http://en.wikipedia.org/wiki/Ley_line

19 Within the animist belief system of Indigenous Australians, a songline, also called dreaming track, is one of the paths across the land (or sometimes the sky) which mark the route followed by localised 'creator-beings' during the Dreaming.

What is this force that is so important for selecting a site and spatializing a building? Force is a condensed flow of force that is generated by the features (material, color, form) of space on earth by its subtle relation to the celestial structure. It, as a concept, can be compared to 'earth radiation'[20], which deals with the belief that there are lines on the surface of the earth that can affect the condition of human life and it is the case that they certainly do. This proves why people in the ancient talked about the notion of 'positive spot' or 'divine site' where their god or spiritual leader must sit. In short, for the ancient, the science of building was an effort to know and manage 'space'[21] in its proper sense through the mode of 'Extrasensory perception'. What follows next is the description of the forces found through years of field studies on the ancient heritages. There are a lot more forces present on earth but the presented are the three main forces that are frequently measured on the sacred sites.

Forces and Ancient Placement

Cylindrical force from above

This is a force that falls perpendicularly from above in the shape of 'cylinder'. 'Above' here refers to the celestial domain which suggests it to be some kind of 'interaction' point between the earth and the celestial. It is supposedly generated from the interplanetary motions and changes. With such characteristics, it is usually compared to acupuncture point or Chakra point in a human body that the spot this force is falling is a kind of optimal point for harmony and balance between the earth and the outer space. It might be seen as 'cosmic pillar' as some of the myths briefly address.

Interesting phenomena from the site investigations on the ancient heritages is the fact that almost every known 'sacred' places are bound with this 'cylindrical force from above'. Sacred places here refers to the places that are known to have functioned as religious or spiritual worship such as temples, sanctuaries, mosques, churches, and shrines. The cause for this correlation is not explicitly known but extrasensory perceptive observation discloses the absolute interaction between the two. One possible theory can be such that because this force falls from above, from the far greater workings of celestial dimension, it might be the channel of important communication and harmonization between the celestial bodies and such ethereal quality embeds more original 'Nature-ness'. The purpose of all religions and spiritual enlightenment is to resemble and finally be 'Nature' and in order to pursuit that path, space must be also in line with the corresponding quality of spiritual sacredness.

The Great Pyramids and the Great Sphinx in Egypt are all bound with the 'Cylindrical force from above'. Other cases where the 'cylindrical force from above' is incorporated includes Stonehenge in UK, The Par-

[20] Earth radiation is a folklore belief that there are lines on the surface of Earth that can affect health. The belief has an old tradition (lay lines, Odic force, Mana, Qi) but was revived by the German authors Manfred Curry and Ernst Hartmann. They described a mystic force field affecting the health of living creatures that supposedly covers the earth at regular intervals and may be detected by dowsing using a divining rod. Source: Wikipedia (http://en.wikipedia.org/wiki/Earth_radiation)

[21] The extent of the term 'space' can be manifold as current debates and discourses on the notion of it are various. The term used in this context indicates 'the three-dimensional extent in which objects have relative position and direction'. The 'objects' in this context imply as from the features of natural geography (mountain, river, field, sea, etc.) and man-made masses structures on earth (buildings, bridges, roads, objects, etc.) to celestial bodies (sun, planets, moons, stars, etc.) in outer space.

thenon in Greece, Dome of the Rock in Israel, numerous megalithic structures across the world and many other sacred places such as temples, cathedrals, mosques and other shrines. Giant Wild goose Pagoda and carved-out statues in Longman Grottoes in China are also bound with the same kind of forces with utmost precision. Detailed use of forces are astonishing.

Cylindrical force from below

Cylindrical force from below' shares almost the same features and qualities as 'Cylindrical force from above' except that this force radiates from under. An earth is also in its own, a sum of subtle aggregates of bodies and forces and in response to the far larger network of celestial realm, it is thought to communicate with it in terms of the channel-like flow of forces. Everything in the universe is connected with each other and they, in a natural way, exchange and act upon one another.

The springs of Moses in Egypt are bound with the cylindrical force from below. All springs at the sites are carefully aligned with the axis of the forces.

Topological force (Similar to Qi as life force in traditional Feng-Shui study)

Every substance is composed of material, colours and form. There has been enormous development and academic research on the category of material and colour but the study on a form-oriented investigation has not received enough attention due to the fundamental difficulties in measuring the multidimensional facets of it with a conventional level of science. In the past, however, there have been many cases and practical applications of utilizing a form and it's radiating forces in wide a range of living. The typical accounts were Yantra[22] and Mandala[23].

Any form that pertains to be in physical structure and shape possesses, in its own particular sense, distinctive features and role in the form of forces. Topological force is significant in this sense because it is generated by such topological features of a certain space in response to the celestial-bound influence.

The natural settings are normally composed of mountains, ridges, valleys, rivers, and plains. In micro scale, there are many vectorial elements within it such as growing plantation and flowing water. Recognizing 'Topological force' is in this respect a carefully engaged process of finding out certain values and significance of the setting by observing the topographical features. Of course, man-made built environment does also impose certain effects on the network of Topological forces with its physicality but the magnitude and degree of influence are, compared to naturally set topography,

22 Yantra is the Sanskrit word for "instrument" or "machine". Much like the word "instrument" itself, it can stand for symbols, processes, automata, machinery or anything that has structure and organisation, depending on context. One usage popular in the west is as symbols or geometric figures. Traditionally such symbols are used in Eastern mysticism to balance the mind or focus it on spiritual concepts. The act of wearing, depicting, enacting and/or concentrating on a yantra is held to have spiritual or astrological or magical benefits in the Tantric traditions of the Indian religions. Source: Wikipedia (http://en.wikipedia.org/wiki/Yantra)

23 Mandala is a spiritual and ritual symbol in Hinduism and Buddhism, representing universe. In various spiritual traditions, mandalas may be employed for focusing attention of aspirants and adepts, as a spiritual teaching tool, for establishing a sacred space, and as an aid to meditation and trance induction. In common use, mandala has become a generic term for any plan, chart or geometric pattern that represents the cosmos metaphysically or symbolically, a microcosm of the universe from an enlightened perspective, i.e. that of the principle deity. Source: Wikipedia (http://en.wikipedia.org/wiki/Mandala)

weak and therefore considered to be minor. 'Topological force' normally comes down along the ridge of a mountain and a hill or flows off the ridges of buildings and other man-made structures.

Some temple compounds and tombs were positioned by reading 'Topological forces' at work. In some cases, the combination of Topological force and Cylindrical force from above are measured at the core point where the Buddha statue sits. Due to its 'flowing' property, Topological force acted as an alignment- line (axis) of spaces in the compounds. The sites are carefully aligned with the axis of the forces.

Future Studies

The evidence of 'Extrasensory perception' in a body is explicitly palpable. Architectural manifestations of the ancient around the world acutely show the direct bond between their method of spatialisation and 'forces' and this nexus cannot be possibly connected without 'Extra sensory perception', an intuitive perceptive mode that is naturally unconcealed when subjective inclinations and predetermined conceptions are receded.

It is definitely not enough to convey the perfect picture of the notions and practicality that have played central in this study in such a short regard and on that account, it is only the start of the infinite voyage into the 'reality'. The domain that the notion of 'All is One, One is All' and the practicality of 'Extrasensory perception' can touch is just infinite. As the reader might have hinted from the contents conveyed, nothing can escape from this infinite domain. In architecture, it is probably the most essential agenda to select a proper site for an intended building. A site must be chosen primarily based on the quality that it holds for the building and the quality is greatly influenced by the forces at work. Within a site, a specific spot where a building with a certain purpose and function should be positioned must be precisely measured with its corresponding proper forces. Design should also be carefully considered and executed based on this basis. If the ideal spot or sites can be verified, dimension and form of the building can also be composed so that it can resonate with the forces at play. 'Ideal' is not different from 'Real'. In fact, the two were never apart. It is our consciousness that fragmented the whole picture. I hope that the contents of this paper might invite the readers to walk on the path of knowing all there is.

Fireplace and Holy Altar in Curiceta at Apuan Alps, Italy

Natalia Tarabella, Paolo Debertolis, Daniele Gullà, Lorenzo Marcuccetti

NATALIA TARABELLA, architect, Super Brain Research Group, Florence, Italy (*[1])
PAOLO DEBERTOLIS, medical doctor, Department of Medical Sciences, University of Trieste (Italy), Super Brain Research Group (*)
DANIELE GULLA', forensic researcher, Super Brain Research Group, Bologna, Italy(*)
LORENZO MARCUCCETTI, historian, Super Brain Research Group, Florence, Italy (*)

ABSTRACT: The Apuan Alps, High Versilia and Garfagnana are part of the Apuan Alps Park, and are rich in petroglyphs and archaeological finds which are, in part, unknown and not precisely datable. These areas have been inhabited since Neolithic times but, the meaning and the reason for signs engraved on stone, is unclear. The Apuan Alps were chosen as a dwelling by people who left many ancestral and Christianity signs including sacred altars, thrones and artefacts carved in stone. This area acts like a stone atlas revealing our past and our roots. The Curiceta site is located inside a thick forest of chestnut trees, in an area where dried stone terraces are perfectly preserved. The first building approached along the path, is the so-called "*fireplace*", a large flat stone that protrudes from the ground and surrounded by a series of aligned stones where, most probably, a fire was lit. Behind the big stone, is a cavity where the smoke could emerge. The lower part of the flat stone features a "handle" carved in the rock, its function is still unknown. Along a short stretch of the uphill path protected by a high dry stone wall, lies the sacred stone altar. This enigmatic structure has revealed many surprises during the tests performed with electronic instruments. This altar is carved from a single block of stone and consists of a backrest and a horizontal supporting surface. From the left side, there are inclined planes which climb down, below these one can find a vertical groove. The altar features the same carved handle found on the "fireplace". Rock altars are very common around the world, for example throughout Southern Italy and the Middle East but, in High Versilia this is the only one example. Archaeoacoustic analysis of the altar found a dominant and powerful frequency present of between -47 and -50db at 25 – 28Hz. A second peak of infrasound at 15-16Hz was also found. This inaudible acoustic characteristic is commonly found at sacred sites, such as the Neolithic temples of Malta (Tarxien – Xaghra Stone Circle). These same vibrations are present near the altar but at a much lower volume. In this case the loudest volume was found directly under the altar decreasing as one walks away from it. In both cases, the most likely source of this frequency is from underground water. The emotional state of eight volunteers was analysed using a TRV camera. 7/8 felt emotionally uncomfortable or uneasy. Based on these results, a hypothesis was formulated on the function of the Curiceta's site. The two stone structures are connected. On the altar, sacrifices were probably, made, with blood flowing along the left side to the groove on the floor. The fireplace, could have been used to burn the bodies or maybe just some organs.

KEYWORDS: archaeoacoustics, Apuan Alps, dolmen, altar, low frequency sound, infrasound

[1] Note. Super Brain Research Group (SBRG) is an international and interdisciplinary team of researchers, researching the archaeoacoustic properties of ancient sites and temples throughout Europe and Asia (www.sbresearchgoup.eu).

Introduction

When the first hominids appeared on Earth, the wild nature in their surrounding environment imposed on them the need to adopt strict rules to guarantee their survival. The advancement of glaciers shaking the northern hemisphere into a frostbite vortex, made those surviving grateful for the shelter and warmth offered by the caves. These ensured an effective thermal insulation by providing shelter from the fury of fierce elements and wild beasts. Like other animals that shared the space, the senses of hearing, sight and smell were acutely developed. Sounds, ultrasounds, vibrations and energy fields were perceived instinctively as part of the natural environment. Inside these caves, people began to erect the first monuments to provide both functionality and sacredness. The caves were equipped with pits dug beneath the stalactites to collect water drops, but they also assumed a ritual charge, as evidenced by many representations and graffiti found that refer to fecundity rites. When man began to build sacred artefacts outside of the rocky environment, the location of such structures was likely chosen according to the vibrations felt and experienced at any specific location. There was a combination of beauty, sacredness and functionality; these three aspects were the foundation which characterized the heritage of mankind, the unity between *man*, the *ecosystem* and the *cosmos*.

At the bottom of the caves of Matera is a cistern devoid of any connecting pipes, yet filled with precious liquid, due to the entire cave working like a water condensation system. A ray of sunshine penetrates the cave and beats against the rocky bottom, thus celebrating the encounter of the solar male principle with the female principle of the earth, which in turn creates water the source of life (Fig. 1).

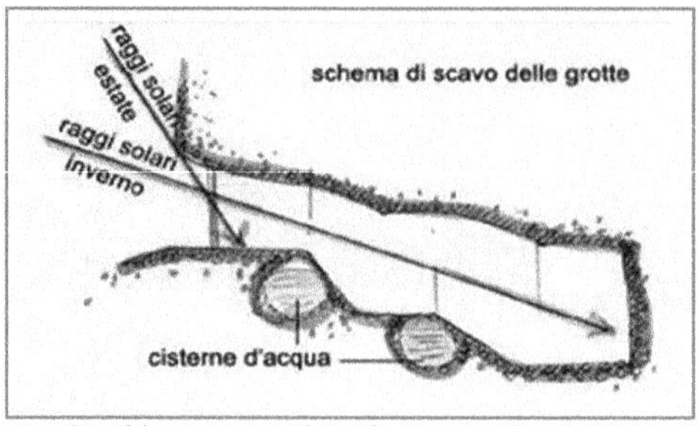

Fig. 1 – Graphic representation of a cave section in Matera (Basilicata) showing the inclination of solar rays in summer (raggi solari estate) & winter (raggi solari inverno).

To consider archaeoacoustics a modern discipline is incorrect, if we consider that man in the past naturally perceived sounds and vibrations of a particular location channelling the energy to favour his body.

In order to hear the voice of sacred structures in the modern age, we need sophisticated instruments to capture their sounds, the interpretation of which can present a challenge.

A spectacular case of such a Bronze Age complex, can be found on the Murgia Materana (Matera-Basilicata, Italy). The site consists of two concentric stone rings crossed by an East-West corridor, which leads to a central hypogeum. The hypogeum is divided into two environments supported by a pillar carved from the rock (Fig. 2, above left). This structure is quite similar to the so-called "*solar monuments*" of the Sahara, whose function remains enigmatic and whose name is attributed to the astral motifs attributed to the stone circles [24] (Fig. 2, above right).

The analogies with other important monuments attest to the fact that its design was connected to water cults. Having a similar aspect to Sardinia with its great sacral complexes from the Metal Age passing through the access corridor to descend a deep stairway into the central hypogeum that features a sacred well that works not because it meets groundwater but because it intercepts the rainwater (Fig. 2, below right).

In Petra, the so-named "*high place*" located on the highest mountain, was a centre of energy and power over life and death, where ablutions and rites with holy waters were celebrated. On its summit lies an altar the shape of which consists of two concentric rings penetrated by a duct which is designed to collect rain water. When this water filled the cisterns it brought the place to life and the ceremonies commenced, filled the cisterns and brought life [24], (Fig. 2, below left). Beauty, sacredness and functionality were united in celebrations of banquets and funeral ceremonies that had an important social value and were permeated by profound symbolic content.

Fig. 2 – Bronze Age Monument in Matera, Basilicata, Italy (above left); Solar Monument in Sahara desert (above right); Holy Pool in Petra, Jordan (below left); S. Cristina sacred well, Sardinia, Italy (below right)

Fig. 3 – Cave of Loltun, Yucatan (above); Cave of Tanaccio, Tuscany, Italy (below).

Apuan Alps: the Atlas of Stone

The Apuan Alps (Fig. 4), the High Versilia and the Garfagnana are part of the Apuan Alps Park, rich in petroglyphs and archaeological finds, many of which are unknown and undeclared. These areas have been inhabited since Neolithic times, but the significance and reason for stone engravings remains unclear. The Apuan Alps were chosen as dwellings by people who left many ancestral and religious signs of their testimony, including sacred altars, thrones and artefacts sculpted from stone [1]. It can be considered a *stone atlas* revealing our past and our roots.

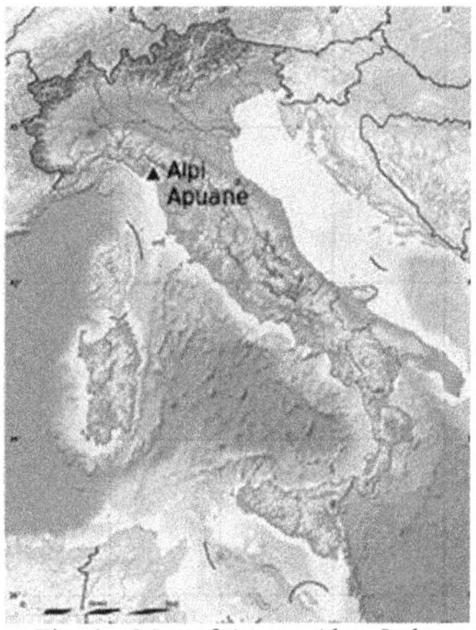

Fig. 4 – Map of Apuan Alps, Italy

Fig. 5 – The Apuan Alps: Nona mount and Procinto

The Site of Curiceta, Seravezza, Tuscany

The Curiceta site is located in a dense chestnut forest, in an area where dry and perfectly preserved stone terraces are found. The first stone building on the path, is the so-called *"fireplace"* (Fig. 6). It is a large flat stone set in the ground with a series of aligned stones where probably, a fire was lit. Behind the big stone, is a cavity where the smoke came out. In the lower part, there is a "handle" carved in the rock, its function is still unknown.

Fig. 6 – The *fireplace* in Curiceta, Seravezza, Tuscany.

Along a short stretch of the path on the hill, lies the sacred stone *altar* (Fig. 7), the second building. This enigmatic structure, has revealed many surprises when tests were performed with electronic instruments. The altar is carved from a single stone block and consists of a backrest and a horizontal supporting surface. From the left side, there are inclined planes which go down. Below which there is a vertical groove. On the altar we find a "carved handle" in the first structure of the so-called "fireplace". Rocky altars are common found throughout the world in sites such as, in Basilicata, Apulia, and Sicily in the South of Italy, the ancient site of Petra in the Middle East. However in High Versilia this is the only one.

Fig. 7 – The altar in Curiceta (Seravezza – Tuscany)

Fig. 8 – Fireplace, detail of the handle (above); altar, detail of the handle (below).

Fig. 9 – Sound recording equipment & set-up at the altar

Fig. 10 – Spectran NF-3010 from German factory Aaronia AG

Fig. 11 – The TRV camera test to value the emotional state of volunteers sitting on the altar.

Materials and Methods

Equipment for the sound recordings consisted of two types of dynamic high-end microphones extended in the ultrasound frequency range, with a digital portable recorder (Tascam DR-680 of TEAC Group, with a maximum sampling rate of 192KHz). Professional studio microphones with a wide dynamic range and a flat response at different frequencies (Sennheiser MKH 8020, response Frequency 10Hz - 60.000Hz) with shielded cables (Mogami Gold Edition XLR) and gold-plated connectors (Fig. 10) were also used.

Before recording a spectrum analyzer (Spectran NF-3010 (Fig. 10) from the German factory Aaronia AG) was used to detect the presence of any electromagnetic phenomena which could influence the results.

Praat program version 4.2.1 from the University of Toronto and Audacity open-source program version 2.1.2 for Windows and Linux were used to analyise the audio recordings.

Thermography was used to analyse the temperature characteristics of the structure by use of a thermal imaging camera (model ThermaCAM SC640 IR Camera by Flir Systems Inc).

A TRV camera (Variable Resonance Imaging camera, known as a Merlin camera in Italy or Defend X system in Japan for industrial use) was used to test the emotional state of 8 volunteers situated on the sacred altar area (anthropologic analysis), a system used in previous research. TRV camera works by valuing the balance of the head and the micro-mobility of the body which is controlled by the vestibular system (inner ear). This system is influenced by the emotional state of the subject (quiet or anxious) and it is possible for a computer camera to perceive these as micro vibrations. By using dedicated software (Vibraimage Pro 8.3) the shape of examined subjects can be coloured to understand their state of mind. This system is used by secret services as a "lie detector" in the field of terrorism. TRV camera has a common CCD backlit, with a three MegaPixel sensor. Its protective anti-aliasing filter was removed to extend its ability to capture light from the infrared (IR) and ultraviolet (UV) bands (the lens has a 25 mm quartz-fluorite optics with a pass band from 200nm to 1800nm).

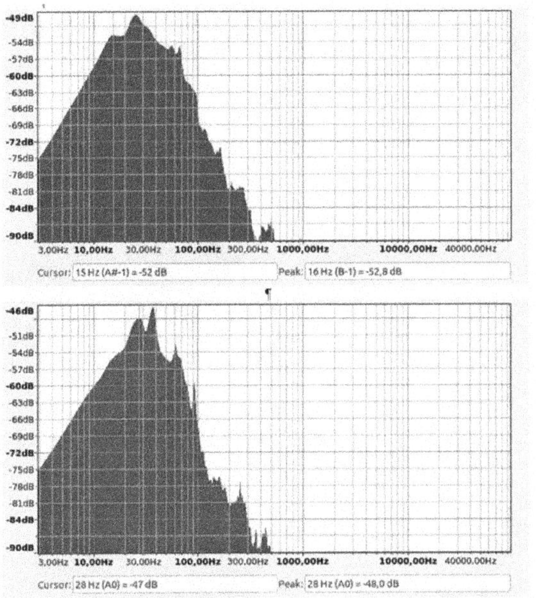

Fig. 12 – The graphic audio analysis at Curiceta's altar by Audacity 2.1.2

Results

Analysis of the microphone recordings revealed a dominant frequency of 25–28Hz that is powerful in volume (-47 to -50db). Another lower peak of 15-16Hz was revealed in the graphic analysis (Fig. 12). To avoid recording mistakes, some analysis was made directly on site by computer (Fig. 16). Near the altar in the fireplace the same vibrations were found, but at a much lower sound level. The source of the sound is directly under the altar with the sound decreasing when walking away from it. The site was full of water and listening by headphones during the recordings the sound of falling water was heard, so the most likely explanation for the source of these low frequency (infrasound) vibrations was from the flow of underground water. Infrasounds have a physiological effect on the body, for example those individuals who consider themselves to be sensitive state they often sense such vibrations as unspecified energy emanating from underfoot,. Infrasound frequencies can also enter the brain without passing through the hearing organ, entraining the brainwave rythm into an Alpha-Theta state.

Some clarification of the characteristics of these results in respect of the measured volume should be mentioned: in that there is a distinction between using decibels to measure sound pressure levels as opposed to signal levels. Sound Pressure Levels are a measurement of air pressure which is caused by sound or noise, this results in physical forces moving against the diaphragm of a microphone and in the acoustic environment this translates to volume. Measurements of this nature are usually expressed as decibels of sound pressure level (dB SPL) and are measured in positive numbers. For example a rock concert can reach 110db or a jackhammer 100db, moreover a person whispering is around 20-30db.

When dealing with signal levels, decibels are used differently. In this case, 0 dB is the highest signal level achievable without any distortion; all signal levels below this are represented as negative numbers. A volume fader may be labeled with a "0", part way up to mark the point at which that fader is neither boosting nor attenuating the signal. The measurements taken at Curiceta altar show a level of -47db which is a medium volume.

Using the thermal imaging camera, we discovered, the altar stone is colder to the rear when compared to the horizontal stone in the foreground, with more than 4 C° of temperature difference (Fig. 13), which is interesting given that this stone is carved from only one piece of rock. It is clear that only a cold flow coming from behind can cause this difference of temperature, so we concluded some sort of cave was located behind the altar.

The results of the eight volunteers tested by TRV camera, revealed that seven out of eight of them experienced a non relaxed state of mind. After few minutes of exposition to the vibrations, we previously measured on the altar, they became anxious and agitated and almost all volunteers felt emotionally uncomfortable, experiencing a sense of fear or feeling like a strange or supernatural event was taking place. The data by TRV camera, which is able to recognize the state of mind of subjects, were really clear in this sense (Fig 14).

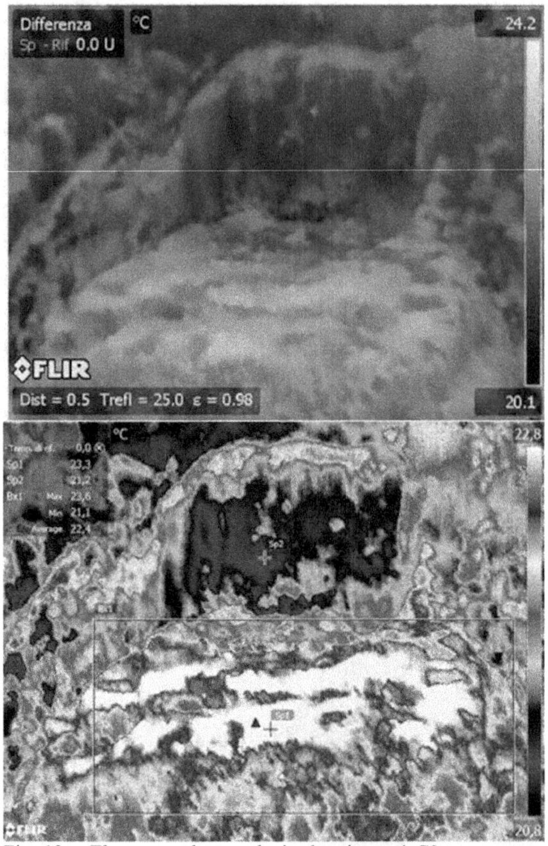

Fig. 13 – Thermographyc analysis showing a 4 C° temperature difference between the horizontal stone in the foreground and the stone behind.

Discussion

The archaeoacoustic study at this site was carried out without any prior archaeological excavations having been undertaken, which actually raised more questions than answered. In both locations where the microphones were placed (about 30 meters apart), the same low vibration frequency signature was detected as a continuous sound. There were no factories or man made activity capable of generating such a frequency in the neighbouring vicinity that we were aware of. The pre-recording clapping tests conducted found the microphones were positioned deep enough as to be scarcely affected by the external noise environment. No sources of electromagnetic fields were found. Those frequencies recorded therefore should be considered as being an accurate representation for this site. According to an anthropologic analysis, we could suppose the two stone structures are connected (altar and fireplace).

Fig. 14 – Some volunteers having their emotional state recorded by TRV camera.

Fig. 15 – To avoid recording mistakes, some analysis was made directly on site by computer.

The results on our volunteers are repeatable, but also incontrovertible as the accuracy of the TRV method has been established in the security field. Why then did almost all volunteers experience a sense of fear? Was there something in the environment affecting their state of mind? Infrasounds can induce feelings of awe or fear and given they are not consciously perceived, it may make people feel like strange or supernatural events are taking place [25]. It is therefore possible to hypothesize that where a concentration of natural low vibrations are present, ancient populations considered these sites to be supernatural or sacred [3].

A similar situation exists at Xaghra Hypogeum, on Gozo Island, Malta, where extremely powerful natural frequencies were found [21]. These are comparable to what was found at Tarxien temples on Malta, but with a slightly longer high frequency range and a small amount of oscillation.

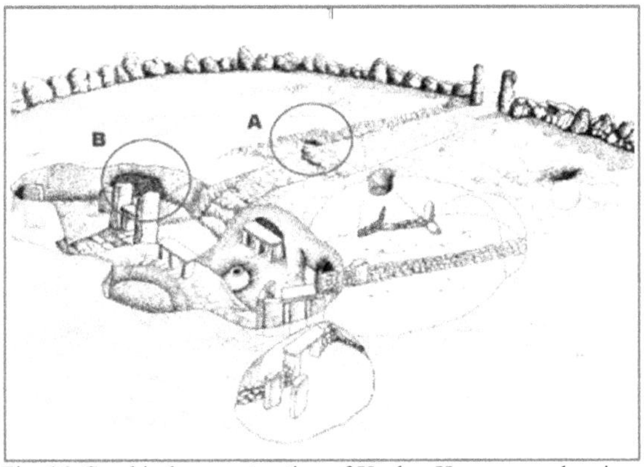

Fig. 16 Graphical reconstruction of Xaghra Hypogeum showing the locations where the microphones were placed (A & B)[21]. Drawing by Natalia Tarabella.

They have a broad peak around 25Hz at -24db. Consideration needs to be given to the fact that Xaghra Hypogeum was carved from the soil making the volume more powerful than at Curiceta altar. However the effect on the mind was totally different, because never discomfort/fear was reported by the people who visited this hypogeum. This raises the following questions: (1) is it possible that the *combination* of frequencies (15-16Hz & 25-28Hz) at the Curiceta site creates a sense of fear? (2) If infrasound does in fact cause feelings of awe or fear as described by other authors, is it possible that this was known about and used in certain rites or ceremonies? These form the basis for stimulating hypothetical questions in which to approach further research.

The study by thermographic camera threw up an interesting result, finding a temperature difference of 4 degrees, that led us to conclude some sort of cave or cavity was located behind the altar. This being the case, why block the cave or cavity with such an altar in front of it? Could this rock actually be acting as some sort of transducer? Is it possible that even stronger vibrations could be found within the cave that might for example be closer to the volume found at Xaghra hypogeum? For now we have no answers, but in future it could be interesting to go on in our research more deeply for having them.

Fig. 17 – Infrared image of Curiceta altar. This image deletes lichens enabling the original structure and the wall behind covering a cave to be more visible.

The altar is carved in a single block of stone and is formed by a backrest and a horizontal supporting surface. From the left side, there are inclined planes which climb down. Below these sits a vertical groove, a perfect channel in which sacrificial blood can flow. The rock altars are very common around the world as, for example, in the South of Italy, in the Middle East but, in High Versilia this is the only example.

Conclusions

The study by thermographic camera threw up an interesting result, finding a temperature difference of 4 degrees. That led us to conclude some sort of cave was located behind the altar. Indeed closer examination revealed a wall of little stones around the altar that look like they were placed there to cover the entrance of the cave. The underground water found at the Curiceta's site is significant because the combination of low frequencies can create an altered state of mind especially during any rituals. We established that where a concentration of natural low vibrations are present, ancient populations considered these sites to be supernatural or sacred and certainly considered as "places of power". We can also suppose, the combination of frequencies is the most likely cause for the discomfort/fear felt by the volunteers. Perhaps Curiceta was used for a number of different ceremonial purposes over the centuries. The shape of the altar draws one to conclude they may well have been used to celebrate sacrifices. The fact that almost all the volunteers, seven on eight, felt emotionally uncomfortable, experiencing a sense of fear or trepidation, lends weight to the argument that the natural frequencies present at this site created the perfect environment in which to conduct sacrificial ceremonies.

ACNOWLEDGEMENTS

The authors are very grateful for the support received by non-profit scientific organization Super Brain Research Group institute (SBRG) for the development of this archaeoacoustic research. The authors are also grateful to Department of Medical Sciences at the University of Trieste (Italy) for supporting this research and in particular to the Director, Professor Roberto Di Lenarda. A sincere thank you to Nina Earl for her support in editing this text.

REFERENCES

[1] E. Calzolari: "Il dolmen del Monte Freddone" in "Il Cielo in Terra ovvero della giusta distanza", 1st edition, Padova University Press, University of Padova, Italy, 2015: pp. 53-62.
[2] P. Debertolis, H.A. Savolainen: "The phenomenon of resonance in the Labyrinth of Ravne (Bosnia-Herzegovina). Resultsof testing" Proceedings of ARSA Conference (Advanced Research in Scientific Areas), Bratislava (Slovakia), December, 3 – 7, 2012: pp. 1133-36.
[3] P. Debertolis, N. Bisconti: "Archaeoacoustics in ancient sites" Proceedings of the "1st International Virtual Conference on Advanced Scientific Results" (SCIECONF 2013), Zilina (Slovakia) June, 10 - 14, 2013: pp. 306-310.
[4] P. Debertolis, N. Bisconti: "Archaeoacoustics analysis and ceremonial customs in an ancient hypogeum", Sociology Study, Vol.3 no.10, October 2013: pp. 803-814.
[5] P. Debertolis, S. Mizdrak, H. Savolainen: "The Research for an Archaeoacoustics Standard", Proceedings of 2nd ARSA Conference (Advanced Research in Scientific Areas), Bratislava (Slovakia), December, 3 – 7, 2013: pp. 305-310.
[6] P. Debertolis, N. Bisconti: "Archaeoacoustics analysis of an ancient hypogeum in Italy", Proceedings of Conference "Archaeacoustics: The Archaeology of Sound", Malta, February 19 - 22, 2014: pp. 131-139.
[7] P. Debertolis, G. Tirelli, F. Monti: "Systems of acoustic resonance in ancient sites and related brain activity". Proceedings of Conference "Archaeoacoustics: The Archaeology of Sound", Malta, February 19 – 22, 2014: pp. 59-65.
[8] P. Debertolis, A. Tentov, D. Nicolić, G. Marianović, H. Savolainen, N. Earl: "Archaeoacoustic analysis of the ancient site of Kanda (Macedonia)". Proceedings of 3rd ARSA Conference (Advanced Research in Scientific Areas), Zilina (Slovakia), December, 1 – 5, 2014: pp. 237-251.
[9] P. Debertolis, D. Gullà, Richeldi F.: "Archaeoacoustic analysis of an ancient hypogeum using new TRV camera (Variable Resonance Camera) technology", Proceedings of the "2nd International Virtual Conference on Advanced Scientific Results" (SCIECONF 2014), Žilina (Slovakia) June, 9 - 13, 2014: pp. 323-329.
[10] P. Debertolis, F. Coimbra, L. Eneix: "Archaeoacoustic Analysis of the Ħal Saflieni Hypogeum in Malta", Journal of Anthropology and Archaeology, Vol. 3 (1), 2015: pp. 59-79.
[11] P. Debertolis, D. Gullà: "Archaeoacoustic analysis of the ancient town of Alatri in Italy", British Journal of Interdisciplinary Sciece, September, Vol. 2, (3), 2015: pp. 1-29.
[12] P. Debertolis, M. Zivić: "Archaeoacoustic analysis of Cybele's temple, Imperial Roman Palace of Felix Romuliana, Serbia", Journal of Anthropology and Archaeology, Vol. 3 (2), 2015: pp. 1-19.

[13] P. Debertolis, D. Nicolić, G. Marianović, H. Savolainen, N. Earl, N. Ristevski: "Archaeoacoustic analysis of Kanda Hill in Macedonia. Study of the peculiar EM phenomena and audio frequency vibrations", Proceedings of 4th ARSA Conference (Advanced Research in Scientific Areas), Zilina (Slovakia), November 9 – 13, 2015: pp. 169-177.
[14] P. Debertolis, D. Gullà, "Anthropological analysis of human body emissions using new photographic technologies", Proceedings in Scientific Conference "The 3rd International Virtual Conference on Advanced Scientific Results (SCIECONF-2015)", Slovakia, Žilina, May 25-29, 2015; Volume 3, Issue 1: pp. 162-168.
[15] P. Debertolis, L. Eneix, D. Gullà: " Preliminary Archaeoacoustic Analysis of a Temple in the Ancient Site of Sogmatar in South-East Turkey", Proceedings of Conference "Archaeacoustics II: Second International Multi-Disciplinary Conference and workshop on the Archaeology of Sound", Istanbul Technical University, Taşkışla Building, Istanbul, Turkey, 30, 31 October and 1 November, 2015: pp. 137-148.
[16] P. Debertolis, D. Gullà: "New Technologies of Analysis in Archaeoacoustics", Proceedings of Conference 'Archaeoacoustics II: The Archaeology of Sound', Istanbul (Turkey), Oct 30-31 Nov 1, 2016, pp. 33-50.
[17] P. Debertolis, D. Gullà: "Preliminary Archaeoacoustic Analysis of a Temple in the Ancient Site of Sogmatar in South-East Turkey. Proceedings of Conference 'Archaeoacoustics II: The Archaeology of Sound', Istanbul (Turkey), Oct 30-31 Nov 1, 2016, pp. 137-148.
[18] P. Debertolis, N. Earl, M. Zivic: "Archaeoacoustic Analysis of Tarxien Temples in Malta", Journal of Anthropology and Archaeology, Vol. 4 (1), June 2016, pp. 7-27.
[19] P. Debertolis, D. Gullà: "Healing aspects identified by archaeoacoustic techniques in Slovenia", Proceedings of the '3rd International Virtual Conference on Advanced Scientific Results' (SCIECONF 2016), Žilina (Slovakia), June 6-10, 2016, pp. 147-155.
[20] P. Debertolis, D. Gullà, F. Piovesana: "Archaeoacoustic research in the ancient castle of Gropparello in Italy", Proceedings in the Congress "The 5th Virtual International Conference on Advanced Research in Scientific Areas" (ARSA-2016) Slovakia, November 9 - 11, 2016: pp. 98-104.
[21] P. Debertolis, N. Earl, N. Tarabella: "Archaeoacoustic analysis of Xaghra Hypogeum, Gozo, Malta", Journal of Anthropology and Archaeology, vol.1 no. 5, June 30, 2017: pp. 1-15.
[22] P. Debertolis, D. Gullà: "Archaeoacoustic Exploration of Montebello Castle (Rimini, Italy)", Art Human Open Acc J 1(1): 00003, DOI: 10.15406/ahoaj.2017.01.00003.
[23] P. Debertolis, D. Gullà, H. Savolainen: "Archaeoacoustic Analysis in Enclosure D at Göbekli Tepe in South Anatolia, Turkey", Proceedings in Scientific Conference "5th HASSACC 2017 - Human And Social Sciences at the Common Conference", Slovakia, Žilina, September 25-29, 2017: pp. 107-114.
[24] P. Laureano: "Giardini di pietra, i Sassi di Matera e la civiltà mediterranea", Bollati Boringhieri, 1993, Torino.
[25] V. Tandy, T. Lawrence: "The ghost in the machine", Journal of the Society for Psychical Research, 62 (851), 1998: pp. 360–364.

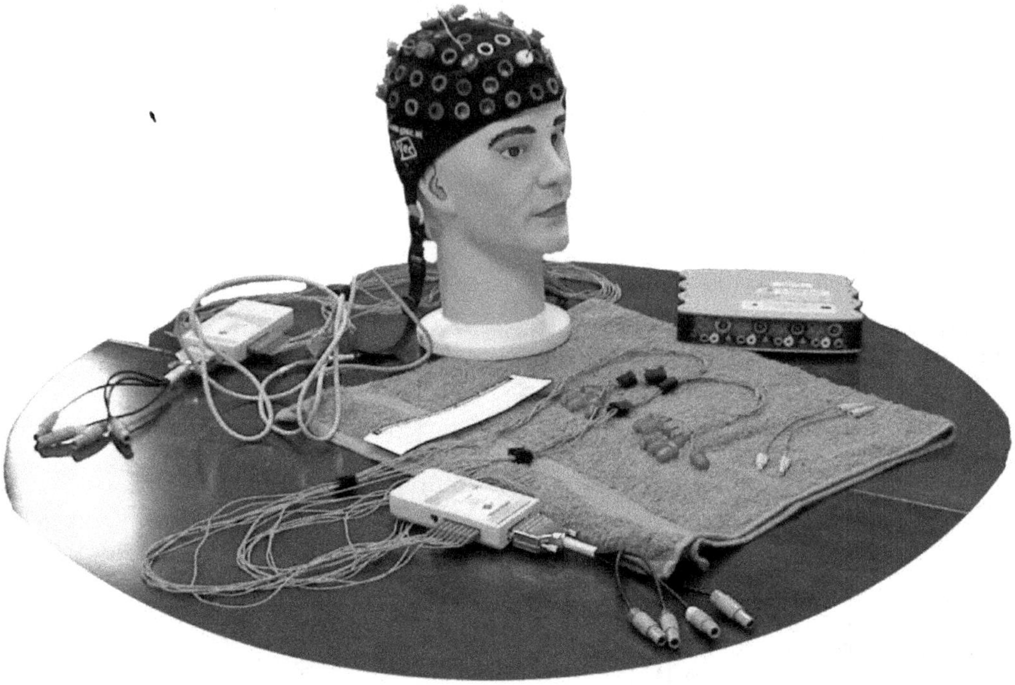

In development for future research.. Image: OTSF

Resonant Form: The Convergence of Sound and Space

Shea Michael Trahan

SHEA MICHAEL TRAHAN, AIA, LEED AP, is a licensed architect and an Associate within the New Orleans based firm of Trapolin-Peer Architects where he directs evidence-based design initiatives. Shea holds degrees in architecture from the University of Louisiana at Lafayette and Tulane University as well as a certificate in Neuroscience for Architecture from the Newschool of Architecture and Design. His research combines aspects of architectural acoustics, neuroscience, and algorithmic design and has been featured in a TEDx presentation entitled *The Architecture of Sound*, an exhibit at the New Orleans Museum of Art, a poster presentation at the 2016 Academy of Neuroscience for Architecture conference, and published within the textbook *Creating Sensory Spaces: The Architecture of the Invisible*.

ABSTRACT: The built environment is a powerful tool for affecting human awareness by embodying experiences which interact with biological rhythms to shift states of consciousness. Through exploration of precedents of architectural sonic phenomenon from throughout history, this body of research aims to identify powerful sonic tools towards such an affect. Looking to the future of designing such spaces, the research delves into algorithmic design and cymatic processes to seek to create forms which are embodiments of sound.

KEYWORDS: neuroscience, cymatics, algorithm

> "Listen! Interiors are like large instruments, collecting sound, amplifying it, transmitting it elsewhere."
> - Peter Zumthor[1]

Human spatial perception is a sensual experience of the world we inhabit. While we experience architecture through all of our senses by varying degrees, the process of design has long preferenced the visual at the expense of other modes of perception. This body of research and proposed design methodology aims to focus on acoustic aesthetics to create spaces which manifest sonic phenomenon that not only elicit psychological responses for inhabitants but also induce physiological shifts toward meditative and/or mystical states. As humans are sonic instruments themselves through use of our hearing and vocal range, the engagement of our sonic qualities within a sensitively designed aural architecture creates the potential for a truly transcendent and immaterial experience. It is in this way that this design proposal strives to enlist the resonant natures of architectural forms to deeply engage and expand the sensory awareness of human spatial perception.

The world which we inhabit is a symphony of oscillating systems, manifesting themselves through a variety of scales. Daily, monthly, and annual cycles are all examples of oscillating systems which impact human symbiosis through their frequency and intensity. Interestingly, the human body is itself a symphony of oscillating systems, each carrying out vital rhythms to maintain

[1] Zumthor, P. *Atmospheres: Architectural Environments, Surrounding Objects*. (Basel; Boston: Birkhauser, 2006): 29.

life, health, and consciousness. Cardiorespiratory rhythms regulate breathing, heart rate, and blood pressure. Pedestrians walking together find that their strides may synchronize in the course of a stroll. The brain conducts its own activities via neural oscillations, or brainwaves.

This inherently rhythmic aspect of human existence permeates the way people experience and interact with the world around them. A characteristic of oscillating systems which is critical to this experiential interaction is the process of entrainment. Through entrainment, two oscillating systems may interact with one another and eventually come to synchronize to a singular frequency.[2] In the case of circadian rhythms, the human body seeks entrainment from the rhythm of the Earth's revolutions.

Entrainment is not specific to biological systems and may be observed in mechanical systems. Non-synchronized metronomes placed on a flexible base will eventually exchange enough interaction to shift phase and become synchronous. However, the fact that a biological rhythm may be definitively changed by an outside frequency is critical to our consideration of architecture. Considering that the EPA estimates that Americans spend up to 93% of their time indoors[3], the quality and characteristics of the spaces we inhabit are very important for wellbeing, even to the extent of shifting biological functions. A team of designers at the Mixed Reality Lab Nottingham showed exactly this in their 2010 installation and experiment entitled ExoBuilding. In the biofeedback installation, participants spent extended durations seated beneath a tensile fabric canopy which stretched away from them as they inhaled and collapsed back near them as they exhaled. The participant's heart rate was also played as a real-time audio track through a subwoofer. What the experiment showed was that participants were able to reach a calmed state similar to mindfulness meditation through the experience of the immersive environment with slower, deeper breathing occurring after a few minutes of experience. Even more compelling was a second experiment in which the participants were similarly introduced to the immersive space as before, but soon after beginning the experiment the biofeedback software took over the rate of movement of the canopy, slowing it by twenty percent. The result was that a majority of participants entrained their respiration to match the slower rate of the installation, and were thus brought to the calmed state quicker by environmental input[4].

Such entrainment can happen in both the brain (brainwave synchronization) and the body (cardiorespiratory coupling), and may be induced by rhythmic visual or auditory stimuli. Indeed, the use of sound for such effects is a powerful tool. The sonic environment has a drastic impact on human wellbeing partially due to the fact that the human brain is hardwired to constantly evaluate the aural landscape. The ear enjoys three times more nerve connections to the brain than the eye does[5] and can decipher a range of sound from 20 Hz to 20,000Hz[6]. This range of frequency is an order of magnitude of 1000 times, equivalent to roughly

[2] Néda Z, Ravasz E, Brechet Y, Vicsek T, Barabsi AL (2000). "Self-organizing process: The sound of many hands clapping". Nature. 403: 849–850.

[3] Klepeis N, Nelson W, Ott W, Robinson J, Tsang A, Switzer P, Behar J, Hern S, Engelmann W (2001). "The National Human Activity Pattern Survey (NHAPS): a resource for assessing exposure to environmental pollutants". Journal of Exposure Analysis and Environmental Epidemiology. Volume 11: 231–252.

[4] Schnadelbach, H. "Embodied adaptive architecture." *Youtube*, 24 September 2016, https://youtu.be/HEkFjx4kxmc.

[5] Crowe B., *Music and Soulmaking: Toward a New Theory of Music Therapy*, Scarecrow Press, 2004.

[6] Helmholtz H., *On the Sensations of Tone*, Dover Publications, 1954.

10 octaves of sound. Considering the range of frequencies detectable by the eye, it equates to an order of magnitude of only 2, or a single octave of frequency, experienced as colors rather than as tones. We know this 'octave' of color as ROYGBIV.

As exhibited, humans are greatly affected by the physical (and specifically sonic) environment. Given the predominance of the built environment in modern human experience, the design of our spaces is coming to be seen as a powerful tool for human well-being. Architecture is the setting in which we experience play, love, rest, learning, struggle, adaptation, and even transcendence. Could architecture go beyond simply containing experience and instead act as a catalyst for transcendent experience?

In search for such a sonic gateway to transcendent experience, two characteristics of spatial sound come to the forefront; reverberation and resonance. Reverberation shall be discussed as the period of time a sound takes to dissipate within a space, while resonance may be understood as the predominant frequency at which a space or object vibrates.

To experience magnificent examples of reverberation one only need visit a gothic cathedral, or perhaps the churches in which Gregorian chant developed. A specific structure is the Baptistry of St. John adjacent to Pisa's iconic Leaning Tower. The Romanesque/Gothic baptistry manifests a rather unique form of reverberation imparting the space with a powerful sonic tool.

Most spaces have reverberation times of a fraction-of-a-second, meaning that once a sound is discontinued the energy of the source sound disappears almost instantaneously. This short moment of continued sound is created by the original sound waves bouncing between the walls of the space until they are eventually absorbed by the materials within the room. The more reflective the materials, the more bounces which are possible (this is the commonly known effect enjoyed by those who sing in the shower). The shape and arrangement of a building also affects the bouncing of waves within a space and can either diffract, focus, delay, or redirect the bouncing sound. While most structures have miniscule reverberation times, many churches and concert venues can approach reverberation times of two seconds.

Due to the particular form and material at Pisa, when a vocalist sings a tone within the Baptistry they are greeted by a reverberation time many times longer than the grandest of concert halls. This allows one to layer their vocal energy, even harmonizing with their own voice as it reverberates within the space for up to twelve seconds. The vocal energies of the singer slowly fill the building's large volume, held in momentary suspension by the geometry of the structure before returning to the listener from another direction.

Studying the Baptistry form and construction, you find that the marble used to construct the space is highly reflective thus bouncing most of the energy from a given sound wave back towards the interior of the room. Working in tandem with this strong reflectance is the arrangement of the walls, columns, and roofs. In plan the Baptistry is a double-layered circle with an exterior diameter of one hundred and sixteen feet. The circular arrangement guarantees that any sound within the space will travel outward radially only to be reflected and re-focused back toward the interior of the room. This refocusing conserves the sonic energy by maximizing the amount of sound returned to the listener.

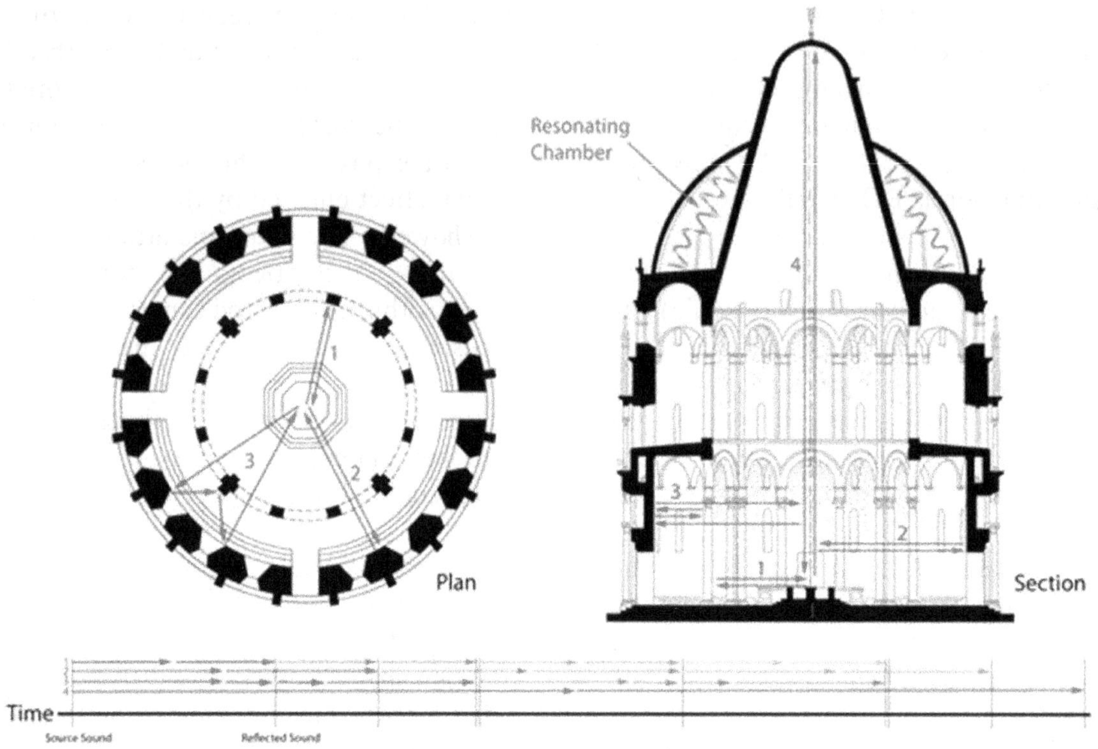

Figure 1 - sonic analysis diagram of Pisa Baptistry

The interior columns which carry the weight of the gallery and ceiling above play a crucial role in the production of the sonic environment as well. Sound waves may take one of three radial paths in plan. Sound waves may be reflected by the interior face of the column and sent back toward the center of the space. This represents the shortest path back to the center and thus it is the fastest. Some of the sound waves will miss the columns and travel to the exterior wall before being reflected back to the interior of the space. These take a slightly longer path (in both distance and time). A third wave path travels past the columns, is reflected by the exterior walls, but rather than returning directly to the interior of the space, these waves are again bounced by the exterior of the column. In this moment the sound wave is deflected back and forth between the column and outer wall, further extending the time between the source sound and its deflected return to the center of the room. In this way the architecture manipulates the acoustic paths of sound waves creating a layering of reflections and a dramatically extended reverberation of the sound.

By studying the Baptistry in section, one can see the original roof (observable as the interior ceiling form) along with the modified design which created a second skin in the shape of a dome around the original roofline. Whether by accident or intent, this double layered spatial arrangement creates a resonating chamber between the two skins. These architectural traits combine to intensify the sonic character of the space. Arguably the most focused architectural example of resonance is to be found within the Hypogeum Hal Saflieni in Paola, Malta. This prehistoric archaeological site contains an underground temple discovered in 1902. A noteworthy element of the temple is a room called the Oracle Chamber. This space is oval shaped in both plan and section with ridges carved into the ceiling of roughly seven feet in height. The ceiling is particularly important as it is elaborately painted in a swirling red paint, potentially a musical annotation. Also notable about the chamber is the presence of a small portal which allows sounds from within the Oracle Chamber to be transmitted and clearly audible to all the rooms of the complex.

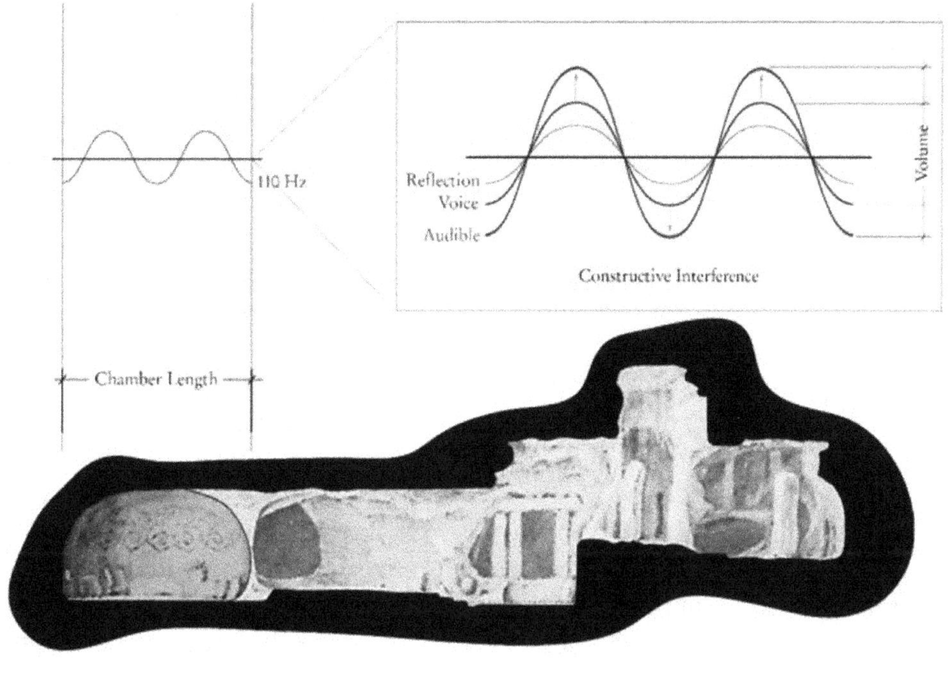

Figure 2: sonic analysis diagram of Hypogeum Hal Saflieni

Acoustic analysis of the Oracle Room has identified a resonant frequency of 110 Hz[1]. Standing in this space one is able to identify the resonant frequency of the room by singing or humming at the lowest range of your vocal register and slowly raising the pitch of the tone. When the resonant frequency is reached at 110 Hz, a multisensory event takes place. The intensity of the sound of your voice within the space increases instantaneously, sounding as though someone has joined in singing. This is caused through constructive interference, created when the sound waves of your voice synchronize. This is a natural form of amplification. (Fig. 2) Additionally, the character of the sound becomes increasingly spatial, observably approaching you from all directions simultaneously and taking on a vibrotactile quality so that you sense the sound as a tingling vibration through the skin. When you discontinue singing the resonant tone, it remains notably audible for a few brief moments before decaying into silence. The experience is at once disorienting (you feel as though your equilibrium is lost) and awe inspiring. This sonic interaction - by which the space informs the performer - inverts the relationship between

[1] Cook I., Pajot S., Leuchter A., *Ancient Architectural Acoustic Resonance Patterns and Regional Brain Activity*, in Time and Mind, Volume 1, Number 1, 2008.
[Editor's note: we now know that the number 110 Hz is not singularly fixed. The acoustics of Hal Saflieni are extremely complex, with more than one frequency triggering high resonance response in the spaces, most lying between 60 Hz and 120 Hz. Thus the chance for any physiological impact is increased, since sensitivity seems also to vary somewhat from one person to another and even at different times in the same individual.]

musician and instrument in that it is indeed the instrument which tunes the singer.

Judged upon the characteristics of the acoustic effects alone, the Hypogeum is an impressive site. More impressive though are the results of a study into the effects of the Hypogeum's particular frequency, 110 Hz, upon the human brain. When the brain is exposed to the resonant frequency of the Hypogeum, the tone causes a shift in the prefrontal cortex from left dominance to right dominance[2]. This shift de-activates the language centers of the brain (focused on rhythm and patterns) and hyper-activates an emotional center often associated with the perception of experiences of spirituality and enlightenment.

This type of physiological reaction to an environment is precisely the sort entrainment discussed earlier. Indeed, it appears that the study of neuroscience as it relates to architecture has become the forefront of this realm of research. Given the strong connection between the sense of hearing and the brain, the sonic realm promises a wealth of potential for creating even more powerful sensory expanding spaces.

The human brain interprets sounds in a variety of ways. Rhythmic sounds of tapping and drumming lead to activation within the left frontal and left parietal lobes of the brain[3]. Tones activate the brain in an entirely different way, activating areas within the right prefrontal cortex which are associated with emotional states and somatosensory perception.

The same areas of the brain which are involved in the perception of sound also play an important role in experiences of enlightenment, or spiritual awareness. In his 2016 book entitled *How Enlightenment Changes the Brain*, Dr. Andrew Newberg reviews decades of brain scan studies conducted on experienced Buddhist meditators, Franciscan nuns, and Sufi mystics. The recurrent results from his and other studies is that at the moment of enlightenment these spiritual practitioners experience a rapid and significant decrease in the frontal and parietal lobes of the brain[4]. This decrease in regional activity within the brain appears to be most heavily felt in the left hemisphere of the frontal and parietal lobes.

For a listener experiencing sound within an immersive environment such as the Hypogeum or the Pisa Baptistry, the initial experience of sound would be met with increased activity within various brain regions: first the primary auditory cortex followed by the secondary auditory cortex. From here the brain begins to send signals to both the parietal and frontal lobes. If the sonic event is significantly immersive and of substantial duration the brain then shifts with rapid decreases in the activity in those areas. This is consequential as the parietal lobe takes in sensory input and creates one's sense of self. A reduction in activity is thus associated with a loss of one's sense of self, creating a new sense of unity or oneness with the universe, God, nature, or consciousness[5]. This notable increase in self-transcendence associated with the drop in parietal lobe activity casts a new light on the neurobiological basis of altered spiritual and religious attitudes[6].

---

[2] Ibid.

[3] Wilkins A., "What Happens to Your Brain Under the Influence of Music", accessed July 17, 2017. http://io9.gizmodo.com/5837976/what-happens-to-your-brain-under-the-influence-of-music

[4] Newberg A., Waldman M., *How Enlightenment Changes Your Brain*, Penguin Random House, 2016.

[5] Ibid.

[6] Urgesi C., Aglioti S., Skrap M., Fabbro F., *The Spiritual Brain: Selective Cortical Lesions Modulate Human Self-transcendence*, in Neuron, 2010.

As the secondary auditory cortex sends signals to the parietal lobe, so too it sends signals to the frontal lobe. The power of sound on the frontal lobe, specifically via the prefrontal cortex, has already been seen in the Hypogeum studies. As in the parietal lobe, given a stimulus of significant duration and intensity, one may experience a sudden drop in activity, predominantly localized to the left frontal lobe. This particular area of the brain is involved in distinguishing the rhythmic patterns of language and is particularly active during focused purposeful concentration. Decreasing activity in this area is notably associated with a sense of surrender as the control mechanisms of the brain give way[7]. Adding to this experience is an increase in the activity of the anterior cingulate cortex, an area of the brain which regulates emotional responses and communicates directly with the self-critiquing tendencies of the frontal lobe, maintaining a balanced control over emotional reactions. As the frontal lobe's control mechanisms decrease and the anterior cingulate cortex increases, the experience of intense emotional responses becomes increasingly possible[8]. "If the Doors of Perception were cleansed, everything would appear to man as it is, infinite." – William Blake[9]

One may be prone to question the value in such an experience, or dismiss it as some sort of fleeting high. Dr. Andrew Newberg addresses this concern directly by stating that a majority of his study subjects have found, "great meaning and purpose to their life", through such enlightenment experiences. He correlates such experiences to defense against depression and anxiety, and cites a study which found that, "a sense of purpose enhances 'coping, generosity, optimism, humility, mature identity status, and more global personality integration'". In an increasingly secular world, Dr. Newberg calls for the need to, "find new tools and experiences to enlighten the minds of the next generation of seekers."[10] This call for a new form of secular ritual is mirrored by the philosopher Alain de Botton. He views ritual as communal acts intended to build community while mediating between the needs of the individual and those of the group[11]. Botton goes even further to specifically tie such ritual to architectural space and calls for a new "Temple of Perspective" as a tool to help contemporary humans cope with the psychological, emotional, and communal needs which religion has traditionally served. "...No less than the church spires in the skyscapes of medieval Christian towns, these temples would function as reminders of our hopes...they would all be connected through the ancient aspiration of sacred architecture: to place us for a time in a thoughtfully structured three-dimensional space, in order to educate and rebalance our souls."[12]

Having explored the scientific power of sound to induce awe and manipulate states of consciousness, and having identified the acoustic phenomenon which might be created through architecture to induce such effects, the question becomes, "Where do we start in the design of such spaces?" It may be argued that in search of design inspiration for spaces intended to affect human experience of frequencies, a designer ought to look to nature's existing vibrational patterns for direction. Specifically related to sound, one branch of physics related to modal vibrational phenomenon is cymatics.

[7] Newberg A., Waldman M., *How Enlightenment Changes Your Brain*, Penguin Random House, 2016.
[8] Ibid.
[9] Blake W., *The Marriage of Heaven and Hell*, Oxford University Press, 1975.

[10] Newberg A., Waldman M., *How Enlightenment Changes Your Brain*, Penguin Random House, 2016.
[11] Botton A., *Religion for Atheists*, Vintage International, 2013.
[12] Ibid.

The study of cymatics reveals the reality that sound is a spatial experience, traveling through media, and that such processes are inherently formal in nature. Such forms are revealed in two-dimensional representations known as cymatic patterns. So it is that this author began the search for a space designed to manifest sound by investigating the possibility of using sound (through cymatics) as an active design agent. Beginning with an algorithm which mapped the cymatic patterns of various tones I have explored hundreds of mandala-like representations of sonic manifestation. Exploiting the symmetrical nature of the cymatic pattern, the algorithm then orbits to explore the spatial implications of the sonic forms in three dimensions. (Fig. 4) Scaled to a precise multiple of the desired tone, such forms could become inhabitable spaces which might offer a tunable sonic chamber similar in performance to the Hypogeum Hal Saflieni. Through layered spatial complexity and material reflectance the Nodal Structures could also offer extended reverberation times much akin to the sonic phenomenon within the Pisa Baptistery

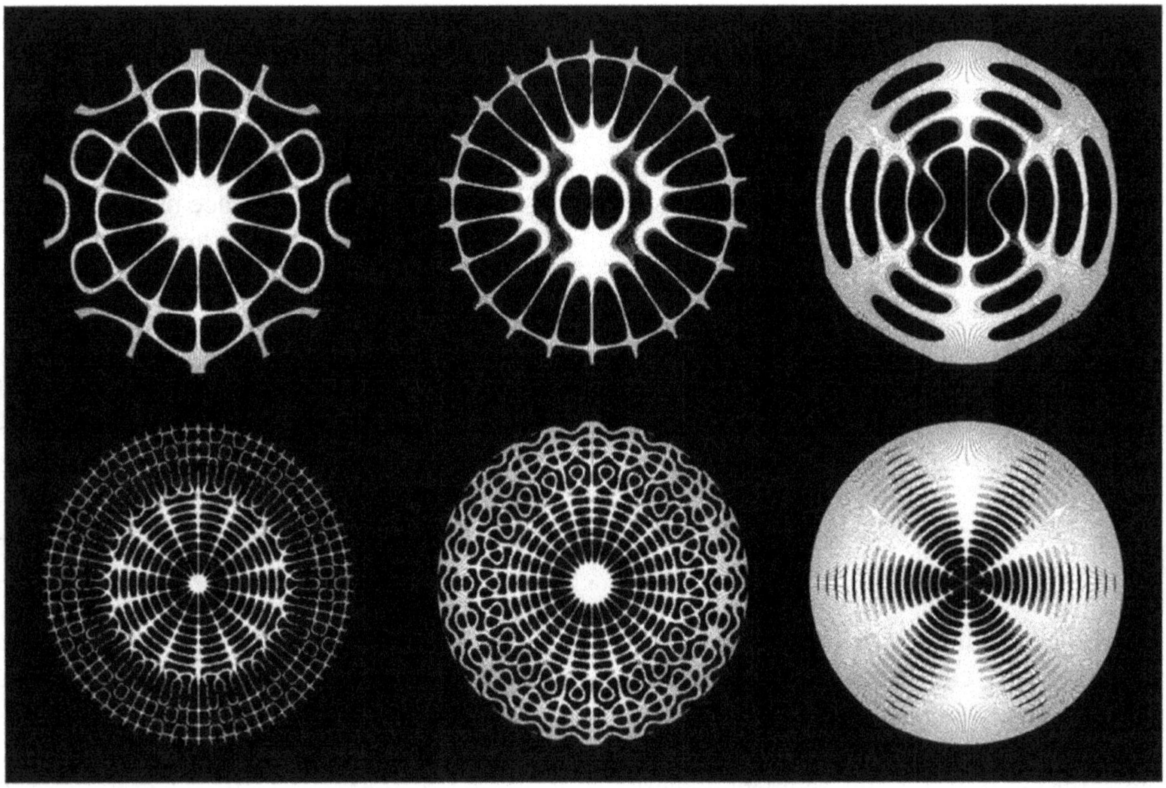

Figure 3 - cymatic explorations of tones

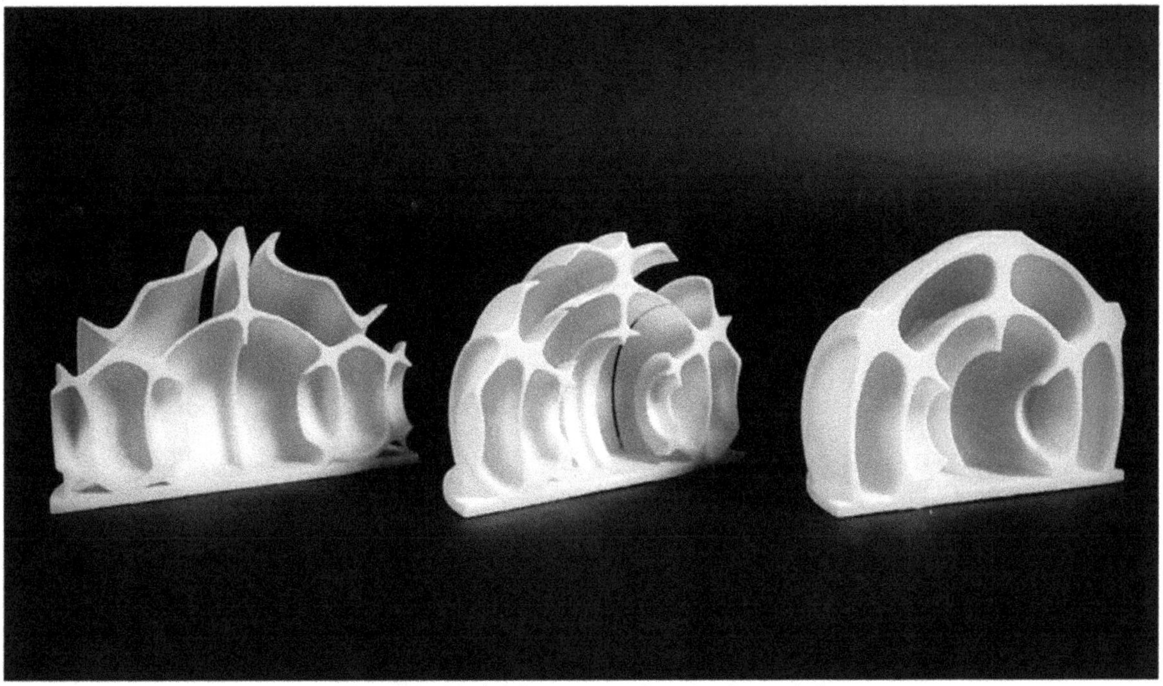

Figure 4 - Nodal Structures of an A minor triad

Such a hyper-resonant and hyper-reverberant space could offer a powerful embodiment of sound in architecture; a temple of sound which not only embodies, but is also manifest of, sound. Users could delight in discovering the resonant frequency as they sing out searching for the constructive interference as the space comes alive in symphony. Extended durations of singing would offer not only an immersive sonic experience, but would also act to slow the breath, only further embodying the potential for meditative states as the brain areas begin to shift in response to the sensory input. (Fig. 5)

Figure 5 – Temple of Sound

As a home for contemporary ritual, the Nodal Structures might offer a new form of secular sacred space, designed to help expand consciousness and prime the brain for enlightenment experiences. Beyond fulfilling the need for such a ritual, these sonic embodiments might offer powerful research opportunities for neuroscientific and physiological experimentation to further explore the power of sound on the human mind and body. Additionally, said space may offer assistance in a variety of sound-based therapies used in treatment for a variety of conditions including depression, PTSD, Alzheimer's, and insomnia. (Fig. 6)

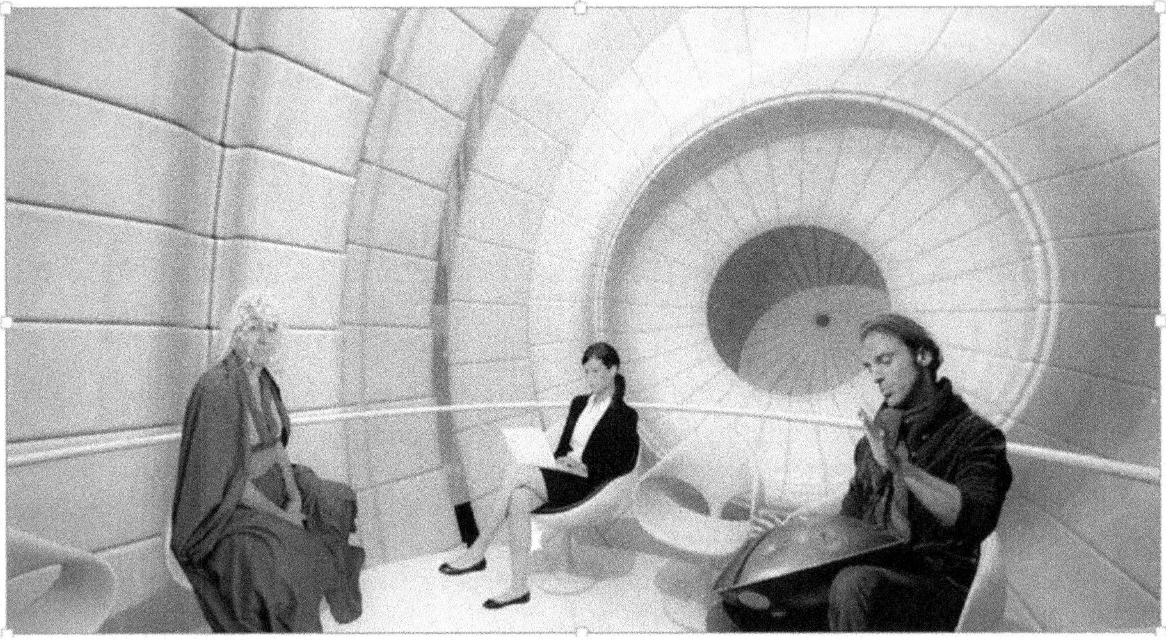

Figure 6 – Sonic Laboratory

Study of the Effect of Different Types of Music in a Large Group of Listeners Within Saint Peter's Church in Cuenca (Spain)

N. Valverde-Gascueña, M.P. Sáez-Pérez, J.P. Ruiz-Fernández

N. VALVERDE-GASCUEÑA (nelia.valverde@uclm.es), PhD Professor, Polytechnic School of Cuenca (University of Castilla-La Mancha, Spain)
M.P. SÁEZ-PÉREZ (mpsaez@ugr.es), PhD Professor, University of Granada (Spain)
J.P. RUIZ-FERNÁNDEZ (juanpedro.ruiz@uclm.es), PhD Professor, Polytechnic School of Cuenca (UCLM, Spain).

ABSTRACT: The aim of this study is to determine the effect caused by listening to different types of music in a large group of listeners within the same site and its influence when subjectively assessing the acoustic quality of that place. For this, the site chosen was one of the most singular locations of the city of Cuenca (Spain), the church of San Pedro. Afterwards, a concert was performed with compositions that spanned three quite different musical genres: electroacoustic music, polyphony with organ and early Christian chant. Immediately after the concert, the audience was submitted to a survey. The study confirmed that the current acoustics of this church, with high reverberation times, are appropriate for vocal music of long notations and slow free time.

Introduction

Sound has constituted a fundamental element of human life from the appearance of the first primitive communities, defining habitable space and determining the place for activities to be held within that space. The acoustic circumstances have evidently influenced this relationship with the environment. Furthermore, as indicated in [1], a close relationship is established between sound and space, which contributes definitively to the perception of a site.

Thanks to the study of oral as well as written traditions, ample information has been compiled concerning the sound contexts of different periods of history, and there are records of acoustic knowledge from early periods [2, 3]. Despite this recognized historical importance, of all sciences that of acoustics has had among the least impact. This has long been focused on fields hardly related to one another, such as acoustic auditoriums and problems of noise, without taking into account such fundamental aspects as human perception and interpretation of sound.

In the 1930s acoustics began to become a new science, due to the development of the technology of electronic equipment for measuring sound and it use as a common tool in field work. From the 1960s on, the evolution in electroacoustic technology has made progress in techniques of measuring psychoacoustics, defined by, among others [4, 5]. This made it possible to relate a series of subjective parameters to other objective parameters derived form *in situ* measurements, which would favour the development of acoustics in fields such as auditory perception [6] and sound spatialization.

The subjects traditionally focused upon in psychoacoustics include those linked to the perception of height, subjective sound, volume and timbre. According to [7], until a short time ago, this discipline dedicated more attention to the behaviour of the auditory system than to the details of cognitive processing. Currently, the areas of research include concepts such as auditory perception [8], the flow of sounds, and the analysis of auditory scenes [9]. The exchange of knowledge between different disciplines has greatly widened the possibilities of acoustic research.

In the present work, the architectonic space under study, in this case the church of San Pedro in Cuenca (Spain), is related to its acoustic properties through three musical works of different genres. After the performance in the church, the listeners were administered a questionnaire asking them to score the works subjectively in terms of the sound quality of the church. The audience had a special part in this research, in that the reception of images, impressions or sensations through their senses revealed different perceptions reflected in the statistical results.

On the other hand, different viewpoints enable the results from the objective methods to be compared with results from subjective methods, both related to the acoustic quality of the church. In this way, some type of equivalence between the two could or could not be established. The aim was to determine which of the three works was most appropriate, according to the listener, for the acoustics of the church. It also enabled a certain correlation with the sensations or images produced by listening to one of the musical works presented.

Among many other authors, [10] shows that broad areas of affinity exist between music and architecture, up to reaching curious points of similarity. The present study has an interdisciplinary basis in that acoustics are presented as an interaction between music and architecture. In fact, [11] indicates the possible connection between the three disciplines investigated.

Study Setting

The venue chosen to conduct the research was the church of San Pedro in Cuenca (Spain), whose architectural design is singular, with an octagonal exterior and a circular interior, generating to fields of sound of extraordinary complexity that make the church special for acoustic research. The church is currently used for Catholic services as well as public concerts.

The church of San Pedro, situated in the highest part of Cuenca (Fig. 1) was originally constructed in the 12th century, shortly after the city was conquered by Alfonso VIII (September 1177). In the 15th century, battles between social classes destroyed the church in 1448, but it was reconstructed in the Gothic style. Three centuries later, (18th century), after the War of Succession (1702-1713), the church was sacked and burned, so that some time later it had to be completely renovated, the work being led by José Martín de Aldehuela, an influential architect of the Baroque in Cuenca [12]. During the Spanish Civil War in the 20th century, the monument again suffered major damage and therefore had to be restored again, this work being finished in 1999 under the direction of the Escuela-Taller Cuenca II y Cuenca III. A few years later (July 4th, 2002), by resolution of the

Dirección General de Bienes y Actividades Culturales de la Consejería de Educación y Cultura de Castilla La Mancha, the church was declared of cultural interest and was catalogued as a monument [13].

Figure 1. Aerial view, exterior view [12] and interior view of the church of San Pedro.

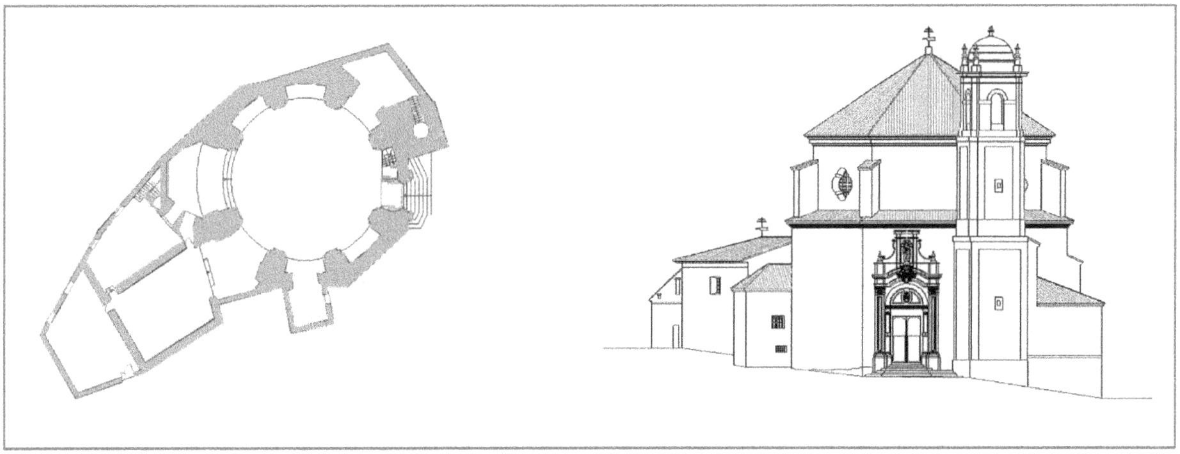

Figure 2. Ground plans and main elevation [14].

As mentioned above, the ground plan was octagonal with a circle inscribed within it. The perimeter of the circle is marked by pilasters connected by Roman arches. In this centralized space, the apse, which is also polygonal, forms an axis ending in the façade, composed of the front door and the tower (Fig. 2). The circular interior space, bearing a pronounced denticulate cornice, and is enclosed by a dome over a drum containing windows with a mixtilinear design.

Most of the materials that adorn the interior of the church (plaster painted on the walls and ceilings, stucco in ornamentation details, marble and granite floors) are acoustically reflectant, except for the wood (wainscoting of the high altar, banisters of the chorus, the balconies and ceiling of the Chapel of San Marcos), whose influence has been considered minor because it covers a minimal surface area. Therefore, these must prevent the excessive loss both of low- as well as high-pitched sounds, which is recommended for a building used for music. However, many other factors cause the high reverberation, such as the shape and volume of the church. These conditions worsen with the marble floor in the central area and the dome crowning the building. Despite being a Baroque-style structure, the ornamentation is spare, and this does not help diffuse the sound well. In addition, there is a lack of carpets, curtains or tapestry that would increase the absorption of the sound and thereby diminish reverberation.

Methodology

The present investigation focuses on the analysis of the results of the acoustic study made as part of previous research on the church of San Pedro in Cuenca (Spain) [14], where its acoustic characteristics were presented through the results of certain acoustic parameters measured *in situ*, with the aim of correlating the objective and subjective results and thereby determining whether it is necessary or not to provide acoustic conditioning and what place within the church is optimal to listen to music.

Afterwards, the psycho-acoustic study was made, this being the main objective of the present research, based on the concert performed inside the church, including the explanation and analysis of the musical compositions chosen for the performance. For the complete documentation, this concert was recorded both in audio and video, incorporating the details related to the equipment used and the technical specifications used, their location in the church, and the diagram of connections between pieces of equipment. Immediately after the concert, the audience was administered a questionnaire to score the acoustics of the building chosen on subjective criteria and based on the musical works performed.

After a description of the methodology and the design of this psycho-acoustic study, and after the format of the questionnaire is presented, the results of the survey are presented for interpretation, and the conclusions are drawn.

Acoustic study

Taking as reference the acoustic study made by [14], and beginning with the minimum characteristic parameters established in [15, 16], apart from knowing the overall dimensions, volume, and surface area of the church (Fig. 3), the acoustics of the building were characterized.

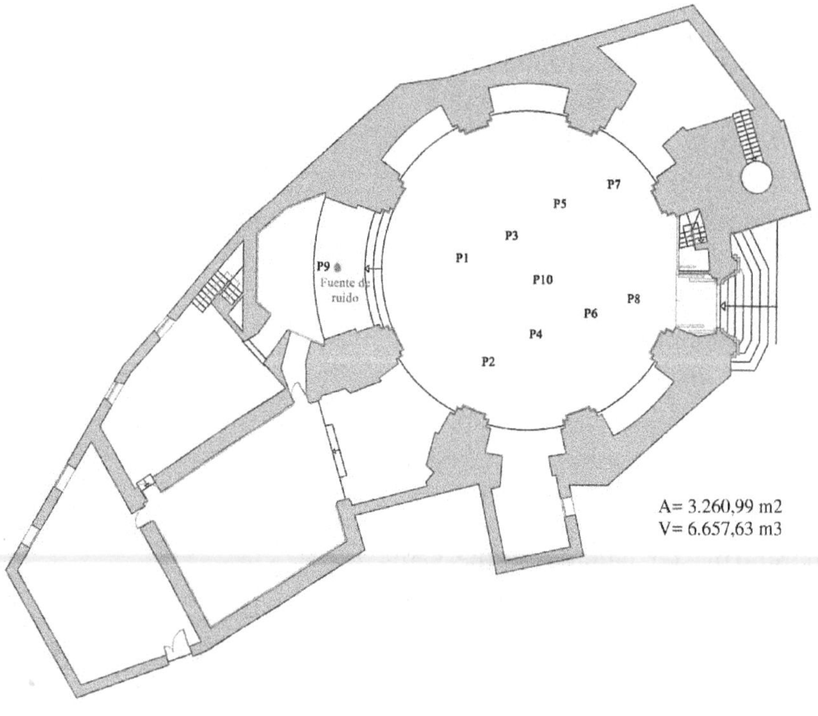

Figure 3. Location of the source of noise and the measurements points (P) for the reverberation time [14].

Table 1 summarizes the mean values found for each of the acoustic parameters considered:

Acoustics parameters	Average values
Reverberation time	5,87 s
Sound pressure level	95,34 dB
Noise level	31,84 dB
Intelligibility (% ALCONS)	15,97%/ 33,48%
Clarity	-6,79 / +6,27
Liveness	5,91 s
Acoustic warmth	1,14
Brightness	0,48

Table 1. Mean values of the acoustic parameters of the church of San Pedro.

According to Table 1, the value for mean reverberation time in the church was 5.87 sec. Taking into account that the optimal times recommended are between 2 and 3 seconds [17], we can interpret this value as high. From a purely acoustic standpoint, this result can be explained by the type of ground floor of the church: octagonal on the exterior and circular on the interior and covered by a central dome, all of which multiplies the focalizing effects. In addition, although the architectural style is Baroque, it does not have excessive ornamentation; on the contrary, the walls are practically bare, resulting in minimal sound absorption.

Nevertheless, this mean reverberation time is suitable in theory for vocal music of long notes and slow free time. On the other hand, it would not be appropriate for ensuring intelligibility of spoken messages. The level of sound pressure was quite uniform, with a mean overall linear level of 95.34 dB from the 110 dB emitted from the source. The overall mean of background noise was 31.84 dB. The spectrum of background noise was good, since it remained under the recommended maximum levels of the NR, NC, and PNC curves advisable for churches, which, according to [18], are 35 dB, 30 dB, and 35 dB, respectively.

The AL*CONS* index is determined from the limit distance, which in our case was 9.80 m. For shorter distances to the limit, the AL*CONS* index is 15.97%; for distances greater than the limit, the AL*CONS* index is 33.48%. Intelligibility of the spoken word within the church is acceptable up to 10%, [16], i.e. up to 7.75 m from the listener; this is not an ideal result, however, as the audience area reaches some 15 m. In the case of the parameter "clarity", which helps appraise the suitability of the church for music, appropriate values usually go from -2 dB to +2dB [16]. However, Table 1 shows that the values found go beyond those recommended [17].

The liveness of the church proved to be 5.91 sec. This means a building of great sharpness, as normal values are 1.5 to 2 sec. [17]. In addition, good acoustic warmth resulted (1.14), as this value is very close to the recommended 1.1 [17]. The brightness of the church (0.48) was very low, compared to recommended values of between 0.8 and 1.0 [17].

Psycho-acoustic study

In the concert entitled "Sounds in the Architecture of San Pedro", which was held in the church of San Pedro for the purposes of this study, three musical works were performed, these adapting in different ways to the acoustics of the church. The works were selected on the need to define significant acoustic parameters and to broaden the artistic spectrum. For this reason, an electroacoustic work was chosen; second, for its creativity, a musical work composed specifically for the church of San Pedro was selected; and the third work chosen was an anonymous work from the 5th and 10th centuries A.D. within the broad category of early Christian chant.

The first work, entitled *Évangile*, belonged to the sixth movement of the work *Requiem*, composed in 1973 on the texts of mass for the deceased by Michel Chion, composer, writer, producer of cinema and video. This is recorded electroacoutic music or "concrete music".

The second work performed in the concert was *Tu es Petrus*, for four mixed voices and organ, composed by Nelia Valverde specifically for this study and for the church of San Pedro, and interpreted by the Cathedral of Cuenca Chapel Music. The text was chosen in an attempt to reflect in the score the works appearing under the balconies that serve as a base for the dome crowning the church, where the phrase appears: "TU ES PETRUS ET SUPER HANC PETRAM AEDIFICABO ECCLESIAM MEAM" (Matt. 16, 18).

The third work programmed was *Alleluia* (*Justus*), an early Christian chant performed by Iégor Reznikoff, a specialist in religious art and old music. It is a monodic work divided into two parts: the first is constructed on the traditional chant of *Alleluia*, and the second is constructed on the text of *Justus Germinabit*. Both were interpreted following the manuscripts existing in the 9th and 11th centuries A.D.

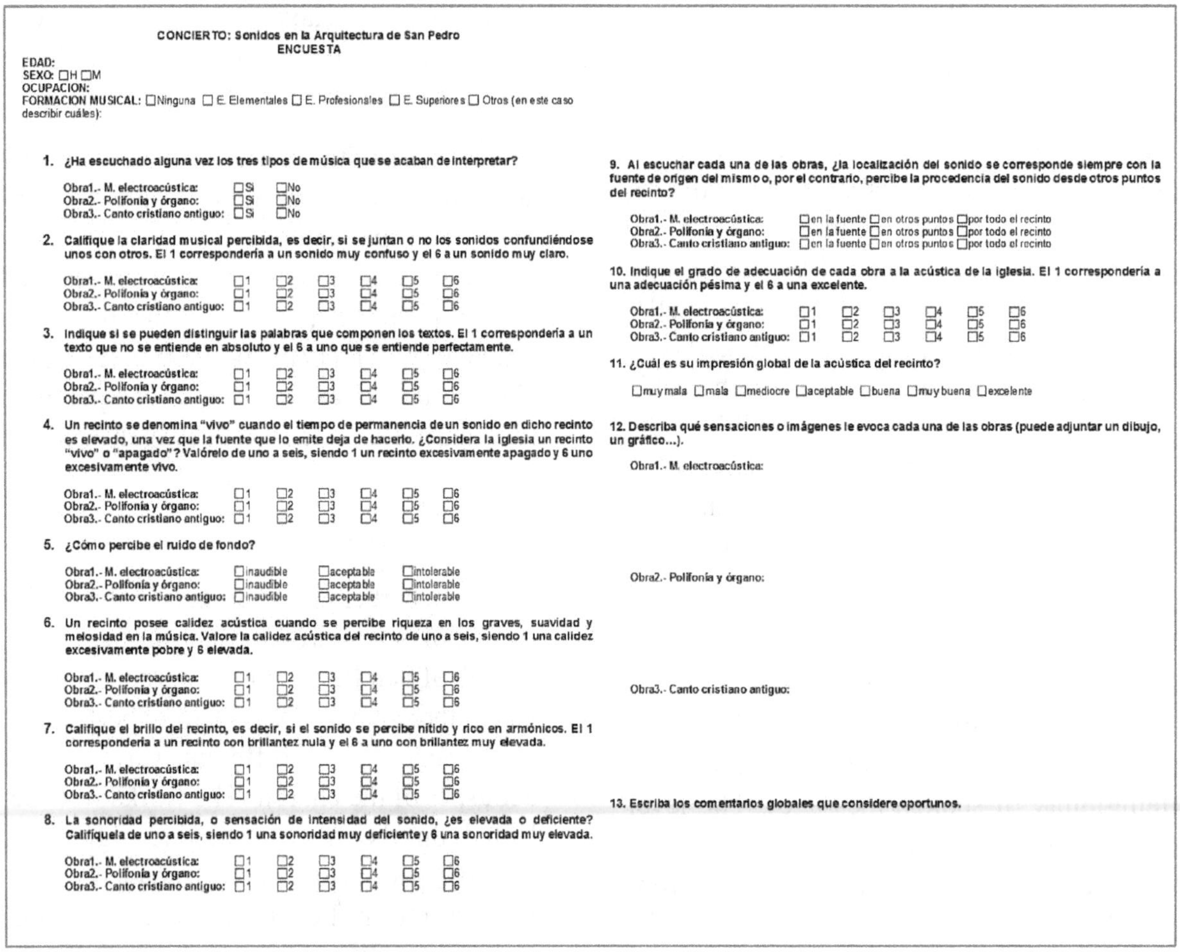

CONCIERTO: Sonidos en la Arquitectura de San Pedro
ENCUESTA

EDAD:
SEXO: ☐H ☐M
OCUPACION:
FORMACION MUSICAL: ☐Ninguna ☐ E. Elementales ☐ E. Profesionales ☐ E. Superiores ☐ Otros (en este caso describir cuáles):

1. **¿Ha escuchado alguna vez los tres tipos de música que se acaban de interpretar?**

   Obra1.- M. electroacústica: ☐Si ☐No
   Obra2.- Polifonía y órgano: ☐Si ☐No
   Obra3.- Canto cristiano antiguo: ☐Si ☐No

2. **Califique la claridad musical percibida, es decir, si se juntan o no los sonidos confundiéndose unos con otros. El 1 correspondería a un sonido muy confuso y el 6 a un sonido muy claro.**

   Obra1.- M. electroacústica: ☐1 ☐2 ☐3 ☐4 ☐5 ☐6
   Obra2.- Polifonía y órgano: ☐1 ☐2 ☐3 ☐4 ☐5 ☐6
   Obra3.- Canto cristiano antiguo: ☐1 ☐2 ☐3 ☐4 ☐5 ☐6

3. **Indique si se pueden distinguir las palabras que componen los textos. El 1 correspondería a un texto que no se entiende en absoluto y el 6 a uno que se entiende perfectamente.**

   Obra1.- M. electroacústica: ☐1 ☐2 ☐3 ☐4 ☐5 ☐6
   Obra2.- Polifonía y órgano: ☐1 ☐2 ☐3 ☐4 ☐5 ☐6
   Obra3.- Canto cristiano antiguo: ☐1 ☐2 ☐3 ☐4 ☐5 ☐6

4. **Un recinto se denomina "vivo" cuando el tiempo de permanencia de un sonido en dicho recinto es elevado, una vez que la fuente que lo emite deja de hacerlo. ¿Considera la iglesia un recinto "vivo" o "apagado"? Valórelo de uno a seis, siendo 1 un recinto excesivamente apagado y 6 uno excesivamente vivo.**

   Obra1.- M. electroacústica: ☐1 ☐2 ☐3 ☐4 ☐5 ☐6
   Obra2.- Polifonía y órgano: ☐1 ☐2 ☐3 ☐4 ☐5 ☐6
   Obra3.- Canto cristiano antiguo: ☐1 ☐2 ☐3 ☐4 ☐5 ☐6

5. **¿Cómo percibe el ruido de fondo?**

   Obra1.- M. electroacústica: ☐inaudible ☐aceptable ☐intolerable
   Obra2.- Polifonía y órgano: ☐inaudible ☐aceptable ☐intolerable
   Obra3.- Canto cristiano antiguo: ☐inaudible ☐aceptable ☐intolerable

6. **Un recinto posee calidez acústica cuando se percibe riqueza en los graves, suavidad y melosidad en la música. Valore la calidez acústica del recinto de uno a seis, siendo 1 una calidez excesivamente pobre y 6 elevada.**

   Obra1.- M. electroacústica: ☐1 ☐2 ☐3 ☐4 ☐5 ☐6
   Obra2.- Polifonía y órgano: ☐1 ☐2 ☐3 ☐4 ☐5 ☐6
   Obra3.- Canto cristiano antiguo: ☐1 ☐2 ☐3 ☐4 ☐5 ☐6

7. **Califique el brillo del recinto, es decir, si el sonido se percibe nítido y rico en armónicos. El 1 correspondería a un recinto con brillantez nula y el 6 a uno con brillantez muy elevada.**

   Obra1.- M. electroacústica: ☐1 ☐2 ☐3 ☐4 ☐5 ☐6
   Obra2.- Polifonía y órgano: ☐1 ☐2 ☐3 ☐4 ☐5 ☐6
   Obra3.- Canto cristiano antiguo: ☐1 ☐2 ☐3 ☐4 ☐5 ☐6

8. **La sonoridad percibida, o sensación de intensidad del sonido, ¿es elevada o deficiente? Califíquela de uno a seis, siendo 1 una sonoridad muy deficiente y 6 una sonoridad muy elevada.**

   Obra1.- M. electroacústica: ☐1 ☐2 ☐3 ☐4 ☐5 ☐6
   Obra2.- Polifonía y órgano: ☐1 ☐2 ☐3 ☐4 ☐5 ☐6
   Obra3.- Canto cristiano antiguo: ☐1 ☐2 ☐3 ☐4 ☐5 ☐6

9. **Al escuchar cada una de las obras, ¿la localización del sonido se corresponde siempre con la fuente de origen del mismo o, por el contrario, percibe la procedencia del sonido desde otros puntos del recinto?**

   Obra1.- M. electroacústica: ☐en la fuente ☐en otros puntos ☐por todo el recinto
   Obra2.- Polifonía y órgano: ☐en la fuente ☐en otros puntos ☐por todo el recinto
   Obra3.- Canto cristiano antiguo: ☐en la fuente ☐en otros puntos ☐por todo el recinto

10. **Indique el grado de adecuación de cada obra a la acústica de la iglesia. El 1 correspondería a una adecuación pésima y el 6 a una excelente.**

    Obra1.- M. electroacústica: ☐1 ☐2 ☐3 ☐4 ☐5 ☐6
    Obra2.- Polifonía y órgano: ☐1 ☐2 ☐3 ☐4 ☐5 ☐6
    Obra3.- Canto cristiano antiguo: ☐1 ☐2 ☐3 ☐4 ☐5 ☐6

11. **¿Cuál es su impresión global de la acústica del recinto?**

    ☐muy mala ☐mala ☐mediocre ☐aceptable ☐buena ☐muy buena ☐excelente

12. **Describa qué sensaciones o imágenes le evoca cada una de las obras (puede adjuntar un dibujo, un gráfico...).**

    Obra1.- M. electroacústica:

    Obra2.- Polifonía y órgano:

    Obra3.- Canto cristiano antiguo:

13. **Escriba los comentarios globales que considere oportunos.**

Figure 4. Format model of the questionnaire administered.

The psycho-acoustic study conducted was based on the concert described, where the audience participated actively in the research by filling out a questionnaire based on previous similar studies [19, 20, 21]. Participants were asked to subjectively qualify the acoustics of the church according to their perception of the different works presented. So that these subjective results would be reliable, the minimum number of participants would be 16 [22], but finally 92 people completed the survey, bolstering the reliability of the results.

The questionnaire used in this study, a model of which is presented in Figure 4, was based on the one formulated for [23]. It consists of 13 questions, several of which are related directly to the church and the acoustic perception in it. Other questions were related directly to the musical perception of the compositions performed, and one was related to the sensations evoked by these works.

The questionnaires were of three types:

1. Score from 1 to 6.
2. Choice of two or more options.
3. Free response.

Results

The data gathered in the survey was grouped according to the type of responses offered. Below the overall result is evaluated.

<u>Questions with scores from 1 to 6.</u>

Table 2 summarizes the results for questions 2, 3, 4, 6, 7, 8, and 10 for each of the works chosen, where the mean value and the standard deviation are indicated.

		Electro-acoustic music	Poli-fony and organ	Early Christian chant
Quest. 2	x average	2,71	4,72	5,63
	deviation	1,33	0,95	0,67
Quest. 3	x average	2,51	4,56	5,20
	deviation	1,34	1,14	1,22
Quest. 4	x average	3,77	4,81	5,22
	deviation	1,47	1,11	0,98
Quest. 6	x average	3,27	4,78	5,46
	deviation	1,50	1,09	0,69
Quest. 7	x average	3,27	4,84	5,24
	deviation	1,59	1,10	0,94
Quest. 8	x average	3,79	5,11	5,28
	deviation	1,50	0,93	0,82
Quest. 10	x average	2,46	4,90	5,56
	deviation	1,40	1,08	0,77

Table 2. Results corresponding to questions 2, 3, 4, 6 7, 8, and 10, on a response scale of 1 to 6.

Below, the results for each of the questions are examined.

Question 2: The musical works in which the greatest clarity is perceived were polyphony with organ and early Christian chant, with very similar values. In this case, the most confusing perception of the sound coincides with the works least known to the listeners.

Question 3: The works of polyphony with organ and early Christian chant in which words are best distinguished.

Question 4: According to the results, until now, certain confusion persisted regarding questions related to the work of electro-acoustic music. On the other hand, those related to the other two works indicate more confidence or conviction in responding. In general, according to the values found, the venue can be considered quite lively.

Question 6: The mean values for the different works indicate that the difference in perception of warmth was appreciated with the electro-acoustic music, implying that the participants considered the church to have poor acoustic warmth. The coincidence in the perception of the high warmth of the church in terms of Christian chant was striking.

Question 7: The results of the brightness of the interior of the church were practically the same as for its warmth. For the works of polyphony with organ and early Christian chant, the room was found to be very bright, while for the electro-acoustic work, the room was considered not to be very bright.

Question 8: The participants perceived a more sonorous quality in the church when listening to early Christian chant than when listing to polyphony with organ. The mean value of the perception of sonority fell notably on listening to electro-acoustic music, where the standard deviation again approached the mean, as occurred in the previous questions.

Question 10: A majority of the participants agreed that, of the three works, early Christian chant was the type of work most acoustically appropriate for the church of San Pedro. Polyphony with organ was also considered appropriate, but less so, while electro-acoustic music was considered to hardly appropriate.

<u>Questions of two or more choices.</u>

Table 3 shows the results in percentages for questions 1, 5, and 9:

		Electro-acoustic music	**Polifony and organ**	**Early Christian chant**
Quest .1	yes	65,22	94,57	71,74
	no	31,52	3,26	25,00
	no answer	3,26	2,17	3,26
Quest .5	inaudible	28,26	31,52	23,91
	acceptable	47,83	60,87	63,04
	unacceptable	13,04	1,09	7,61
	no answer	10,87	6,52	5,43
Quest .9	from the source	46,74	18,48	18,48
	in other points	23,91	30,43	15,22
	all over the room	18,48	44,57	64,13
	no answer	10,87	6,52	2,17

Table 3. Results in percentages corresponding to questions 1, 5, and 9.

As in the preceding case, the result for each question are examined:

Question 1: The results show that most of the audience had in fact listened to these types of works before. As we suspected at the outset, the electro-acoustic music was the least known, followed by early Christian chant.

Question 5: According to the results, it is evident that the background noise in the church was considered acceptable. In the appraisal of polyphony with organ, practically no participant considered the background noise intolerable, while in relation to the electro-acoustic and early Christian chant, the percentages of intolerability reached 13% and 8%, respectively. Practically the same percentage in the three works indicates that nobody perceived background noise.

Question 9: Again, the works of polyphony with organ and early Christian chant were grouped inasmuch the participants perceived sound emanating throughout the entire church interior, as opposed to the electro-acoustic music, for which the listeners invariably located the origin of the sound with its source.

Table 4 shows results in percentages corresponding to question 11:

Quest. 11	very bad	bad	poor	acceptable	good	very good	excellent	no answer
%	0,00	1,09	1,09	6,52	38,04	42,39	8,70	2,17

Table 4. Results in percentages corresponding to question 11.

Question 11: Most of the listeners classified the overall acoustics of the church as "very good". Almost all of the responses clustered between "acceptable" and "excellent" (with only two individuals who characterized it as mediocre or bad) and thus were considered representative.

<u>Questions of free response</u>

Question 12: Not all the participants answered this question and, furthermore, practically all did so with words. It appears that the group submitted to the survey did not dare to express their emotions through a drawing. However, this not prevent the observation or analysis of sensations or images that the three compositions evoked for each of the listeners. The images that each musical composition evoked were practically the same for all the groups analysed.

Question 13: The overall comments insisted on the predilection for early Christian chant, on not habitually listening to electro-acoustic music, and the differences in perception between acousmatic genre and pieces performed live.

Discussion and Conclusions

In general, the work of early Christian chant received the greatest acceptance, followed by the work of polyphony with organ, and finally the electro-acoustic music. The chant registered the least dispersion in the results and therefore the greatest coincidence in the responses of the participants. By contrast, the responses to the electro-acoustic music revealed much confusion in what was perceived.

According to the initial results, the Christian chant proved to the composition most appropriate to the acoustics of the church according to the perception of the listeners. The lack of having habitually listened to negatively influenced the perception of the electro-acoustic music. In general, the questionnaires reflected that a church was not the appropriate place for this type of music. Furthermore, the lack of musicians in the acousmatic genre also influenced the perception of the majority with respect to the other works performed live.

For the great majority of the listeners, the setting strongly influenced the perception of the music.

This study includes technical issues as well as artistic ones, interrelated by means of a psycho-acoustic analysis in a specific space for a concert with different musical instruments.

From the results as well as their analysis and discussion, the following conclusions were drawn:

1. The acoustics of the church of San Pedro were pleasing to the audience for the musical performances presented. The subjective acoustic appraisal of the church proved to be in general "very good". Despite the

measurement of the objective acoustic parameters and that there were certain orientations for assessing of the acoustics of any venue according to the values thus assigned, it was confirmed that the listener of the concert are the final judges of the acoustics of that place.

2. The results indicate that the musical composition most appropriate to the perception and acoustic conditions of the church is old Christian chant. This is due in part to the high reverberation of the church that aids in the execution of the chant. However, it is also because the audience adjusts its ear to be able to perceive the advantages of this situation, according to the commentaries on the questionnaire.

3. The difference between the broadcasting of recorded electro-acoustic music through speakers and the live performance of the other two works notable influenced the perception of the listeners. This was noted in that the electro-acoustic music was transmitted without any performer contrasted with the scene of polyphony with organ, where a large number of people appeared dressed for the occasion, following the movements of the director. This was also true with the performance of the early Christian chant, where one person chanted while moving, searching the entire church for the place that had the best acoustics for this purpose. It was confirmed that, just as the aesthetic impact that the listeners witnessed entering the room can strongly influence their judgement of the acoustic quality of the church, so the visual impression received during the performance of a given musical work can also have an impact. Therefore, in part, the electro-acoustic music received more negative opinions than did the other two works. In any case, this revealed a predilection for vocal music in the church of San Pedro.

4. For this particular case, in a general way, certain equivalencies can be established between the results of the acoustics of the church determined from the objective *in situ* measurements and the results of the subjective acoustic evaluations provided by the listeners. The objective results coincided in general with the subjective perceptions in the parameters of liveliness, acoustic warmth and background noise. On the contrary, the objective results that did not coincide with the subjective results of the listeners, the latter being higher, were musical clarity, the intelligibility of the words and brightness (see table 5).

Parameters	**Objetive results**	**Subjetive Results**
Liveness	5,91 s	quite liveness
Acoustic warmth	1,14	good
Noise level	31,84 dB	acceptable
Clarity	-6,79 dB / +6,27 dB	almost high
Intelligibility	ALCONS = 15,97 % - 33,48 %	high
Brightness	0,48	high

Table 5. Comparison of the objective and subjective results. Source: own preparation

5. After interpreting the information compiled from the questionnaires, we can state that these images or sensations evoked by each of the works were practically the same for all the respondents. The electro-acoustic music evoked above all the following sensations: nervousness, chaos, uneasiness, suffering, fear, confusion, disorder, tension and darkness. Some of these sensations, such as confusion, were reflected in general in the subjective assessments made in the context of this general music. The work of polyphony with organ evoked order, harmony, exaltation, grandeur, power, fullness,

intensity, sonority, beauty, colour, organization. The continuous references to sonority reaffirmed the results of the subjective evaluation with respect to this work. The early Christian chant suggested peace, tranquillity, relaxation, beauty, meditation, reflection, mysticism, profoundness, stability, peacefulness, equilibrium. Perhaps all this spirituality and tranquillity had an influence on the coincidence of the evaluations in relation to the chant, which was the general trend in all the questions posed in the questionnaire.

In this sense, the question related to the sensations evoked by the works for each person who responded to the questionnaire added a further viewpoint to the questionnaire designed by [23], enriching the conclusions of this line of research.

6. Given the results of this study, the best place to listen to music in the church of San Pedro, considering the location of the performers at the altar, would be situated at any point within a radius of 7 to 8 m from the source of the sound to the listener, preferably in the central area of the church, in the direction of the projection of the sound. The audience situated in this area, especially, perceived greater musical clarity and greater intelligibility of the words (in the case of voices). The possible background noise perceived attenuated when the listener was far from the front door of the church, and the intensity of the music was perceived without problems throughout the church.

The results of this study reveal the close relation between the sound quality of the venue, the characteristics of the works performed there, and the perception by the listeners. The psycho-acoustic analysis confirmed the decisive influence of the acoustic conditions of the church on the musical perception of the audience.

In summary, the current acoustics of the church, with high reverberation times, is adequate for vocal music of long notation and slow free time.

Given that the subjective evaluations vary in large part even for the same person, it would be impossible to repeat exactly the same results with the same works but with different listeners, although belonging to the same groups according to musical training. Nor is it likely that another person with the same data can reach the same conclusions. What can be guaranteed is that any other researcher could verify the validity of the data compiled in this survey and critically assess the analyses and interpretations provided in this study.

REFERENCES

[1] PALMESE, Cristina y CARLES, José Luis: "Acústica y arquitectura: El marco acústico y su evolución", in Revista Scherzo, Año XX, Núm. 193, Enero 2005.

[2] REZNIKOFF, Iégor: *Le Chant du Thoronet: splendeur du grégorien pour soliste dans la plus belle résonance romane*. SM 12 16.62, Paris, 1989/2005 (CD).

[3] REZNIKOFF, Iégor: "The evidence of the Use of Sound Resonance from Palaeolithic to Medieval times". *Archaeoacoustics*, C. Scarre & G. Lawson, ed. University of Cambridge, Cambridge, 2006.

[4] ZWICKER, Eberhard & FASTL, Hugo: *Psychoacoustics. Facts and Models*. Springer. Berlin, 1999.

[5] HOWARD, David M. & ANGUS, James: *Acoustics and Psychoacoustics*. Focal Press. Oxford, 2001.1

[6] SCHAFER, R. Murray: *Le paysage sonore*. Lattès. Paris, 1979 (trad. del original *The Tunnig of the World*, por Sylvette Gleize, A. Knopf Inc., New York, 1977.

[7] TRUAX, Barry: *Handbook for Acoustic Ecology*. CD-ROM. Cambridge Street Publishing. CSR-CDR 9901, 1999

[8] GOTZENS, Antonia Mª y MARRO, Silvia: *Prueba de valoración de la percepción auditiva: explorando los sonidos y el lenguaje*. Masson. Barcelona, 2001, p. 12

[9] Ears: ElectroAcoustic Resource Site (on line) http://www.ears.dmu.ac.uk/spip.php [21 September 2017]

[10] GARCÍA DE PAREDES, José María: *Paseo por la arquitectura de la música*. Discurso del Académico electo José María García de Paredes. Real Academia de Bellas Artes de San Fernando. Madrid, 1986.

[11] CÁRDENAS, Inmaculada: Música & Arquitectura. S/I: U.M./Unió Músics, 2001 (CD).

[12] TROITIÑO VINUESA, Miguel Ángel: Arquitecturas de Cuenca: el paisaje urbano del casco antiguo, Servicio de Publicaciones de la Junta de Comunidades de Castilla-La Mancha, p.17 (1995)

[13] BOE nº 235, 1 de Octubre de 2002

[14] VALVERDE-GASCUEÑA, Nelia: (2009). Sonidos en la Arquitectura de San Pedro. Un estudio acústico desde la perspectiva técnica y artística de la creación sonora en la Iglesia de San Pedro en Cuenca (España). (Thesis unpublished). Universidad de Castilla-La Mancha, España.

[15] RECUERO, Manuel: Acondicionamiento Acústico, Paraninfo, Madrid (2013)

[16] SENDRA, J.J.; ZAMARREÑO, T.; NAVARRO, J.; ALGABA, J.: El problema de las Condiciones Acústicas en las Iglesias: Principios y Propuestas para la Rehabilitación, I.U.C. Construcción, Sevilla (1997)

[17] CARRIÓN ISBERT, Antoni: Diseño Acústico de Espacios Arquitectónicos. Edicions UPC, Barcelona (1998)

[18] RECUERO, M.; GIL, C.: Acústica Arquitectónica, Paraninfo, Madrid (1993)

[19] SENDRA, J.J.; NAVARRO, J.: La Evolución de las Condiciones Acústicas en las Iglesias: del Paleocristiano al Tardobarroco, I.U.C. Construcción, Sevilla (1997)

[20] BERANEK, Leo L.: Music, Acoustics and Architecture. New York: Krieger (1979)

[21] EDWARDS, R.M.: A subjective assessment of concert hall acoustics. Acustica 39, 183-195 (1974)

[22] HAWKES, R.J.; DOUGLAS, H.: Subjective acoustic experience in concert auditoria. Acustica 24, 135-150 (1971)

[23] Delta-Nordtest, Measurements and Judgments of Sound in relation to Human Sound Perception, Report AV 1461/01, Nordtest-Delta (2001)

[24] BARRON, Mike: Auditorium Acoustics and Architectural Design, Spon Press (2010)

An Interpretative Phenomenological Analysis of the Impact of Frequency-Dependent Sound on Human Thought and Feeling in a Bronze Age Cave in Morayshire, North-East Scotland

Michelle Walker

MICHELLE WALKER is a Postgraduate Student from The Archaeology Institute of the University of the Highlands and Islands, Orkney.

ABSTRACT: Ten voluntary participants were asked to listen to drumming frequencies of 90, 100, 110 and 120 hertz (Hz) for two minutes per frequency in a Bronze Age cave in Morayshire, North-East Scotland in order to assess the impact of frequency dependent sound on human thought, feeling and imagery. After which time participants were given, and asked to complete, a semi-structured qualitative questionnaire containing 10 questions relating to any thoughts, feelings and imagery they may have experienced during the individual frequencies. An Interpretative Phenomenological Analysis was utilized for the analysis of the data. Previous research undertaken by the Super Brain Research Group in 2007 (SBRG) and Princeton Engineering Anomalies Research (PEAR) in 1996 found that in Neolithic temples and hypogea (underground temple or tomb) in Europe had significant resonance properties and concluded that "These ancient structures possessed resonant acoustical properties that may have contributed to their functional purpose."

Introduction

Interpreting and analyzing the past calls for an interdisciplinary approach. One may study documents which are classified as History and Literature; others study oral traditions and folklore which may be classified as ethnography, and the study of material culture which lands in the realms of archaeology. In addition, others utilize a biological methodology which includes a wide variety of areas, for example DNA, Palynology and Dendrochronology. The analysis of linguistics and ancient languages is another method by which the past can be interpreted (Zubrow, 2014), thus creating a very rich and diverse knowledge base for analyzing all aspects of our ancestry.

How do we listen to the past? Archaeoacoustics is a relatively new interdisciplinary approach that analyzes sound from and within ancient sites and structures. Archaeoacoustics is in a pre-paradigmatic stage. There are no generally accepted paradigms, frameworks, theories, methodologies or models as yet to work within. These may be structured and developed through a systematic crossing over of paradigms from, perhaps, copious academic fields. Many practitioners from this developing field do not call themselves archaeoacousticians, primarily because they originate from various academic fields of history, archaeology, the sciences, engineering, and neurophysiology to name a few.

Previous research undertaken by the Super Brain Research Group (SBRG) in 2007 and Princeton Engineering Anomalies Research in 1996 (PEAR) found that Neolithic temples and hypogea in Europe had interesting and significant resonance properties. Within six different Neolithic temples in England and Ireland, an acoustic resonance around 110Hz was discovered. PEAR concluded "*These ancient structures possessed resonant acoustical properties that may have contributed to their functional purpose*". Furthermore, a research group from the University of California, Los Angeles (UCLA) in 2008, undertook research that looked at the effect and correlation of the frequencies found at the English and Irish temples by PEAR on brain activity using electroencephalogram (EEG). In particular they found that the pattern of asymmetric activity over the prefrontal cortex shifted from one of higher activation on the left side at most frequencies to right-sided dominance at 110 Hz. These findings are compatible with relative deactivation of language centres along with a shift in prefrontal activity that may be related to emotional processing. These results demonstrate that the acoustic properties of ancient structures may influence human brain function. A wider study of these interactions should be undertaken. In addition, from the findings of these studies Cook *et al* in 2008 hypothesised: "The resonances of the chamber cavities might have been intended to support human ritual chanting. There is the possibility that tones at these frequencies may specifically affect regional brain activity."

Cook *et al* in 2008 postulated that Archaeoacoustics has been employed to analyze variable sound frequencies, many of which may alter human physiological and emotional states. Furthermore, Debertolis in 2013 suggested that the documented effects of these acoustic properties may have been realized by the inhabitants of some ancient sites and their knowledge of the process used to enhance their rituals.

Hal Saflieni on the Mediterranean Island of Malta is an underground cemetery which was used for burial in Maltese prehistory spanning 4000 to 2500 BCE. The acoustic properties of this site were discovered in 1920 by William Arthur Griffiths. In one chamber, a curved projection was observed specially carved out of the back of the cave and acting as a sounding-board, showing that the designers had a good practical knowledge of sound-wave motion. Ritual chanting into a notable niche became the highlight of countless visits so much so that the bottom of the niche has become worn and discoloured from having the chanters' hands resting on it.

Debertolis *et al* 2014 carried out a study of acoustic resonance at ancient sites and related brain activity to assess the effects of resonance phenomena on the human body. Healthy volunteers underwent examination by EEG and were subjected to listening to tones between 90 and 120Hz. These tones are similar to the resonant sounds found at some Neolithic structures in England, Ireland, Italy and Malta. They discovered that each volunteer has their own individual frequency of activation that can be significantly different from 110Hz, but was always between a range of 90 to 120Hz.

Moreover, Gaona *et al* in 2014 investigated the archaeoacoustics of El Castillo in Northern Spain and found that the highest frequencies in the cave were again at 108 and 110Hz. The archaeological studies of this cave revealed the presence of prehistoric ritual activity associated with early shamanism.

Sculptors Cave, Covesea, North East Scotland

Sculptor's Cave, Covesea, is located near the base of impressive cliffs where the Moray Firth meets the sea in the North-East of Scotland. The National Grid Reference number is 1750 7072. The cave interior, is 13.5m x 20m in area and up to 5.5m high, reached via twin passages over 11m long, from a large entrance canopy. The site – inaccessible at high tide – produced a range of objects of Late Bronze Age date including several bronze arm-rings, ring money, a swan's neck pin, pottery and worked bone, along with a very large quantity of human remains, predominantly those of sub-adults. (Benton 1931; Shepherd 2007). The presence of skull and mandible fragments in the entrance passages suggests the display of heads at the entrance to the cave. There is evidence to suggest that one juvenile frontal bone was deliberately de-fleshed, hinting at practices which involved the curation of human remains (Armit *et al* 2011).

The very nature of the cave that was, and still is, difficult and dangerous to access is likely to have made this a highly appropriate location for the undertaking of ritual practices situated between land and sea, the upper world and the underworld. Was this a liminal location in which rites of passage, transforming children to adults or the living to the dead taken place? It may also have provided an access point to the gods or spirits of the underworld (ScARF, 2011).

The Rationale

One may postulate from the secondary research findings that were outlined in the introduction, that the acoustic resonance properties within prehistoric sites in Scotland may have had an impact on the emotional state of human beings during ceremonies and rituals. It may be plausible to suggest that resonant acoustic properties exist within prehistoric sites in Scotland. Perhaps the reasoning behind the structural similarities of ancient sites is to assist with the ceremony, ritual and deposition of human remains of both adults and children in prehistoric times, as were discovered by Benton in the Sculptors Cave excavation in 1929 – 1931. Moreover, the methodology employed in this piece of research lies within the significant lack of phenomenological and qualitative data in the developing field of archaeoacoustics. Phenomenology is the study of human experience and consciousness in everyday life from a first person perspective; a process of allowing a phenomenon to reveal itself by experiencing or re-experiencing movements through an ancient site or landscape (Smith *et al*, 2009).

Aims and Objectives

The aims and objectives of the research project were to investigate the impact of drumming for two minutes on the thoughts, feelings and imagery of 10 voluntary human participants at frequencies of 90, 100, 110 and 120Hz within a Late Bronze Age cave in Morayshire, North-East Scotland, UK. The qualitative data was collected by means of a 10 question semi-structured questionnaire from participants. The questionnaires contained questions pertaining to any imagery, thoughts or feelings they may have experienced during the drumming. These were then analysed using the methodology of Interpretative Phenomenological Analysis (IPA) for any commonality and/or recurring themes amongst participants whilst listening to drumming at the aforementioned frequencies.

IPA as a Methodological Approach to Archaeoacoustics

Interpretative phenomenological analysis (IPA) is a qualitative approach to research with an idiographic focus, which means that it aims to offer insights into how a given person, in a given context, makes sense of a given phenomenon (Smith *et al*, 2009).

The aim of IPA is to explore in detail the participant's personal perception of the topic under investigation. Thus the approach is phenomenological, in that it is concerned with an individual's personal perception and/or experiences of the research being undertaken, as opposed to an attempt to produce an objective statement of the event or events. Moreover, IPA also recognises that the research process is a dynamic one. The researcher endeavours to obtain access and get close to, the participants physiological and emotional states while participating within the research. However, the researcher cannot do this directly. Access depends on, and is complicated by, the researchers' own conceptions. Indeed, these are required in order to make sense of the personal perceptions and experiences of the participant. The analysis itself is the interpretative work, which the researcher rigorously carries out at each of the five stages of analysis.

IPA was the chosen qualitative methodology within the research project as the participant's thoughts, feelings and imagery could be recorded and analyzed whilst listening to variable frequencies in a Late Bronze Age Cave. This data may give us a glimpse of how Bronze Age People may have been physiologically and emotionally affected by variable brainwave frequencies influencing their ceremonies, rituals and/or burials within the Late Bronze Age Cave. In addition, contemporary Neuroscience suggests that the human auditory system is the only human sensory system that hasn't changed for over a millennia (Fannin, 2009). It may be suggested that the frequencies that alter our brainwaves today may have altered our ancestor's brainwaves in a similar way. Furthermore, utilising the phenomenological qualitative methodology within archaeoacoustics can add pertinent qualitative data to a significantly low level of qualitative data within the field.

The Five Stages of Analysis

(i) Stage one of the analysis, thematic coding, the data from 40 questionnaires which had been written by the participants' was rigorously read and re-read, initial notes, interpretation and comments were recorded onto the questionnaires.

(ii) Stage two, the semantic content and language was examined on an exploratory level and initial noting was undertaken.

(iii) In Stage three, the data was analyzed for emerging themes such as Physiological factors.

(iv) Stage four, the emergent themes were then broken down into categories.

(v) Stage five, the categories were then recorded and participants' responses were analyzed for any occurrence of commonality.

The Findings

The analysis of the qualitative data found that five categories had emerged from the themes. Each category illustrated a written response experienced during the drumming. None of the participants had recorded experiencing any imagery during the drumming at variable frequencies. Eighty-percent of the participants experienced physiological changes to their body whilst listening to the drumming at two-minute intervals. The physiological factors were: increased breathing, increased heart rate and a mild feeling of anxiety, a feeling of anticipation that something was about to occur. Interestingly, the participants recorded these physiological experiences whilst they were listening to the drumming frequencies of 120Hz the highest frequency of drumming. Other studies outlined above found neurological changes at 110Hz and others' between 90 and 120Hz. This may suggest that the higher frequencies affect human physiological factors too.

Moreover, all participants experienced 'internal dialogue' during the four variable frequencies. The internal dialogue was variable between participants' and ranged from dialogue pertaining to menial chores to conversations they were going to have or had had during the last 24 hours.

In addition, all participants reported a feeling of a 'lapse of time'. Participants felt that a shorter period of time had elapsed than the reality of the two minutes that had passed during each frequency of drumming. The participants' concept of time seemed to have been altered whilst listening to the variable drumming frequencies within the Bronze Age cave.

Participants recorded a multitude of detailed experiences onto a 10 question semi-structured questionnaire that included physiological insights into their experiences. In light of the qualitative data analysis, experiences of increased breathing and heart rate were identified along with a mild feeling of anxiety and a feeling of apprehension. These experiences were constant and didn't fluctuate whilst listening to the drumming for two minutes at the frequency of 120Hz. Contemporary research in Physiology suggests that the physiological changes in the participants are linked to a feeling of anxiousness, anticipation and the 'flight or fight response' (Burns, 1999). It appeared that the participant's normal state of being was affected by the drumming at 120Hz.

It may be suggested that the inhabitants of this late Bronze Age cave may also have been physiologically affected by these drumming frequencies and were indeed preparing themselves for a ceremony, ritual or burial within the cave. Evidence from Silvia Benton's 1931 excavation report points towards the cave being utilised for burials as over 6,000 human remains were yielded from the excavation. Over 60% of these were mandibles from sub-adults. In addition, Armit in 2011 suggested in his research paper 'Death, Decapitation and Display' that the cave was a place for deposition and the curation of human remains.

Interestingly, in addition to the physiological changes, participants also reported experiencing a skewed perception of time. Whereby participants felt that a shorter time had passed whilst listening to the drumming than the reality of the two minutes. Eriksson *et al in 2009* reported that 'disturbed' time perception is one of the hallmarks of an Altered State of Consciousness (ASC). Therefore, it may be suggested that Bronze Age people experienced an altered state of consciousness whilst preparing and/or conducting ceremonies, rituals and/or burials within Sculptors Cave.

Suggestions for Future Research

Interesting experiences were recorded by participants whilst listening to variable drumming frequencies within this research. Although there were some intense physiological changes recorded, none of the participants reported experiencing any adverse effects whilst listening to the drumming. In conclusion, drumming at variable frequencies within ancient sites and recording the data qualitatively can be an interesting and fruitful domain for future research in the field of archaeoacoustics. Its value as a supplement to quantitative methodologies needs to be investigated and further evaluated in order to establish a place in the future of archaeoacoustic methodologies.

Final Remarks

It is well documented that set and setting can heavily influence the experiences of participants whilst participating in such a research project. The experiences recorded within this phenomenological research were considered real and genuine and were of particular interest. The intention of this study was to investigate any thoughts, feelings and imagery that were experienced by participants whilst listening to variable drumming frequencies and to increase the relatively low level of qualitative data within archaeoacoustics.

BIBLIOGRAPHY

Armit, I. (2013). Schulting, R., Knusel, C.J., Shepherd, I.A.G. *Death, Decapitation and Display? The Bronze and Iron Age Human Remains from the Scultors Cave, Covesea, North East Scotland.* Proceedings of the Prehistoric Society, 77 251-278.

Benton, S. (1931) The excavations of the Sculptors Cave, Covesea, Morayshire. Proceedings of the Society of Antiquaries of Scotland, 65 177-216.

Burns, D.D. (1999). *The Feeling Good Handbook.* 2nd Edition. London: Plume.

Cook, I.A. Pajot, S. K. and Leuchter, A.F. (2008) Ancient Architectural Acoustic Resonance Patterns and Regional Brain Activity. *Time and Mind: The Journal of Archaeology, Consciousness and Culture,* 1(1).

Debertolis, P., Tirelli, G., and Monti, F., (2014). Systems of Acoustic Resonance at Ancient Sites and Related Brain Activity: Preliminary Results of Research. In: L.C Eneix, eds., *Archaeoacoustics: The Archaeology of Sound.* Florida: OTS Foundation. pp.59-66.

Fannin, J.L. (2009). *Understanding Your Brainwaves.* PhD. University of London.

Gaona, J.M. Rouleau, N. Caswell, J.M. Tessaro, W.E. Burke, R.C. and Schumacher, D.S. (2014). *Archaeoacoustic Investigation of a Prehistoric Cave Site: Frequency Dependent Sound Amplification and Potential Relevance for Neurotheology.* NeuroQuantology, 12(4).

Princeton Engineering Anomalies Research (1996) Consciousness and Anomalous Phenomena. [online] Princeton: Princeton University. Available from: http://www.princeton.edu/~pear/pdfs/1995-consciousness-anomalous-physical-phenomena.pdf [Accessed 16 August 2016].

ScARF, (2011). *Case Study: Sculptors Cave, Covesea.* [online] Edinburgh. Available from http://www.scottishheritagehub.com/content/case-study-sculptor's-cave-covesea [Accessed 13 February 2016].

Smith, J.A., Flowers, P. and Larkin, M., (2009). *Interpretative Phenomemological Analysis: Theory, Method and Research.*

Super Brain Research Group (2007) *Archaeoacoustics in Archaeology.* [online] Florence: University of Trieste. Available from: http://www.sbresearchgroup.eu/index.php/en/research-papers [Accessed 23 September 2016].

Watson, A. and Keating, D., (1999). Architecture and sound: an acoustic analysis of megalithic monuments in prehistoric Britain. *Antiquity,* **73**, (280), pp.325-336.

Zubrow, E,B,W (2014). The Silence of Sound: A Prologue. In: Eneix, Zubrow and Douglas (eds.) Archaeoacoustics: The Archaeology of Sound. Publication of the 2014 Conference in Malta.

Can We Hear the Sounds of Archaeological Sites ...The Way Our Ancestors Heard Them?

Steven J. Waller

STEVEN J. WALLER, Ph.D. in Biochemistry/Biophysics, University of Virginia 1977; has conducted archaeoacoustic research since 1987; advocates preserving natural soundscapes of archaeological sites. Email: wallersj@yahoo.com

ABSTRACT: The principles of Auditory Scene Analysis modeling are applied to the study of Archaeoacoustics, with an emphasis on differing cultural interpretations of various acoustic phenomena. Examples are given in which auditory perceptions are shown to be heavily influenced by prior expectation according to the world view of the listener. A given sound can be interpreted ambiguously – e.g., what was that noise: thunder, or stampeding hoof beats, or reverberation? This serves as a caution that precise mechanical acoustic measurements and detailed documentation of the sound characteristics at archaeological sites are only a beginning -- what happens between the eardrums is equally, if not more, important. Neural representations of auditory stimuli in the brain can be studied in the laboratory or in the field with modern human volunteers, but are these really representative of the mental "visions" evoked by sounds that were mysterious to ancient humans? Case studies are presented to illustrate the value of including ethnographic information, including myths and legends, in the study of sound at archaeological sites to gain insight into the reactions our ancestors had to various sounds they heard. One interesting case is the possible use of parrot petroglyphs to symbolize the repetition of words that can be heard in the form of echoes at Pony Hills, Petroglyph National Monument, Mesa Verde, Willow Springs, Chaco Canyon, and Bandelier National Monument rock art sites.

Introduction

This paper is a continuation of the talks the author presented at the first two Archaeoacoustics conferences, entitled "Auditory Illusions in the Soundscapes of Rock Art and Stonehenge" (Waller 2014), and "Recent Advances in Archaeoacoustics: Echo 'Spirits' and Thunder 'Gods' Predicted at Archaeological Sites." (Waller 2016). This present paper emphasizes the importance of not just recording and analyzing sounds at archaeological sites, but in giving weight to the various cultural interpretations of sounds throughout history. A given sound does not just mechanically produce a given response. As a trivial example, if someone walks up behind you and says "Hi", you will react very differently depending on the circumstances -- you might simply reply "Hi" if you planned to meet a friend, or you might be startled half to death yelling "Ahhhhh!" and jumping out of your skin if you were working late at night in an empty building thinking you were alone. The exact same pressure waves hitting your eardrums in these two instances could be measured as identical, but can under different circumstances produce extremely different responses. As shown in studies of auditory scene analysis (Bregman 1994, Gordon and Rosenblum 2004; Rosenblum and Robart 2007) in which people try to interpret their surroundings based on listening, the response to a sound is determined not just by the characteristics of the sound itself, nor even by the mechanics of hearing and neural stimulation, but is heavily influenced by prior experiences and expectations of the listener. This principle will be shown now to be applicable to studies of the archaeology of sound, including presentation of a Black Box model for Archaeoacoustics, and case studies of the interrelationships between archaeological sites, sound reflection, and mythology.

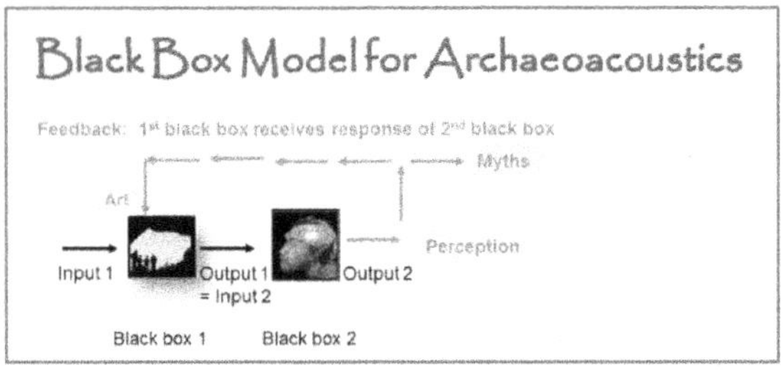

Figure 1. Black box model for archaeoacoustics. Illustration of an example situation in which Input 1 is an impulse sound, Black box 1 is a cave chamber, Output 1 is the impulse response (e.g., echoes) that also serves as Input 2 for coupled Black box 2 which in this case is the ancient mind in prehistoric times. Output 2 (which depends on the many prior inputs that constitute the past experiences of the human brain, including in this case beliefs in supernatural beings) consists of perceptions/interpretations including: ECHO SPIRITS, DISEMBODIED VOICES, PORTALS TO THE SPIRIT WORLD RESPLENDENT WITH SOUND, STAMPEDING HOOFED THUNDER GODS, etc., plus additional tangible Output 2 reactions in the form of oral myths and artwork describing and portraying these auditory perceptions of supernatural beings. Notice the feedback loop: Black box 1 (cave) can receive input (art) as a result of the output (perception) of Black box 2 (artist's brain) that was in turn a response to the output (sound reflection) from Black box 1.

Theoretical Framework for Archaeoacoustics: Black Box Model

The technique of Black Box modeling can be applied to Archaeoacoustics. As noted in Wikipedia, "In science, computing, and engineering, a black box is a device, system or object which can be viewed in terms of its inputs and outputs (or transfer characteristics), without any knowledge of its internal workings. Its implementation is "opaque" (black). Almost anything might be referred to as a black box: a transistor, an algorithm, or the human brain." As a simplified example, a cave chamber could be considered to be a black box: it can be studied by making a sound (inputting an audio impulse), and listening for any sound effects that follow (determining the output as impulse response: I.R). This I.R. is the resulting effect that the chamber has on the sound -- i.e., determining the output in terms of additional sounds produced due to reflection -- without needing to know the specific details of the chamber's size, shape, coefficient of absorption, etc. So the black box (cave chamber) can be characterized by the input (audio impulse) and output (impulse response, or I.R.). In a further level of complexity, a number of archaeoacoustic studies have been conducted in which the acoustic response from an archaeological site is input into audio equipment that provides quantitative measurements as output (e.g., Waller, Lubman and Kiser 1999; Fazenda 2017). The audio equipment is another type of black box, in which the user does not need to necessarily understand the inner workings of the hardware and software in terms of circuits and algorithms in order to gain benefit from the output data. This then is an illustration of a system comprised of coupled black boxes: the output of the first black box becomes the input of the second black box. Thus the output of the second black box is an indirect function of the first black box. This can be represented in a simplified way as follows.

<u>Example A</u>

Cave chamber = black box 1 (characteristic behavior of the chamber is the way in which sound is altered)

Input 1 = audio impulse; output 1 = Impulse response (I.R.)

Audio equipment = black box 2

Input 2 = I.R. (= output 1 from black box 1, which goes into black box 2);

Output 2 = quantitative measurement parameters including: dB, Delay time, RT60, Hz, D50, C50, etc.

Note that the output 2 in this case is only as good as the quality of the equipment and design of the software, which in itself is a result of many scientific inputs and could be either state-of-the-art or contain design flaws.

Example B

Let us now examine the situation in which the black box 2 is the scientist's mind; the output 2 will depend on the many prior inputs that constitute the past experiences of the brain, including in this case the extent of scientific knowledge of acoustic theory and practical audio training.

Cave chamber = black box 1

Input 1 = impulse; output 1 = I.R.

Scientist's human brain = black box 2

Input 2 = I.R. (= output 1 from black box 1, which goes into black box 2);

Output 2 = Perceptions/reactions including: Sound reflection, Distortion, Reverberation, Coloration, Definition, Clarity, Audibility, etc.

Example C

In this final black box example, let us examine the situation in which the black box 2 is the ancient mind in prehistoric times; the output 2 will depend on the many prior inputs that constitute the past experiences of the human brain, including in this case beliefs in supernatural beings.

Cave chamber = black box 1

Input 1 = impulse; output 1 = I.R.

Pre-scientific ancient artist's brain = black box 2

Input 2 = I.R. (= output 1 from black box 1, which goes into black box 2);

Output 2 = Perceptions/reactions including: *ECHO SPIRITS, DISEMBODIED VOICES, PORTALS TO THE SPIRIT WORLD RESPLENDENT WITH SOUND, STAMPEDING HOOFED THUNDER GODS, PLACES OF POWER, MAGIC PILLARS, QUETZAL CALLS, ANGELS*, etc., plus additional Output 2 = oral myths and artwork describing and portraying these auditory perceptions of supernatural beings.

Thus, as shown in these examples, a sound (input 1) produced at an archaeological site (black box 1) can have acoustical characteristics (output 1 impulse response) that, depending on how and when these resulting sounds (impulse response that serves as input 2) were processed (black box 2), may have different results (output 2) ranging from quantitative data, to scientific descriptions, to supernatural interpretations. This black box model of Archaeoacoustics also includes feedback: black box 1 can receive input from the output of black box 2 that was in turn a response to the input/output from black box 1, forming a loop (see Figure 1). An example of this would be noise from pounding on rock producing a percussive echo, which is interpreted as hoof beats and evokes the perception of hoofed animals, inspiring more pounding in order to hear more hoof beats and perhaps in an effort to communicate with the unseen hoofed animals in the rock, and consequently the marks from the repeated percussive strikes possibly being formed into the shape of the hoofed animals, thus resulting in a pictorial image of the auditory perception that occurred at the echoing site. Acoustics can in such a way help explain why rocky sound-reflecting environments were decorated in the past with images that correspond to mythical beings.

Case Study:
Rock Art Sites with Parrot Designs

There are many rock art sites in the American Southwest with bird-shaped images that have been identified as parrots or macaws. These include:

Petrified Forest National Park* (AZ)

Willow Springs* (AZ)

Mesa Verde National Park* (CO)

Petroglyph National Monument* (NM)

Chaco Culture National Historical Park*(NM)[1]

Bandelier National Monument* (NM)

La Cieneguilla (NM)

Pony Hills* (NM); (see Figure 2)

Black Mesa (NM)

Rio San Jose (NM)

Zuni Reservation (NM)

Hueco Tanks* (TX)

Hovenweep National Monument (UT)

* sites tested acoustically by the author; these all were found to be echoing.

Parrots and Macaws are a recurring rock art motif (Figure 2). These birds are revered in myths. The Acoma origin story includes the search for the perfect echo; an integral part of the story is the people must choose which of two eggs will hatch a parrot and which one will hatch a crow. There is an ancient belief that the human soul enables parrot to repeat words and talk like humans, i.e., a supernatural explanation for the vocal mimicry. The author has noted the presence of strong echoes at the rock art sites he visited with parrot designs. Might parrot petroglyphs symbolize repeats of human speech heard in the form of echoes? This would be another case in which a given sound could be perceived in different ways: the repeat of an echo interpreted as the repeat from a parrot.

Other Rock Art Sites Recently Characterized As Having Strong Echoes:

The following rock art sites were tested for acoustics between the time of the 2nd and 3rd Archaeoacoustics conferences (February 2014 to October 2015). As predicted by the Rock Art Acoustics theory, all these sites were found to possess strong sound reflection characteristics, in the form of echoes and/or reverberation:

OREGON: Tumalo; Picture Gorge

WYOMING: Dinwoody (see Figure 3)

CALIFORNIA: Palo Verde

NEW MEXICO: Hembrillo Canyon

SPAIN: Cantabrian Caves of El Castillo, Las Monedas, La Cullalvera, and Covalanas.

PORTUGAL: Vale do Côa (in collaboration with António Batarda)

Testable Hypothesis of a Connection between Acoustics and Pottery in Maya Caves

At the second Archaeoacoustics conference in late 2015, the author presented a new theory along with another example of a testable archaeoacoustics hypothesis. Holley Moyes' research (Moyes 2006; Moyes et al. 2009) describes evidence for ritualistic use

[1] Legend has it that Crooked Nose (One who wears a turquoise feather) reshaped the cliff face to enhance the projection of sound

Figure 2. Example of a parrot design at an echoing rock art site: Pony Hills, New Mexico.

Figure 3. The Dinwoody site in Wyoming: an example of a rock art location recently tested and found to possess strong echoing characteristics, as predicted.

of Maya caves for rain-making ceremonies including uneven distribution of pottery. A literature search by the author revealed that Maya mythology describes the belief that "Each thunderclap is a separate Chaak in action, breaking a jar open and letting the rain fall" (Guillermoprieto 2013). The author theorized a possible connection between acoustics and pottery since thunderous reverberation in the caves could have been misperceived as thunder that was produced by the same thunder gods who caused thunderstorms in the sky by breaking pots. A testable hypothesis is that there would be found a correlation between cave acoustics and pottery placement.

In June of 2016, the author travelled to Belize and conducted systematic acoustic tests in several Maya caves in collaboration with Holley Moyes. Preliminary analysis of the acoustical data shows a correlation between reverberation time and the location of certain forms of pottery. Thus the hypothesized archaeoacoustical connection between sound and pottery in Maya caves, as suggested by Maya cosmology, is tentatively supported (detailed results to be co-authored and published elsewhere). These preliminary results are presented herein as another example of how an appreciation for the ways in which ancient people perceived sound as recorded in myths (e.g., thunder as pots breaking) can inform archaeoacoustic studies.

Discussion

Auditory perceptions are heavily influenced by prior expectation according to the world view of the listener. A given sound input can be interpreted ambiguously and result in a variety of responses – e.g., a loud broad-spectrum noise that decays over time could be characterized as a long RT60, perceived as thunderous reverberation, a thunderstorm, breaking pots, or the thundering hoof beats of stampeding animals as described in thunder god myths. This serves as a caution that precise mechanical acoustic measurements and detailed documentation of the sound characteristics at archaeological sites in modern descriptive terms are only a beginning -- what happens between the eardrums is equally, if not more, important. Neural representations of auditory stimuli in the brain can be studied in the laboratory under controlled conditions with present day human volunteers, but are these results really representative of the mental "visions" evoked by sounds that were mysterious to ancient humans? This underscores and illustrates the value of including ethnographic information, including myths and legends, in the study of sound at archaeological sites. We need not only to hear these sounds, but to attempt to hear them through the ears of our ancestors, to gain insight into the reactions their minds had to various sounds they heard, as described in the variety of myths and legends passed down to us through countless generations.

https://sites.google.com/site/rockartacoustics

ACKNOWLEDGEMENTS

The author thanks António Batarda, Richard Braun, and Holley Moyes for collaboration with acoustical studies.

REFERENCES

"Black Box" Wikipedia. https://en.wikipedia.org/wiki/Black_box (accessed December 31, 2017).

Bregman, A.S. 1994. Auditory Scene Analysis: The Perceptual Organization of Sound. Cambridge, Mass.: MIT Press.

Fazenda, Bruno, Chris Scarre, Rupert Till, Raquel Jiménez Pasalodos, Manuel Rojo Guerra, Cristina Tejedor, Roberto Ontañón Peredo, Aaron Watson, Simon Wyatt, Carlos García Benito, Helen Drinkall, and Frederick Foulds, 2017, The Journal of the Acoustical Society of America 142, 1332; doi: 10.1121/1.4998721, accessed online January 21, 2018 at: http://dx.doi.org/10.1121/1.4998721

Gordon S.G. & L.D. Rosenblum. 2004. Perception of acoustic sound-obstructing surfaces using body-scaled judgments. Ecological Psychology, 16:87-113.

Guillermoprieto, Alma 2013. Secrets of the Maya Otherworld. National Geographic, accessed online December 30, 2015 at http://ngm.nationalgeographic.com/2013/08/sacred-cenotes/guillermoprieto-text

Moyes, Holley. 2006. "The Sacred Landscape as a Political Resource: A Case Study of Ancient Maya Cave Use at Chechem Ha Cave, Belize, Central America." PhD dissertation, Department of Anthropology, State University of New York at Buffalo.

Moyes, Holley, Jaime J Awe, George A Brook, and James W Webster. 2009. The Ancient Maya Drought Cult: Late Classic Cave Use in Belize. Latin American Antiquity 20, no. 1: 175-206.

Rosenblum, L.D. & Robart, R.L. 2007. Hearing silent shapes: Identifying the shape of a sound-obstructing surface. Ecological Psychology 19: 351-366.

Waller, Steven J., 2016 Recent Advances in Archaeoacoustics: Echo "Spirits" and Thunder "Gods" Predicted at Archaeological Sites. Proceedings of Conference "Archaeoacoustics: The Archaeology of Sound", Istanbul, Turkey, 29 OCT - 02 NOV 2015. (2016 Print).

Waller, Steven J., 2014 "Auditory Illusions in the Soundscapes of Rock Art and Stonehenge" in Archaeoacoustics: The Archaeology of Sound: Publication of Proceedings from the 2014 Conference in Malta, eds. Eneix, Linda C. and Ezra B. W. Zubrow.

Waller, Steven J., Lubman, David and Kiser, Brenda, 1999, "Digital Acoustic Recording Techniques Applied to Rock Art Sites" in, American Indian Rock Art, Ridgecrest, California, Vol. 25:179-190, American Rock Art Research Association, San Miguel, California

Conference participants explore an audio/visual installation by Steve Waller. Image: OTSF

The International Association for Archaeoacoustics: a Prolegomenon

Ezra Zubrow

EZRA B. W. ZUBROW, Ph.D., F.S.A. President of UUP Buffalo Center Chapter, SUNY Distinguished Service Professor, Professor Anthropology University at Buffalo & Toronto (SO), Department of Anthropology University at Buffalo, 380 Academic Core Millard Fillmore Ellicott Complex, Buffalo, New York, 14261-0026. email: ezubrow@gmail.com

CALL: Archaeoacoustics is at a crossroads. This is a call to create an International Association for Archaeoacoustics. At best it will be the rallying cry for scholars, scientists, engineers, and other interested people to begin the process. At worst, it will be a consideration to reflect upon.

Introduction:

Although it may seem paradoxical, the more one learns, the more complex it becomes. What initially seems simple, turns out not to be so. This is particularly true for what has come to be called archaeoacoustics. We know that sound has permeated all of hominid and human existence for millions of years. It has been and is one of the essential senses.

Archaeoacoustics is the study of past sound. It is a very inclusive field that draws upon acoustics, music, ethnomusicology, history, archaeology, anthropology and other disciplines. Archaeacoustics includes the notation of past sounds, the instruments of past sounds, the environments of past sounds, the motivation for the creation of past sounds, and many other aspects of past sounds. In this call, we want to be inclusive and include all such studies and even others that we even have not considered. For example neuroscience offers great potential through the use of electroencephalography for monitoring of hearing for different instruments and sounds in different environments.[1]

We know from skeletons that our hominid predecessors had the ability to make sounds similar to ours. Given that the oldest Homo sapiens sapiens now date back to almost 200,000 years ago conservatively and maybe as far back as 300,000 years there has been a long time for sound making to evolve. We can make strong interpretations from ethnographic analogy. Given that all known all societies around our globe, including the most isolated tribal groups, have language and music in a variety of forms, one may infer that it was a constituent of the ancestral society's tool kit prior to the great radiation of humans around the world. There is evidence it has been existence for at least 50,000 years. It probably was invented in Africa and evolved into a fundamental constituent of humanity.[2]

Secondarily, we find sound makers and "musical instruments" such as the bone flutes from the Geisenklösterle cave dating

[1] Fernando Coimbra, Geosciences Centre, University of Coimbra Personal Communication February 26, 2018.

[2] I was part of the expedition at Djebel Qafzeh that dated early Homo sapiens.

to 42,000 or 43,000 years. The ones from the Hohle Fels cave supplement them. There is evidence of stone blades being used during the Paleolithic to strike particular sounds[3]. Other evidence includes lithiphones[4], the "ringing stones" and the selection of prehistoric caves and other locations for their "sound characteristics"[56].

Today and in the past, a society's language and music are influenced by all aspects of culture including the economic and ecological domain, access to technology, and ideological issues. Thus, how music is played and listened to, and the attitudes towards musicians and composers vary among regions and periods.[7]

The ability to record past sound has a considerable history. There have been systems of indirectly notating sound. These include early writing systems and musical notation systems. The earliest writing systems dating to the 3rd millennium record words but the actual sounds that were pronounced at the time disappear in the mists of time and are not known today[89]. Music is known from the same period approximately[10]. But our knowledge dates complete compositions about a millennium later. Fragments of Hurrian songs exist in cuneiform. The oldest examples of more or less complete musical composition are the Hurrian Hymn h6 dating to 1400 BC in cuneiform and the Seikilos epitaph written in old Hellenistic Greek and dating to the time of Christ.

However, the ability actually to preserve past sound is quite recent in comparison to the above. Édouard-Léon Scott de Martinville, a French printer and bookseller who lived in Paris, is generally acknowledged to have invented the first device to record sound based on even earlier work by Young, Weber, Duhamel, and Koenig[11]. This device called the phonautograph was patented in France on 25 March 1857, patent #17,897/31,470. However, it only recorded sound. It made a visual record. However, there was no playback ability. It preserved sound but one could not hear it.

Twenty years later on July 18, 1877 Thomas Edison invented the phonograph. It was a successful recording and reproducing ma-

[3] Ezra Zubrow and Elizabeth Blake "*The Origin of Music and Rhythm*" in Archaeoacoustics edited by Chris Scarre and Graeme Lawson McDonald Institute Monographs McDonald Institute for Archaeological Research University of Cambridge 2006v p.117-126

[4]Lya Dams *Palaeolithic Lithophones: Descriptions and Comparisons*. Oxford Journal of Archaeology 1985 4/1 pp 31-45

[5]Maja Hultman *Soundscape Archaeology: Ringing Stone Research in Sweden* 2013 Time and Mind: The Journal of Archaeology, Culture and Consciousness Taylor and Francis 7/1: pp3-12

[6] Paul Devereaux *Sacred Geography* Gaia (Octopus Publishing Group) London pp 160

[7] Matthew 17 (Umaomolhe 17) Fresh Cube News (Noticias Fresquinhas do cubo) *The emotion of music evolution, the emotional wave of sound.* http://umaomolhe.com/the-emotion-of-music-evolution-the-emotional-wave-of-sound/ 2/18/2016

[8] Roger Woodward *Writing Systems* International Encyclopedia of Social and Behavioral Sciences (2nd Edition) Editor-in-Chief James Wright, Elsevier 2016 pp 776-779

[9] We know that at the end of the Bronze Age in the beginning of 3rd millennium BC there were languages being written down in the Middle East. These would include Sumerian, Hurrian, Hattic and Elamite languages. By the second millennium there were Cretan Hieroglyphic, Linear A and Linear B, Ugaritic cuneiform, Akkadian, Phoenician 'alphabet' or 'abjad', Aramaic and Hebrew alphabets, and early alphabetic systems in Greece, Turkey, and Italy. The CREWS project, an ERC grant housed at the University of Cambridge studies the Contexts and the Relations of Early Writing Systems (2016-2021). It looks at not only the systems themselves but the cultural context in which they were adapted.

[10] Supposedly, the "oldest known song was written in cuneiform, dating to 3400 years ago from Ugarit. It was deciphered by Anne Draffkorn Kilmer, and was demonstrated to be composed in harmonies of thirds, like ancient gymel and also was written using a Pythagorean tuning of the diatonic scale.

[11] G. Brock-Nannestade and J. –M Fontaine Early use of the Scott-Koenig phonoautograph for documenting performance Acoustics 2008 Paris http://webistem.com/acoustics2008/acoustics2008/cd1/data/articles/001974.pdf

chine. It meant that not only one could record past sounds but that one could reproduce the sound for posterity. Not only could sounds be heard later but also it could be stored for generations.[12]

As the field has evolved with new discoveries, the data of archaeoacoustics, the methodologies, and the theory have increased logarithmically. This is particularly true as other disciplines have imported and exported theory, methodologies and data.

This paper argues the time for the archaeoacoustic dilettante, the amateur, is clearly in decline. The amateur, similar to the dinosaur is no longer adapted to the modern environment. During the first half of the nineteenth century the amateur was of considerable importance. The person who was avocationally interested in a specific area of endeavor, or even the "gentlepeople" who enjoyed science in a local scientific or philosophical society once important is now less so. The Industrial Revolution led to the rapid increase in the importance of the professional scientist. Amateur scientific interest and the number of amateur societies have decreased in number but their significance as a major source of scientific knowledge has decreased much more.[13][14][15]

Details of the Call

For all of the above – the substance, the new techniques, the interest, and the responsibility, the time is ripe for the field of archaeoacoustics to consider the benefits of establishing a broad, vigorous, engaged scientific and professional society. This society I have tentatively entitled The International Association for Archaeoacoustics.

The benefits are:

1. legitimizing the field,
2. advocacy and education for the public regarding the field,
3. setting up standards,
4. oversight of the appropriate practice,
5. safeguard the public interest,
6. represent the interests of the field and the interests of its membership in other disciplines professional societies and organizations,
7. help with the development of funding opportunities,
8. to be a learned society,
9. to publish peer reviewed materials in appropriate print and electronic outlets,
10. to have meetings and conferences,
11. to help archaeoacoustic researchers obtain employment,
12. to help younger scholars.

Let us consider each in detail.

<u>Legitimizing the Field</u>

This means creating a field where the research within is not spurious. Today, hat the relationship between the public and scientists is important. The dispersal and uptake of information, is more complex and subtler.

There are lots of professional societies that impinge on archaeoacoustics. For example looking at archaeological societies and associations. There are literally dozens of professional archaeological associations. They include continental and sub continental associations such as the European Archaeological Association, Society of East Asian

---

[12] Thomas A Edison *The Perfected Phonograph* 1888 The North American Review 146/379 pp 641-650

[13] Mims, Forrest M. "*Amateur Science -- Strong Tradition, Bright Future.*" *Science*, vol. 284, no. 5411, 1999, pp. 55–56. *JSTOR*, JSTOR, www.jstor.org/stable/2899127.

[14] H.K. Burgess, L.B. DeBey, H.E. Froehlich, N. Schmidt, E.J. Theobald, A.K. Ettinger, J. HilleRisLambers, J. Tewksbury, J.K. Parrish, *The science of citizen science: Exploring barriers to use as a primary research tool*, Biological Conservation, 2017. v208, pp 113-120

[15] H.S. Torrens, *Notes on 'The Amateur' in the development of British geology,* Proceedings of the Geologists' Association, 2006 v117/1/pp1-8

Archaeology or the Society of American Archaeology: National associations such as such as the Canadian Archaeological Association (Association Canadienne d' Archelogie or the Indian Archaeological Society; temporal associations such as Paleoanthropology Society or the Society of Medieval Archaeology: substantive archaeological societies such as the Society of Archaeological Sciences and the International Association of Landscape Archaeology, the Nautical Archaeological Society and its counterpart the Maritime Archaeological association. Each brings members of related interests together in a variety of manners, provides professional standards for its members and provides a public statement that its members are as legitimate as any other professional group.

Without a professional organization we justifiably are considered illegitimate by other researchers. Not only do they belong to such organizations but also their entire careers are structured through the prism of such organizations. Our research interests and we who study archaeoacoustics are as legitimate as any of these other societies and should get the advantages of being recognized as a legitimate field of research interest.

Advocacy and Education to the Public Regarding the Field

A professional society will provide a central place to advocate and to educate the public regarding the field. Frequently, they are a crucial component of the legislative or executive process as these associations are consulted regarding their areas of expertise and specialization as governments seek to develop and implement sound public policy. For example, the aforementioned International Association of Landscape Archaeology lobbies throughout the world for the preservation of prehistoric landscapes. Similarly, if it existed an International Association for Archaeoacoustics would be lobbying for the preservation of prehistoric soundscapes including places with ringing stones or caves with particular echo characteristics etc. Advocacy always is more effective when a group collaborates rather than there only are individuals even if they have the same agendas and priorities. Legislators respond to numbers.

There are other advantages that professional societies have. They have access to professionals and courses regarding best practice in lobbying and advocacy. They are able to teach their members not only how to advocate effectively for issues that impact archaeoacoustics, but to create campaigns across both space and time -multi regional and multi -generational. Because their members feed them information, they will be aware of the most current issues at both the state and national levels.

In terms of educating the public, a professional society frequently provides many services. First, it provides information that is up to date and professionally recognized as being relevant. Secondarily, when information is controversial it will give equal weight to the arguments of both sides and provide a balanced evaluation. Third, an organization is more knowledgeable than an individual because it has the breadth of all of its members. Finally, it is better in contextualizing the profession and our research results in the public realm.

In short, an organization such as the International Association for Archaeoacoustics will be the public face of our profession.

Setting Up Standards

Standards capture the benchmarks against which performance within the profession can be measured and they are vital in improving outcomes since the standards are directly connected to performance. This is true for the individual members of the profession and for the profession as a whole. They demand a minimum threshold above which all work must be performed.

Standards may be processual, phenomenal, or definitional. Processually, one will expect members of the society to follow the approved standardized operating procedures. Phenomenally, there will be standard descriptions or measurements for particular phenomena. Definitional, standards enhance communication. Different practitioners will know exactly what each other means for the professional standards provide dictionaries with standardized definitions. For example when GIS began to be used such terms as location, distance, vector, and point had to be standardized as well as path, road, highway, and freeway as well as hamlet, village, town, and city. The definitions had to hold across countries, languages, software, platforms etc. Another of my favorite examples is that during the construction of the Chunnel from 1988 to 1994, the English and the French firefighting associations had to agree on definitions of types of and scopes of fires so that firefighting teams from each side would know precisely what the other was trying to communicate.

The creation of standards will force the discipline to capture the vision of the field of archaeoacoustics as difficult problems are resolved by the senior policy makers and are informed by best practices in a region and further afield. Beyond that, such standards will facilitate the maintenance of established competencies of professionals and will bring to the profession a much-needed degree of professional accountability. Ultimately, not only are archaeoacousticians the stakeholders but they also are the direct beneficiaries.

Standards are particularly important in today's internetted and socially medially connected world. For example, there are predatory organizations, conferences, and journals that will publish fake studies and make them look like real science. They help unqualified people to get credentials and the unsuspecting do not realize that they will make anything appear as published legitimate discoveries for a fee[16].

Finally, given that standards are necessary inevitably they will be instituted. In most professions the members prefer to set up their own rather than having them imposed on them. So one reason for such a body is that it would set up agreed upon standards rather than having them imposed by either a governmental of business entity.

Of course this process would take a while and would have to be inclusive of the practitioners. Most professional societies create drafts and then have them voted upon by their membership. In addition, they create committees to review and update their standards as a continuous process.

Oversight of the Appropriate Practice:

Another function of the International Association of Archaeoacoustics would be to oversee practice. Today, there is no structured oversight for archaeoacoustics. There is a limited type of "peer review" dependent upon where research money is obtained and upon the outlet where one publishes one's results.

Oversight is related but not the same as standards. It is the mechanisms by which

[16] Tom Spears "When Pigs Fly: Fake Science Conferences about for Fraud and Profit. Ottawa Citizen, March 10, 2017 http://ottawacitizen.com/news/local-news/when-pigs-fly-fake-science-conferences-abound-for-fraud-and-profit

one discovers problems with standards being maintained. Generally speaking there are three types of oversight. One is oversight by government through regulating bodies. Second is by competition or steering by market, academic, or research forces. Third is "mutuality" or professional oversight by self-regulation.

Similarly, there are three types of warnings that indicate that oversight might be insufficient. They are first —a dramatic deviation from the expected outcome. The second is a reasonable outcome despite what appears from outside to be a faulty or questionable process or methodology. This is indicative that the practice is not in control and may deteriorate further causing very adverse events or results. The third category of warning is when there is a reasonable outcome with an acceptable process or methodology that is inefficient, expensive, and unproductive but that might be further improved beyond the prevailing levels.

Errors are inevitable. However, oversight will not only reduce them but what is far more important it provides a continuous process of recognizing when errors occur and begin the pathway to correction. Furthermore, the ultimate goal of oversight by such a body is to create a genuine "learning organization" in which there can be taught a "culture of evidence", canons of scholarly inquiry, peer accountability, and dissemination of best practice to minimize the necessity of oversight.

<u>To Safeguard the Public Interest</u>

In order to safeguard the public interest, a professional body such as the International Association for Archaeoacoustics needs to interact with the government in order to protect the public interest. Clearly, the protection of soundscapes, prehistoric artifacts regarding sound, and the locations of prehistoric buildings, caves, and sites surely fall under the public interest. However, there may be much more. As we know the United Nations now divides heritage into tangible and intangible heritage. This includes the practices, representations, expressions, knowledge, and skills involved with past sound making. Recognizing the intangible heritage in archaeoacoustics would include aspects of ethnomusicology and folk music as well as storytelling, knowledge of instrument making and potentially many other types of intangible information.

There must be integrity of the archaeoacoustic community in order to gather the trust of the public and to influence the political regulators of society. Archaeacousticians have important values. We already know how critical these values are when issues are in high-risk situations. For example, what should be and how should the relationship between the public and private sector be implemented when the public interest in heritage is balanced against the private interest in development.

Archaeoacousticians need to lobby. Lobbying can contribute to good decision-making and improve governments' understanding of policy issues regarding our insights into the origin and continuation of sound in our societies. We need to provide valuable insights and data as part of open consultation processes. Yet, lobbying can also lead to unfair advantages for vocal vested interests if the process is hazy and standards are lax.

We need to create a level playing field by granting all stakeholders fair and equitable access to the relevant public policies surrounding heritage[17]. Gaining balanced perspectives on issues leads to informed policy debate and formulation of effective policies. We need to lobby for public hearings on archaeoacoustic issues. We should have integrity in our lobbying and relate it to the wider heritage policy and regulatory frameworks. Furthermore, the society clearly should label our lobbyists so that they are recognized as being the spokesmen for the profession. We should provide an adequate degree of transparency to ensure that public officials, citizens and businesses can obtain sufficient information on lobbying activities. We should be inviting all stakeholders – including civil society organizations, businesses, the media and the general public – to participate in our lobbying activities and thus become more effective. Our lobbyists should act with professionalism and transparency[18]. In particular they should involve key legislative and international actors in implementing a coherent strategy and practice to achieve governmental compliance with our goals.

Not only should we be lobbying with our municipal, provincial, state and national governments. We should be lobbying, participating, and involved at the international level. We should participate in ICCROM, the International Centre for the Study of the Preservation and Restoration of Cultural Property. This UN organization based in Rome is an intergovernmental organization dedicated to the preservation of cultural heritage worldwide. Similarly, we should be lobbying ICOMOS, the International Council on Monuments and Sites. The Parisian professional association that works for the conservation and protection of cultural heritage places around the world and is the scientific advisor on World Heritage Sites to UNESCO. And the IUNC and the International Center for the Conservation of Nature located in Gland Switzerland. As far as I know no other archaeoacoustician has interacted with these organizations in any way and that is a sorry state.

As a professional body the developing of a conflict of interest policy is very important. It needs to correlate with the political, administrative and legal structure of a country's public life. One need only consider the "revolving door" phenomenon, involving an increased movement of staff between the public and private sectors[19]. It has raised concerns over pre- and post-public employment conditions and its negative effects on trust in the public sector (i.e. the misuse of "insider information", position and contacts).

In short, not only do we have a responsibility to both "safeguard" and the develop the public interest; to take positions on heritage and its safety, to take a position on public versus private ownership (any privatization raises interest of safeguarding the public interest); develop norms and regulations for this and upcoming generations; and lobby for creating a "statutory guardian" charged with safeguarding the public interest.

To Represent the Interests of the Field and the Interests of Its Membership in Other Discipline Organizations

There are many purposes for forming relationships with other professional organiza-

[17] European Committee on Legal Co-operation (CDCJ) – *Study on the feasibility of a Council of Europe legal instrument on the legal regulation of lobbying activities* https://rm.coe.int/16805c482c section 4.42.

[18] The Organization for Economic Co-operation and Development *Transparency and Integrity in Lobbying: The 10 Principles for Transparency and Integrity in Lobbying* Principle number 5 *https://www.oecd.org/gov/ethics/Lobbying-Brochure.pdf*

[19] Jordi Blanes i Vidal, Mirko Draca, and Christian Fons-Rosen *Revolving Door Lobbyists* American Economic Review. 102/7 December 2012 pp 3741-48

tions. Relationships with other organizations help to maintain the identity of archaeoacoustics as well as expanding the diversity of the field.

Members of other disciplines frequently are unfamiliar with archaeoacoustics, our theories, our methods and our data. There are several results. First, the potential for interdisciplinary research is lost. Second, there is a chance that the other disciplines will do redundant research that archaeoacoustics already has done. Third, they may try to stake out particular areas of research and prevent archaeaocousticians from entering those subject areas.

On the other hand, having an archaeoacoustic association fosters inter-organizational relationships. To the extent that professions can share ideas about how productive interrelationships can be accomplished all will benefit regardless of their affiliation. Likewise all professional organizations share the common responsibility of promoting the values of professionalism for the benefit of all who work in this discipline.

One size does not fit all. So these relationships should and do vary from organization to organization. The relationship with the European Association of Acoustics and the Society for Ethnomusicology perhaps might be and should be different. Similarly, the mechanism to be used to form the linkages would differ widely from group to group. Whether there are liaisons, a written agreement of common understanding, or formal exchange of members of each other's executive board have all been used and are situation dependent.

Every discipline is judged by both professional peers both inside and outside of the field. They judge both the fields and the individuals within the fields. The result is that physics is respected as a "solid stem field" while "transnational studies" is seen as an advocacy profession. To put it mildly, there is much work to be done in the area of "image enhancement: for archaeoacoustics"[20]. If we are not careful we will end on the dunghill of academic disciplines with alchemy, phrenology, and other such fields.

In short, the professional public still lacks an accurate knowledge about the discipline of archaeoacoustics. Interacting with other professional organizations on an equal footing will advocate interdisciplinarity, create a more free exchange of information about each other,

This goal is easy to begin to implement. One needs to begin by simply participating in each other's conferences, publish in each other's journals, and try to do joint research.

<u>To Help With the Development of Funding Opportunities</u>

Funding for archaeoacoustics is very limited or not available all together. Generally, there are five research markets: academic, corporate, policy, professional and public markets. Of these, archaeoacoustics has a very small presence in the academic and public markets. Furthermore, our field is going against the national and international trends in monetary support. Academic research is increasingly perceived and evaluated from a utilitarian and economic perspective, emphasizing its ability and responsibility to promote economic growth and competitiveness, and, correspondingly, its need to expand and intensify linkages and collaboration with industry. It is the rise of academic capitalism and "entrepreneurial research"[21]. These are not the areas with

[20] Paul Pitkoff *Image Enhancement: An Integral Part of Good Practice* Law Practice Management v20/is2 (3/1994) 58-60

[21] Sheila Slaughter and Gary Rhoades *Academic Capitalism and the New Economy*

which our field is concerned. Rather we are part of the "general knowledge" aspects of science and the social sciences. Instead of being based on the national or international relevance of research, archaeoacoustic research is far more determined by the participants and their research partners and the development of broad teaching and social priorities regarding heritage.

Many national and international funding bodies will sector money by discipline. For example, there is thirty million dollars for climate change research, ten million dollars for social geography, five million for archaeology, fifty million for astronomy, etc. etc. The disciplinary associations lobby both the funding bodies and the legislatures to put aside so much funds for their discipline. Without an association, archaeoacoustics will continue to have no funding "set aside". Presently, we are not even a priority within say an "acoustics" "set aside".

Furthermore, private donors who have an interest in a discipline will go to the association for recommendations for where and to whom to donate. This is particularly true for donors who do not have large staffs to find and evaluate the kind of research that fits closest to their patron's interests. An association has both the knowledge and the evaluative depth to provide insight and expertise. In addition to across the domain expertise, they can help create long term and more institutionally planned giving strategies.

To Be a Learned Society

An archaeoacoustic association also should be a "learned society". As such a society rather than espousing "market driven ideals with proprietary, commissioned and authoritarian ideals, it favors the Mertonian ideals of "universalism", "skepticism", and professional "disinterestedness".

Here, but seldom elsewhere in our country, we find an unchanging devotion to the truth for its own sake, and a simple conviction, such as a plain man may grasp, of the value of the truth.

Organized to promote the discipline of archaeoacoustics membership may be open to all or may require possession of some qualification or may be an honor conferred by election.

As a "learned society" it is important to accumulate, store, and make accessible knowledge related to the discipline. This includes providing the sources of the knowledge as well as arranging them so that they can be preserved and retrieved with minimal effort. Furthermore, learned societies answer questions and inquiries received from internal as from outside the society. They provide outreach through serving the international, national and local communities through organizing lectures and publishing materials. In addition they exchange, gift, and cooperate with other societies.

Although learned societies have a certain antiquarian and "musty air" about them. They should not be discounted. They have remarkable "staying power" and longevity. For example, Académie des Jeux floraux, a literary association (founded 1323) in Toulouse still exists. They frequently are very modern having taken off with globalization, the internet, and social media. The ease for learned societies to create "virtual communities" for interaction and scientific collaboration today is remarkable. Members of these online academic communities, grouped by areas of interests, use for their

Markets, State, and Higher Education 2009 John Hopkins University Press Baltimore. pp.384

communication shared and dedicated listservs to research, to educate and to reify values.

To Publish Peer Reviewed Material in Appropriate Print and Electronic Outlets

One of the functions of a professional association such as the International Association for Archaeoacoustics is publishing peer reviewed material. Furthermore, it is important for it to remain up to date on "communication developments". For in the same way print has been augmented by electronic outlets, electronic outlets will be augmented by new "communication technologies and an association must adapt to the technological changes facing communication.

There is no absolute measure to evaluate the quality of journals or academic monographs and books. Historically informally and now far more formally journals, monographs, and books are not rated equally. Certain presses have better reputations than others. For example, Cambridge University Press, Harvard Press, Oxford Press, and the University of Chicago press are among the first tier of academic presses. While University of Colorado, SUNY Press, and the University of New Mexico Press would be in the second tier. Peer reviewed are considered better than non-peer reviewed. However, not all peer-reviewed journals are rated equally. Not always, but in general those journals which are peer reviewed, as part of a professional society are considered better than those peer reviewed that are not part of a professional association[22].

Of course, each association decides in what languages they wish to publish. However, I would recommend English because of its present dominance in the scientific world[23].

To Have Meetings and Conferences

One of the great advantages of a professional society is the potential for them to hold meetings and conferences. Meetings and conferences set the stage for researchers to get together. They share information. Above all they should serve as a sort of "information bureau." They are the academic "rapid response team". For meetings and conferences make available information pre publications. Also they may otherwise go unnoticed. Future plans for the profession are suggested, debated and implemented. Sometimes they are significant as symbols; sometimes as keeping open channels of communication.

Meetings and conferences are a convenient place to meet professionals from other places and to develop both individual and institutional networks. These networks may be formal or informal. It is one locus for creating new projects as well as bringing together members of older projects to evaluate progress.

There is an equalizing and democratizing character to professional conferences. As long as one is a member one has equal rights as does every member. The equal status means all members have the right to apply for officer positions, to submit speech presentations to the scientific committees of the meeting as well as to the editor of the society's journals.

[22] There are two types of peer review. Occasionally they are confused. There is individual peer review that involves evaluation of the quality of a paper. There also is agency review that assesses the general patterns of the discipline.

[23] As early as the millennium almost 90% of the 5,000 articles published every day is were published in English. Furthermore, the vast majority of articles indexed in international databases are written in English and The Web of Science/International Science Index has a distinct preference for journals which publish in English, stating that they focus "on journals that publish full text in English or at very least, their bibliographic information in English.

Other functions are undertaken at society conferences and meetings. They include lobbying for needs, election of officers, vendor displays and frequently employment interviews.

An important advantage of meetings and conferences supported and administered through professional societies such as the International Association of Archaeoacoustics is that there is a guarantee of continuity in the association's annual conferences that may not be possible with other supporting institutions.

Ultimately in a dynamic and rapidly changing field conferences and meetings play a strong role in informing the professional opinions of individuals and experts. They are the center of presentation, assessment, and reassessment and thus the development of the field.

To Help Archaeoacoustics Researchers Get Employment

As has already been pointed out the field of archaeoacoustics is growing. Today, one must be able to take advantage of employment diversification. Archaeoacoustci researchers have and will continue to have a wide variety of positions. Some are fully employed as archaeologists, acousticians, musicians, ethno musicians, historians etc. Others do archaeoacoustic contingent employment of a variety of types. This would include part time, extra service, or simply avocational employment.

An advantage of a professional society is that one may take advantage of employment diversification by expediting information about employment opportunities. In many disciplines there is a tendency to centralize employment opportunities. The society can help standardize both job expectations and qualifications for some types of employment. The over qualification and under qualification issue are an important aspect of job expectations. Furthermore, the professional society can analyze detailed employment trends and direct its members toward those areas. This is particularly important because there are so many markets that archaeoacoustics touch. For example, one needs to examine the archaeology, the anthropology, the acoustic, the music, the art, and other related markets. And to additionally complicate the situation, there is a blurring of discipline and sector boundaries.

Help Younger Scholars

There are some members of the archaeoacoustic community who like the general society believe in the "trickle down model" of research, apprentice employment, and economics. These suppose that senior scholar beneficence and employment will trickle down to younger scholars. This model is particularly inappropriate and insufficient for a fast growing field where younger and older scholars are making new discoveries and applications rapidly and equally.

A more institutionalized approach would be far better. The professional society could create skill courses in particularly useful areas---for example "field recording", or "acoustic equipment" etc. etc. They could create "development" courses to help young scholars analyze both their skills and weaknesses and match them to employment opportunities. There need to be vital forums for the discussion, development, and dissemination of authoritative knowledge related to traditional technologies, while also focusing more on emerging and disruptive technologies.

Mentorship programs developed through the society would provide all the benefits of the "trickle down model" with none of the costs created by lack of access or lack of diverse mentors as in the "apprentice system".

In today's marketplace hiring is not simply a process of skills sorting. Rather, it is a process of "cultural matching" between candidates, evaluators, researchers, NGO's and firms. Employers seek candidates who were not only competent but are also culturally similar to themselves in terms of leisure pursuits, experiences, and self-presentation styles[24]. This process of "cultural matching" is far easier in a professional society environment than in a non-regulated, "everyone on their own", competitive environment. Similar to a union or open shop older members help younger ones in a variety of manners including cultural indoctrination. In fact, it must be a culture of respect for younger employees that invests in enhancing capabilities and empowering both younger individuals and groups of younger individuals building a concept of membership in the profession.

Today's scientific and humanity's communities are not only global but are diverse. A global community cultivates interactions not only across the generations but also across space, which greatly enhances the ability to help young members of the profession. Young scholars need to be dynamic, nimble, flexible and diverse.

Conclusion

We end where we began making a complete circle. This is a general call to all individuals and groups of people interested in archaeoacoustics. No matter where in the world you are located. No matter what aspect of past sound you are interested in. No matter what your profession is. The time has come to create a professional organization. Whether it is a Society of Archaeoacoustics or an International Association for Archaeoacoustics does not make a difference. It should be inclusive not exclusive. It should be dynamic and not static. If created, it will help the field collectively and each of us individually.

NOTE: Subsequent to Prof. Zubrow's presentation at the Portugal conference, several participants volunteered to help launch such an organization.

The International Society for the Study of Archaeoacoustics
is now a legal entity with its own website www.Archaeoacoustics.org.

ARCHAEOACOUSTICS

[24] Lauren A. Rivera *Hiring as Cultural Matching: The Case of Elite Professional Service Firms* American Sociological Review 2012 77(6) 999–1022

The **ARCHAEOACOUSTICS I, II,** and **III** international conferences have been initiated and made possible by

The OTS Foundation

A United States of America not-for-profit 501 (c)(3) educational foundation

OTSF.org

Back Cover

Sound Is an Integral Piece of the Archaeological Puzzle
by
Steve Waller

Winner of the fine art competition:

Illustrating Archaeoacoustics

Sponsored by

ARTSHIP Foundation

and

OTSF

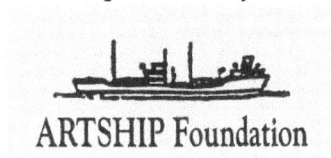

EDITORS

LINDA C. ENEIX works with archaeologists, anthropologists, architects, engineers, historians and other specialists toward a multi-disciplinary understanding of the Neolithic phenomenon. She has lectured on Malta's megalithic monuments for the Smithsonian Associates, has been consulted for film, television and print media in the USA and Europe, and has appeared on international television. Eneix spearheaded initiatives for conservation management and educational outreach in Malta. Her research related to archaeoacoustics, combined with an architectural appreciation of the oldest freestanding buildings on earth have enabled fresh insight into interpretation of motivation and original function. She is also responsible for the design and operation of educational programming in Malta for Road Scholar's Adventures in Lifelong Learning.

MICHAEL W. RAGUSSA is a teacher of History and Sociology, recently retired. A lifelong interest in interpretation and cognative anthropology manifests itself in his active support of open dialogue between researchers in a range of professions. He is a member of several NFP boards and community advisory panels, and is the proud grandfather of three student scientists.

www.ingramcontent.com/pod-product-compliance
Lightning Source LLC
Chambersburg PA
CBHW061354210326
41598CB00035B/5985